D1574032

Edited by
Genserik L.L. Reniers, Kenneth Sörensen,
and Karl Vrancken

**Management Principles
of Sustainable Industrial Chemistry**

Related Titles

Azapagic, A., Perdan, S. (eds.)

Sustainable Development in Practice

Case Studies for Engineers and Scientists

2011
Softcover
ISBN: 978-0-470-71872-8

Yasuda, N. (ed.)

The Art of Process Chemistry

2011
Hardcover
ISBN: 978-3-527-32470-5

Leimkühler, H.-J. (ed.)

Managing CO_2 Emissions in the Chemical Industry

2010
Hardcover
ISBN: 978-3-527-32659-4

Centi, G., Trifiró, F., Perathoner, S., Cavani, F. (eds.)

Sustainable Industrial Chemistry

2009
Hardcover
ISBN: 978-3-527-31552-9

*Edited by Genserik L.L. Reniers, Kenneth Sörensen,
and Karl Vrancken*

Management Principles of Sustainable Industrial Chemistry

Theories, Concepts and Industrial Examples for Achieving Sustainable Chemical Products and Processes from a Non-Technological Viewpoint

WILEY-VCH

WILEY-VCH Verlag GmbH & Co. KGaA

The Editors

Prof. Genserik L.L. Reniers
Universiteit Antwerpen
City Campus, Office B-434
Prinsstraat 13
2000 Antwerpen
Belgien

Prof. Kenneth Sörensen
University of Antwerp
Operation Res. Group ANT/OR
Prinsstraat 13
2000 Antwerpen
Belgien

Prof. Karl Vrancken
University of Antwerp
Dept. Bio-Engineering
Boeretang 200
2400 Mol
Belgien

All books published by **Wiley-VCH** are carefully produced. Nevertheless, authors, editors, and publisher do not warrant the information contained in these books, including this book, to be free of errors. Readers are advised to keep in mind that statements, data, illustrations, procedural details or other items may inadvertently be inaccurate.

Library of Congress Card No.: applied for

British Library Cataloguing-in-Publication Data
A catalogue record for this book is available from the British Library.

Bibliographic information published by the Deutsche Nationalbibliothek
The Deutsche Nationalbibliothek lists this publication in the Deutsche Nationalbibliografie; detailed bibliographic data are available on the Internet at <http://dnb.d-nb.de>.

© 2013 Wiley-VCH Verlag GmbH & Co. KGaA, Boschstr. 12, 69469 Weinheim, Germany

All rights reserved (including those of translation into other languages). No part of this book may be reproduced in any form – by photoprinting, microfilm, or any other means – nor transmitted or translated into a machine language without written permission from the publishers. Registered names, trademarks, etc. used in this book, even when not specifically marked as such, are not to be considered unprotected by law.

Print ISBN: 978-3-527-33099-7
ePDF ISBN: 978-3-527-64951-8
ePub ISBN: 978-3-527-64950-1
mobi ISBN: 978-3-527-64949-5
oBook ISBN: 978-3-527-64948-8

Cover Design Grafik-Design Schulz, Fußgönheim, Germany
Typesetting Laserwords Private Limited, Chennai, India
Printing and Binding Markono Print Media Pte Ltd, Singapore

Contents

Preface *XIII*
List of Contributors *XV*

Part I **Introductory Section** *1*

1 **Editorial Introduction** *3*
Genserik L.L. Reniers, Kenneth Sörensen, and Karl Vrancken
1.1 From Industrial to Sustainable Chemistry, a Policy Perspective *4*
1.2 Managing Intraorganizational Sustainability *5*
1.3 Managing Horizontal Interorganizational Sustainability *5*
1.4 Managing Vertical Interorganizational Sustainability *6*
1.5 Sustainable Chemistry in a Societal Context *6*

2 **History and Drivers of Sustainability in the Chemical Industry** *7*
Dicksen Tanzil and Darlene Schuster
2.1 The Rise of Public Pressure *7*
2.1.1 The Environmental Movement *8*
2.1.2 A Problem of Public Trust *9*
2.2 Industry Responded *10*
2.2.1 The Responsible Care Program *10*
2.2.2 Technology Development *12*
2.2.3 Corporate Sustainability Strategies *14*
2.3 An Evolving Framework *15*
2.3.1 New Issues and Regulations *15*
2.3.2 Sustainability as an Opportunity *16*
2.3.3 Recent Industry Trends *16*
2.4 Conclusions: the Sustainability Drivers *18*
References *18*

3	**From Industrial to Sustainable Chemistry, a Policy Perspective** 21
	Karl Vrancken and Frank Nevens
3.1	Introduction 21
3.2	Integrated Pollution Prevention and Control 22
3.2.1	Environmental Policy for Industrial Emissions 22
3.2.2	Best Available Techniques and BREFs 23
3.2.3	Integrated Pollution Prevention and Control in the Chemical Sector 24
3.3	From IED to Voluntary Systems 25
3.4	Sustainability Challenges for Industry 26
3.4.1	Introduction 26
3.4.2	Policy Drivers for Sustainable Chemistry 27
3.4.3	Transition Concept 28
3.5	Conclusion 30
	References 31
4	**Sustainable Industrial Chemistry from a Nontechnological Viewpoint** 33
	Genserik L.L. Reniers, Kenneth Sörensen, and Karl Vrancken
4.1	Introduction 33
4.2	Intraorganizational Management for Enhancing Sustainability 36
4.3	Horizontal Interorganizational Management for Enhancing Sustainability 37
4.4	Vertical Interorganizational Management for Enhancing Sustainability 38
4.5	Sustainable Chemistry in a Societal Context 39
4.6	Conclusions 40
	References 41

Part II   Managing Intra-Organizational Sustainability 43

5	**Building Corporate Social Responsibility – Developing a Sustainability Management System Framework** 45
	Stefan Maas, Genserik L.L. Reniers, and Marijke De Prins
5.1	Introduction 45
5.2	Development of a CSR Management System Framework 47
5.2.1	Management Knowledge and Commitment (Soft Factor) 49
5.2.2	Stakeholder Knowledge and Commitment (Soft Factor) 49
5.2.3	Strategic Planning – the Choice of Sustainable Strategic Pillars (Hard Factor) 50
5.2.4	Knowledge and Commitment from the Workforce (Soft Factor) 50
5.2.5	Operational Planning, Execution, and Monitoring (Hard Factor) 51
5.3	Conclusions 52
	References 52

6	**Sustainability Assessment Methods and Tools** 55	
	Steven De Meester, Geert Van der Vorst, Herman Van Langenhove, and Jo Dewulf	
6.1	Introduction 55	
6.2	Sustainability Assessment Framework 56	
6.3	Impact Indicators and Assessment Methodologies 59	
6.3.1	Environmental Impact Assessment 62	
6.3.1.1	Emission Impact Indicators 62	
6.3.1.2	Resource Impact Indicators 68	
6.3.1.3	Technology Indicators 71	
6.3.1.4	Assessment Methodologies 72	
6.3.2	Economic Impact Assessment 75	
6.3.2.1	Economic Impact Indicators 76	
6.3.2.2	Assessment Methodologies 76	
6.3.3	Social Impact Assessment 77	
6.3.3.1	Social Impact Indicators 78	
6.3.3.2	Assessment Methodologies 79	
6.3.4	Multidimensional Assessment 79	
6.3.5	Interpretation 81	
6.4	Conclusions 81	
	References 82	

7	**Integrated Business and SHESE Management Systems** 89	
	Kathleen Van Heuverswyn and Genserik L.L. Reniers	
7.1	Introduction 89	
7.2	Requirements for Integrating Management Systems 90	
7.3	Integrating Management Systems: Obstacles and Advantages 92	
7.4	Integrated Risk Management Models 95	
7.4.1	FERMA Risk Management Standard 2003 95	
7.4.2	Australian/New Zealand Norm AS/NZS 4360:2004 96	
7.4.3	ISO 31000:2009 97	
7.4.4	The Canadian Integrated Risk Management Framework (IRM Framework) 98	
7.5	Characteristics and Added Value of an Integrated Model; Integrated Management in Practice 100	
7.6	Conclusions 103	
	References 103	

8	**Supporting Process Design by a Sustainability KPIs Methodology** 105	
	Alessandro Tugnoli, Valerio Cozzani, and Francesco Santarelli	
8.1	Introduction 105	
8.2	Quantitative Assessment of Sustainability KPIs in Process Design Activities 107	
8.3	Identification of Relevant KPIs: the "Tree of Impacts" 111	
8.4	Criteria for Normalization and Aggregation of the KPIs 121	

8.5	Customization and Sensitivity Analysis in Early KPI Assessment *123*
8.6	Conclusions *128*
	References *128*

Part III   Managing Horizontal Interorganizational Sustainability *131*

9   Industrial Symbiosis and the Chemical Industry: between Exploration and Exploitation *133*
Frank Boons

9.1	Introduction *133*
9.2	Understanding Industrial Symbiosis *134*
9.2.1	Industrial Symbiosis Leads to Decreased Ecological Impact *135*
9.2.2	Industrial Symbiosis Requires a Highly Developed Social Network *136*
9.2.3	The Regional Cluster Is the Preferred Boundary for Optimizing Ecological Impact *136*
9.3	Resourcefulness *137*
9.4	Putting Resourcefulness to the Test *138*
9.4.1	Petrochemical Cluster in the Rotterdam Harbor Area *138*
9.4.2	Terneuzen *139*
9.4.3	Moerdijk *141*
9.5	Conclusions *142*
	References *144*

10   Cluster Management for Improving Safety and Security in Chemical Industrial Areas *147*
Genserik L.L. Reniers

10.1	Introduction *147*
10.2	Cluster Management *148*
10.3	Cross-Organizational Learning on Safety and Security *150*
10.3.1	Knowledge Transfer *150*
10.3.2	Overcoming Confidentiality Hurdles: the Multi-Plant Council (MPC) *151*
10.3.3	A Cluster Management Model for Safety and Security *152*
10.4	Discussion *157*
10.5	Conclusions *158*
	References *159*

Part IV   Managing Vertical Inter-Organizational Sustainability *161*

11   Sustainable Chemical Logistics *163*
Kenneth Sörensen and Christine Vanovermeire

11.1	Introduction *163*
11.2	Sustainability of Logistics and Transportation *165*

11.3	Improving Sustainability of Logistics in the Chemical Sector	*166*
11.3.1	Optimization *167*	
11.3.2	Coordinated Supply Chain Management *170*	
11.3.3	Horizontal Collaboration *171*	
11.3.4	Multimodal, Intermodal and Co-Modal Transportation *174*	
11.4	Conclusions *178*	
	References *179*	

12 Implementing Service-Based Chemical Supply Relationship – Chemical Leasing® – Potential in EU   *181*

Bart P.A. Van der Velpen and Marianne J.J. Hoppenbrouwers

12.1	Introduction *181*
12.2	Basic Principles of Chemical Leasing (ChL) *182*
12.3	Differences between Chemical Leasing and Other Alternative Business Models for Chemicals *186*
12.3.1	Classical Leasing *186*
12.3.2	Chemical Management Services *186*
12.3.3	Outsourcing *187*
12.4	Practical Implications of Chemical Leasing *187*
12.4.1	Strengths and Opportunities for the Supplier *189*
12.4.2	Strengths and Opportunities for the Customer *190*
12.5	Economic, Technical, and Juridical Aspects of Chemical Leasing *191*
12.5.1	An Example *191*
12.5.2	Barriers to the Model *191*
12.5.3	Analysis of the Legal Requirements Impacting Chemical Leasing Projects *193*
12.5.3.1	The Importance of Contracts *193*
12.5.3.2	Competition Law and Chemical Leasing *194*
12.5.3.3	REACH and Chemical Leasing *195*
12.5.3.4	Legal Aspects, a Bottleneck? *196*
12.6	Conclusions and Recommendations *197*
	References *198*

13 Sustainable Chemical Warehousing   *199*

Kenneth Sörensen, Gerrit K. Janssens, Mohamed Lasgaa, and Frank Witlox

13.1	Introduction *199*
13.2	Risk Management in the Chemical Warehouse *200*
13.2.1	Hazard Identification *200*
13.2.2	Quantifying Risk: Probabilities and Consequences *205*
13.2.3	Mitigation Strategies *209*
13.2.3.1	Minimize Risk *209*
13.2.3.2	Transfer Risk *211*
13.2.3.3	Accept Risk *213*

13.2.4	Control and Documentation	*213*
13.3	Conclusions	*214*
	References	*214*

Part V Sustainable Chemistry in a Societal Context *215*

14 A Transition Perspective on Sustainable Chemistry: the Need for Smart Governance? *217*
Derk A. Loorbach

14.1	Introduction	*217*
14.2	A Transitions Perspective on Chemical Industry	*219*
14.3	A Tale of Two Pathways	*223*
14.4	Critical Issues in the Transition Management to Sustainable Chemistry	*225*
14.5	Governance Strategies for a Transition to a Sustainable Chemistry	*227*
14.6	Conclusions and Reflections	*230*
	References	*231*

15 The Flemish Chemical Industry Transition toward Sustainability: the "FISCH" Experience *233*
Luc Van Ginneken and Frans Dieryck

15.1	Introduction	*233*
15.1.1	Societal Chemistry	*233*
15.1.2	The Belgian and Flemish Chemical and Life Sciences Industry in a Global Context	*233*
15.1.3	The Challenge of Sustainable Development for the Chemical Industry in Flanders	*234*
15.2	Transition of the Chemical Industry in Flanders: the "FISCH" Initiative	*236*
15.2.1	Setting the Scene: the "FISCH" Feasibility Study	*236*
15.2.2	Outcome of the Study – Goals and Overall Setup of "FISCH"	*237*
15.2.2.1	Vision, Mission, and Setup of FISCH	*237*
15.2.2.2	FISCH in a Flemish and European Context	*241*
15.2.2.3	Added Value of "FISCH" and Spillover Effects	*242*
15.2.3	Putting It All into Practice: Implementing "FISCH"	*243*
15.3	Concluding Remarks and Lessons Learned	*244*
	Acknowledgments	*245*
	References	*245*

16	**The Transition to a Bio-Based Chemical Industry: Transition Management from a Geographical Point of View** *247*	
	Nele D'Haese	
16.1	Introduction *247*	
16.2	Composition of the Chemical Clusters in Antwerp, Ghent, Rotterdam, and Terneuzen *249*	
16.2.1	The Rhine–Scheldt Delta *249*	
16.2.2	Past and Present of the Petrochemical Industry in the Ports of Antwerp, Ghent, Rotterdam, and Terneuzen *250*	
16.3	Regional Innovation Projects to Strengthen the Transition to a Bio-Based Chemical Industry *254*	
16.3.1	First Step: Substitution of Fossil Resources by Bio-Based Feedstocks Making Use of Vested Technologies *254*	
16.3.2	Second Step: Development of a New Technological Paradigm for the Production of Second-Generation Bio-Based Products *257*	
16.3.3	Third Step: Closing Material Loops *258*	
16.4	Conclusions *259*	
	References *262*	

Part VI Conclusions and Recommendations *265*

17 Conclusions and Recommendations *267*
Genserik L.L. Reniers, Kenneth Sörensen, and Karl Vrancken

Index *269*

Preface

Chemical products make an irreplaceable contribution in every aspect of our modern day lives. Chemical processes and products play an essential role in industrial sectors as diverse as agriculture, automotive, clothing, communication, construction, food, health, leisure, mobility, plastics, space, transport, and so on. We can easily observe that our advanced society depends on the wealth-creating aspects of industrial chemistry.

Nonetheless, societal expectations and the depletion of natural resources are pushing toward chemical processes becoming cleaner, more efficient, less consuming, safer, and more secured. The ecological footprint of chemical products needs to be decreased.

Sustainable chemistry being concerned with the development of sustainable chemical products and processes and thereby integrating economic, environmental, and social performance, can provide an answer to these major challenges.

To achieve sustainable industrial chemical processes and products, companies, research centers, and academia tend to focus mainly on technological solutions such as cleantech, green technology, process intensification, new catalysts, new membranes, ecofining, and so on. However, nontechnological approaches are essential as well to succeed in adequate sustainable chemistry. Integrated management systems, cluster management, business models, measuring criteria and methods, sustainable supply chain management, chemical leasing, transition management, societal expectations, and so on are all important nontechnological aspects of sustainable chemistry. To date, most of the know-how and expertise on nontechnological issues is developed on individual company or academia basis and in a fragmented way. An overview of management principles, theories, concepts, and so on from a nontechnological holistic (People, Planet, and Profit) perspective has, to the best of the Editors' knowledge, not yet been discussed in one book volume.

The objective of writing a book from a managerial viewpoint consists in leveraging the search for truly sustainable chemical products and processes, and to disseminate the available knowledge to captains of industry and to leaders of the public sector, as well as to company management (within all organizational levels and from all different departments, and disciplines). It is crucial for the vision of sustainable chemistry to be realized that not only novel technology is conceptualized and developed but also that innovative management models, intraorganization models, and interorganization models are elaborated, promoted, and implemented within the chemicals using industries.

We are convinced that a clear interdisciplinary approach within technological areas, supported by cross-cutting managerial actions, is required for truly successful tackling of these new chemistry challenges and paradigms.

Antwerp
May 10, 2012

*Genserik L.L. Reniers, Kenneth Sörensen,
and Karl Vrancken*

List of Contributors

Frank Boons
Erasmus University Rotterdam
Department of Public
Administration
Post Box 1738
3000 DR Rotterdam
The Netherlands

Valerio Cozzani
Alma Mater Studiorum –
University of Bologna
Dipartimento di Ingegneria
Chimica Mineraria e delle
Tecnologie Ambientali (DICMA)
via Terracini 28
40131 Bologna (BO)
Italy

Jo Dewulf
Research Group ENVOC
Ghent University
Coupure Links
9000 Ghent
Belgium

Nele D'Haese
Flemish Institute for
Technological Research (VITO)
Boeretang 200
2400 Mol
Belgium

Frans Dieryck
essenscia vlaanderen
Diamant Building
A. Reyerslaan 80
1030 Brussels
Belgium

Marianne J.J. Hoppenbrouwers
Universiteit Hasselt
Centrum voor Milieukunde
Campus Diepenbeek
3590 Diepenbeek
Belgium

Gerrit K. Janssens
University of Antwerp
Operation Research Group
ANT/OR
Prinsstraat 13
2000 Antwerp
Belgium

Mohamed Lasgaa
University of Antwerp
Operation Research Group
ANT/OR
Prinsstraat 13
2000 Antwerp
Belgium

and

Groenewout
Nijverheidssingel 313
4800 DG Breda
The Netherlands

Derk A. Loorbach
Dutch Research Institute for
Transitions (DRIFT)
Erasmus University Rotterdam
PO Box 1738
3000 DR Rotterdam
The Netherlands

Stefan Maas
Univsersity of Antwerp
Antwerp Research Group on
Safety and Security ARGoSS
Chemistry Lab
City Campus
Koningstraat 8
2000 Antwerp
Belgium

Steven De Meester
Research Group ENVOC
Ghent University
Coupure Links
9000 Ghent
Belgium

Frank Nevens
Flemish Institute for
Technological Research (VITO)
Boeretang 200
2400 Mol
Belgium

Marijke De Prins
HUB, Stormstraat 2
1000 Brussels
Belgium

Genserik L.L. Reniers
Univsersity of Antwerp
Antwerp Research Group on
Safety and Security ARGoSS
Chemistry Lab
City Campus
Koningstraat 8
2000 Antwerp
Belgium

Francesco Santarelli
Dipartimento di Ingegneria
Chimica
Mineraria e delle Tecnologie
Ambientali (DICMA)
Alma Mater Studiorum –
Università di Bologna
via Terracini 28
40131 Bologna (BO)
Italy

Darlene Schuster
Institute for Sustainability -
American Institute of Chemical
Engineers (AIChE)
3 Park Avenue 19 Fl
New York, NY 10016-5991
USA

Kenneth Sörensen
University of Antwerp
Operation Research Group
ANT/OR
Prinsstraat 13
2000 Antwerp
Belgium

Dicksen Tanzil
Golder Associates Inc.
500 Century Plaza Dr. Ste. 190
Houston, TX 77073-6027
USA

Alessandro Tugnoli
Dipartimento di Ingegneria
Chimica
Mineraria e delle Tecnologie
Ambientali (DICMA)
Alma Mater Studiorum –
Università di Bologna
via Terracini 28
40131 Bologna (BO)
Italy

Luc Van Ginneken
Flemish Institute for
Technological Research (VITO)
Boeretang 200
2400 Mol
Belgium

Kathleen Van Heuverswyn
HUBrussel
Stormstraat 2
1000 Brussels
Belgium

Herman Van Langenhove
Research Group ENVOC
Ghent Univesity
Coupure Links
9000 Ghent
Belgium

Bart P.A. Van der Velpen
Royal HaskoningDHV
Schaliënhoevedreef 20D
Hanswijkdries 80
2800 Mechelen
Belgium

Geert Van der Vorst
Research Group ENVOC
Ghent University
Coupure Links
9000 Ghent
Belgium

Christine Vanovermeire
University of Antwerp
Operations Research Group
ANT/OR
Prinsstraat 13
2000 Antwerp
Belgium

Karl Vrancken
Flemish Institute for
Technological Research (VITO)
Boeretang 200
2400 Mol
Belgium

and

University of Antwerp
Department Bio-Engineering
Groenenborgerlaan 171
2020 Antwerpen
Belgium

Frank Witlox
Ghent University
Department of Geography
Krijgslaan 281, S8
B9000 Ghent
Belgium

and

University of Antwerp - ITMMA
Kipdorp 59
B2000 Antwerp
Belgium

Part I
Introductory Section

1
Editorial Introduction

Genserik L.L. Reniers, Kenneth Sörensen, and Karl Vrancken

There has been an ever-growing worldwide interest in sustainability in all industrial sectors since the Rio declaration two decades ago (UN, 1992). Especially in industries using chemicals, topics related to sustainability are gaining importance by the year. Sustainability should be seen as an ideal. It is an objective of perfection that will never be completely achieved. It is a target of continuous improvement. It should be a business imperative. The interconnectedness of organizational actions and decisions should have an impact on the social, ecological, and economic sustainability of the community in which it operates. To achieve this ideal, and all its accompanying aims, technological as well as a nontechnological innovations and operations should be strived for and implemented. This book specifically deals with the nontechnological path that should be taken within the chemical industry to achieve sustainability in business needs.

However, these are rather vague concepts. All this wisdom about sustainability, the awareness, and information, does not suggest concrete actions and tactics needed to change an organization for the better. This book describes how to significantly enhance the sustainability of chemical plants from the management's perspective.

By taking into consideration the needs for nontechnological advancements toward sustainability, the present book, whose structure is illustrated in Figure 1.1, aims at covering all aspects and all principles leading to truly sustainable industrial chemistry from a managerial perspective. The first introductory section provides a description of the history and importance of sustainability in the chemical industry and of the evolution in managerial themes and models leading to a steady transition toward sustainability. The second section discusses the management system requirements and the needs to build corporate social responsibility within one plant, and provides tools and methods to measure sustainability within a chemical company or a part thereof. The third section investigates the managerial needs to improve cross-plant management and collaboration at the same level of the supply chain, for moving toward ever more sustainable chemical products and processes. The fourth section provides insights into some innovative managerial approaches with respect to collaboration and cooperation between organizations not situated on the same level of the supply chain, leading toward so-called vertical

Management Principles of Sustainable Industrial Chemistry: Theories, Concepts and Industrial Examples for Achieving Sustainable Chemical Products and Processes from a Non-Technological Viewpoint,
First Edition. Edited by Genserik L.L. Reniers, Kenneth Sörensen, and Karl Vrancken.
© 2013 Wiley-VCH Verlag GmbH & Co. KGaA. Published 2013 by Wiley-VCH Verlag GmbH & Co. KGaA.

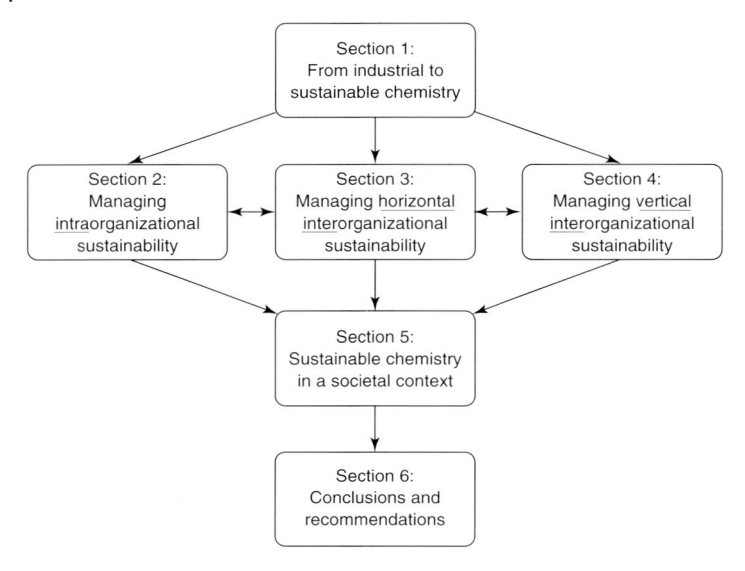

Figure 1.1 Structure of the book.

interorganizational sustainability. The fifth section presents and elaborates on the societal context of sustainable chemistry.

The following paragraphs offer an outlook of the 13 contributions that constitute the various sections of the book. In order to provide an introduction to the various chapters, a description of the main themes that are dealt with in each one is given.

1.1
From Industrial to Sustainable Chemistry, a Policy Perspective

This first, introductory, section contains three contributions. The first one, History and Drivers of Sustainability in the Chemical Industry, provides a brief description of the chemical industry's path toward sustainability. The incentives and drivers for a step-by-step advancement, from the Responsible Care® program to the various corporate sustainability initiatives, are listed out and expounded.

The second contribution of this section, From Industrial to Sustainable Chemistry, a Policy Perspective, clarifies the policy developments that could be observed over the past decades in relation to chemistry on an industrial scale. The contribution clearly demonstrates that there has been a shift in focus over the last two decades from strict rule-driven regulations and authorities toward performance-based and stakeholder-based governance. This shift has initiated and empowered a shift of industry – such as the chemical industry – toward new managerial and governance approaches.

The third contribution of this introductory section, Sustainable Industrial Chemistry from a Nontechnological Viewpoint, briefly discusses what is understood in this book by "sustainable chemistry" and what constitutes a "nontechnological viewpoint." The foundations are laid for the further chapters by elaborating on the different managerial topics for achieving sustainable chemistry in a simple, nontechnological manner.

1.2
Managing Intraorganizational Sustainability

The second section of the book is composed of four chapters. The first one, *Building Corporate Social Responsibility* – Developing a Sustainability Management System Framework, deals with the creation of a conceptual sustainability management system, mainly on the basis of the umbrella guideline ISO 26000. The proposed coherent and systematic framework contains five inherent and consecutive features of sustainability. The current overload of standards makes organizations uncertain how to translate the idea of sustainability optimally into a management system, and this section provides an answer to this organizational need.

The second chapter of this section, Sustainability Assessment Methods and Tools, discusses a sustainability assessment framework and impact indicators and assessment approaches from both a uni- and a multidimensional perspectives. The chapter argues that harmonization and standardization of knowledge in three dimensions (environment, economic, and social) should be pursued for the chemical industry.

The third contribution of this section, Integrated Business- and SHESE Management Systems, takes a closer look at the added value of integrated management systems and the required steps to successfully implement an integrated management system approach. The chapter provides arguments for treating sustainability as a holistic, organization-wide objective, to be achieved by an integrative generic framework that leaves space for specificities wherever and whenever needed.

The last contribution is concerned with the identification of relevant impact categories and suitable KPIs for sustainability performance. How the KPIs should be interpreted and aggregated is explained, amongst others. The method elaborated in this contribution helps decision makers in the design for sustainability within chemical process plants.

1.3
Managing Horizontal Interorganizational Sustainability

The third section of the book is contains two chapters. The first chapter, Industrial Symbiosis and the Chemical Industry: between Exploration and Exploitation, explains industrial symbiosis and compares different chemical clusters from the Netherlands in this regard. The advantages and hurdles of realizing cross-plant collaboration initiatives to advance environmental symbiotic linkages are discussed.

The second contribution in this section, Cluster Management for Improving Safety and Security in Chemical Industrial Areas, proposes a framework and an approach for chemical plants situated within the same chemical cluster, to transfer knowledge, know-how, and best practices, and a more intensive collaboration on safety and security topics.

1.4
Managing Vertical Interorganizational Sustainability

The fourth part of this book has three contributions. The first contribution, Sustainable Chemical Logistics, investigates the status of sustainability in chemical logistics, and argues that organizational aspects have an important role to play in this area. Furthermore, different ways to improve sustainability of chemical logistics are discussed: optimization in logistics, coordinated supply chain management, horizontal collaboration, and intermodal transportation.

The second contribution, Implementing Service-Based Chemical Supply Relationship – Chemical leasing® – Potential in EU?, explains "chemical leasing" as a new business model that aligns economic incentives in the chemical supplier–user relationship toward reduced material use on the one hand and waste prevention on the other. The contribution clarifies this novel business concept and shows its innovative nature and possible role in "servicizing" the chemical supply chain. Furthermore, the synergy that exists between chemical leasing and several relevant legal frameworks, such as REACH, is addressed.

The third contribution deals with the needs as regards sustainable warehousing. It is evident that adequate risk management policies and -procedures and risk treatment strategies need to be in place in warehouses. The different factors important in this regard, are given and clarified. The chapter further discusses sustainable inventory management and vendor management inventory, and their importance.

1.5
Sustainable Chemistry in a Societal Context

The fifth section of the book is based on three contributions. The first one, A Transition Perspective on Sustainable Chemistry: the Need for Smart Governance, offers an exploratory transition perspective on challenges and changes going hand in hand with sustainable chemistry. The author argues and explains that a transition toward sustainable industrial chemistry is not so much a technological challenge as it is an institutional, economic, and political challenge.

The second chapter, The Flemish Chemical Industry Transition toward Sustainability: the "FISCH" Experience, discusses the peculiarities and the obstacles and hurdles of developing an initiative in the Flanders' region in Belgium, to advance the chemicals-using industries toward becoming a sustainability-driven and an innovation-driven industrial sector. Factors to be taken into account when developing a similar initiative are given.

The third contribution, The Transition to a Bio-based Chemical Industry: Transition Management from a Geographical Point of View, analyzes the regional characteristics and their influence on bio-based innovation. The chapter discusses the hard and soft influential factors in this regard, and four cases are examined: the port regions of Antwerp, Ghent, Rotterdam, and Terneuzen.

2
History and Drivers of Sustainability in the Chemical Industry
Dicksen Tanzil and Darlene Schuster

This section provides a historical look on the emergence of sustainability issues and awareness in the chemical industry and how the industry has responded to them, especially over the last 50 years. It describes industry's initial reactive response to the rising public and regulatory pressures, and how this has morphed into the more proactive stance taken by leaders in the chemical industry today in managing environmental, social, and economic issues. The history also illustrates how addressing sustainability issues helps business to better manage risks and capture opportunities for new markets and innovations.

2.1
The Rise of Public Pressure

At the birth of the modern chemical industry some 200 years ago with the beginning of mass production of chemicals such as acids and bleaching powder, what we now consider as "sustainability issues" were hardly on anyone's mind. Natural resources were thought to be plentiful, the environment was for industries to exploit, and workers' and community safety was little more than an afterthought. This mind-set had stayed for most of those 200 years. While one can point to measures taken by some early chemical manufacturers that benefited the environment or safety, such as reducing products released to rivers or in the workplace, those examples are few and far between.

Through its history, the chemical industry has certainly made important contributions to society. Fertilizers and other agricultural chemicals increase crop production, synthetic polymers make various new industrial products possible, and mass production of medicines saves lives – just to name a few. Nevertheless, the advancement of the chemical industry was accompanied with growing environmental concerns. Early examples include documented cases of water pollution from chemical plants in the early twentieth century, which led to the 1935 listing of the chemical industry as among the most polluting industries in the United States by the country's National Resources Committee (Geiser, 2005). Yet, the state and federal governments in the United States remained slow in enacting environmental or public health policies in spite of the growing concerns.

2.1.1
The Environmental Movement

Many would point to the 1960s as the turnaround, with the chemical industry becoming a primary target of a growing environmental movement (Hoffman, 1999). Many attributed the rise of the public pressure on the chemical industry to the publication of the book *Silent Spring* by Rachel Carson (1962) and the controversy that followed (e.g., Gottlieb, 1993). *Silent Spring* meticulously presented the adverse environmental effects from the indiscriminate use of the chemical pesticide DDT and became an immediate best seller in the United States. Beyond questioning the safety of synthetic pesticides, the book brought up concerns on the widespread use of synthetic chemicals without fully understanding their potential impacts to the environment and human health. The discovery of 5 million dead fish in the lower Mississippi River later that year, which was attributed to the insecticide endrin, further exacerbated the public concern. Pesticide manufacturers and others in the chemical industry reacted strongly and negatively to the book and the public concern (Natural Resources Defense Council, 1997). The reaction, however, appeared to have largely backfired and further elevated the issues to a high-profile national discourse on the potential environmental and public safety impacts of synthetic chemicals.

The rising public pressure associated with the environmental movement of the 1960s resulted in the many new environmental bills brought to the floor of the US Congress. The *National Environmental Policy Act* (NEPA) was passed by the US Congress in 1969, and signed by President Nixon on 1 January, 1970. The United States Environment Protection Agency (USEPA) was formed shortly after. It was followed by the proliferation of environmental regulations passed by the US Congress. The *Clean Air Act, Occupational Safety and Health Act, Clean Water Act, Safe Drinking Water Act, Consumer Product Safety Act, Resource Conservation and Recovery Act*, and *Toxic Substances Control Act* were all passed between 1970 and 1976, often with strong bipartisan support in the US Congress.

Many European countries and Japan enacted similar regulations during the same period (Desai, 2002). These regulations affected the chemical industry as well as many other industries. Among these regulations, the *Toxic Substances Control Act* and the similar 1979 Sixth Amendment to the Dangerous Substances Directive of the European Community were particularly directed to the chemical industry. These regulations address the intrinsic hazards of chemical products and provide government agencies with the authority to demand health and safety data on chemical products and restrict the use of chemical substances so as to reduce "unreasonable risks" to the public and the environment (Geiser, 2005).

2.1.2
A Problem of Public Trust

A series of industrial incidents and controversies in the late 1970s and early 1980s further elevated the public awareness on the environmental and public safety risks posed by industries in general. These include the Amoco Cadiz oil spill off the coast of Brittany, France, in 1978 and the Three Mile Island nuclear incident in Pennsylvania, United States, in 1979. Three incidents and controversies involving chemical products and processes particularly stood out in their impact on the public perception of the chemical industry: a train derailment in Canada, the Bhopal chemical disaster in India, and the Love Canal controversy in the United States.

The train derailment incident occurred in December 1979 in Mississauga, a major business and residential suburb of Toronto, Canada. While the chemical industry was not directly responsible, the transportation incident drew additional public attention on the environmental and societal impacts of chemical products and the industry that makes them. The train derailment resulted in the rupture of several tankers carrying chlorine, propane, styrene, toluene, and caustic soda and a fireball explosion that rose to a height of 1500 m visible 100 km away (City of Mississauga, undated). Because of concern of a possible spread of toxic chlorine gas cloud, 218 000 residents were evacuated, making it the largest peacetime evacuation in North America at the time.

The Union Carbide incident in Bhopal, India, ignited even greater global public controversy due to its massive impact. Just after midnight on 3 December, 1984, water contamination of a tank of methyl isocyanate (MIC) in Bhopal, India, initiated a series of events that led to a catastrophic toxic release, killing more than 3000 residents and injuring over 100 000. According to Indian Government estimates, the incident resulted in an immediate death toll of over 2500 people. More recent government estimates puts the long-term mortality at of 14 400 people and permanent disabilities to about 50 000 people due to exposure to the MIC toxic cloud (Lapierre and Moro, 2001). Other independent estimates put the figures higher. However, for sure, the Bhopal disaster constituted one of the worst industrial disasters of all time.

Along with these high-profile incidents, other controversies related to chronic chemical exposure also posed problems to the chemical industry. Most infamous among these is the Love Canal controversy toward the end of the 1970s. Residents of the Love Canal neighborhood of Niagara Falls, New York, were found to have unusually high rates of miscarriages and birth defects as well as toxin content in breast milk, which were attributable to the long-term exposure to hazardous chemicals released from a nearby decades-old chemical waste dump. The Love Canal controversy led to the passage of the 1980 *Comprehensive Environmental Response Compensation and Liability Act* (CERCLA, or the "Superfund" Act) in the United States. Among others, the "Superfund" Act assigns liability for the release of hazardous chemicals from a waste site and provides a trust fund for the cleanup of contaminated areas when no responsible party can be identified.

This series of incidents and controversies resulted in the lack of public trust in the chemical industry especially in the United States and other developed economies. Lingering in the public mind were questions about the safety of chemical products and operations, and more importantly, on whether the chemical industry is providing the public with an accurate and complete picture on the risks associated with its products and operations. These concerns on transparency have stayed (SustainAbility, 2004) and are associated with the declining public opinion on the chemical industry in the United States and Europe from 1970 to the 1990s (Boswell, 2001; Milmo, 2001).

2.2
Industry Responded

As environmental regulatory framework developed in the 1970s, the chemical industry in the United States and in most other developed economies began to institute internal processes to assure compliance with the new environmental laws and regulations. In fact, a survey of chemical industry literature of that period indicated a high degree of confidence in the industry that all the regulatory requirements could be met through technological innovation (Hoffman, 1999).

However, following the chemical incidents of the late 1970s and early 1980s, many leaders in the industry realized that regulatory compliance and technology were not sufficient to address the increasing public pressure. Additional voluntary programs were necessary to re-earn public trust and protect the industry's societal license to operate. These include industry-wide and corporate efforts engaged by members of the industry and their partners, as described below.

2.2.1
The Responsible Care Program

Following the Mississauga train derailment incident, the chemical industry in Canada faced tremendous increase in public and regulatory pressure that threatened the survival of the industry. In the words of Jean Bélanger, president of the Canadian Chemical Producers' Association (CCPA), by the early 1980s the chemical industry "risked losing its public license to produce in Canada." Bélanger and other chemical industry leaders in CCPA intuitively understood that the issue was that of credibility and public trust. This was later confirmed by a series of polls that showed the prevailing public perception that the chemical industry knew about the risks associated with its products, but did not care to share the information with the public (Bélanger, 2005).

By 1985, a simple one-page "Statement of Policy on Responsible Care" that was originally prepared by CCPA in 1979 had evolved into a full-fledged Responsible Care® program. The program was intended to gain the trust of the public in communities near chemical plants and throughout Canada. It was developed on three fundamental principles that were deemed necessary to earn the public trust (Bélanger, 2005):

- **Doing the right thing** – to do what is right and ethical even when it is not required by the regulations, including accurately presenting the risks of the industry's products and operations;
- **Caring about the products from cradle to grave** – recognizing that the industry's responsibilities do not stop at the plant gate, but extend to the products' use and disposal, including working with supply chain partners and consumers on the proper handling, use, and disposal of the products;
- **Being open and responsive to public concerns** – being transparent and accountable not only to the public but also among industry peers, as a problem with one member company or operation could damage the credibility of the chemical industry as a whole.

The Responsible Care program includes a set of codes, verification processes, visible performance measurement, and advisory panels. The community advisory panels (CAPs) were probably the most revolutionary element of the program at the time. The CAPs establish a dialogue channel between chemical companies and the communities surrounding their operations, and enable community members to express their concerns and work with the chemical companies to maximize the benefits to both the chemical companies and the communities (Hook, 1996).

Following the Bhopal incident of 1984, the value of the Responsible Care program became clear also to US chemical manufacturers. In 1988, the Chemical Manufacturers Association in the United States (now the American Chemistry Council) formally adopted the Responsible Care program and principles.

Today, the Responsible Care program and principles have been adopted and implemented in 60 countries and regions throughout the world. Not all chemical companies are part of these national/regional chemical trade groups that have adopted Responsible Care. Most small chemical producers and a few large ones are notably absent from this commitment. However, in terms of production volume, the companies committed to Responsible Care represent about 90% of global chemical production.

In 2003, the global chemical industry acting through the International Council of Chemical Associations (ICCA) undertook a strategic reexamination of the Responsible Care program, which resulted in the new Responsible Care Global Charter document (ICCA, International Council of Chemical Associations, 2004). The Responsible Care Global Charter further extended the scope of the program to focus on new challenges to the chemical industry, including the growing public dialogue over sustainable development; public health issues related to the use of chemical products; and the need for greater industry transparency (Yosie, 2005).

The Responsible Care program provides an industry-wide platform for managing environmental, health, and safety (EH&S) issues that are implemented by individual chemical companies. However, many chemical companies have also applied additional processes internally to respond to various sustainability issues, including technology development and strategic processes discussed below.

2.2.2
Technology Development

As mentioned earlier, during the proliferation of environmental regulations in the 1970s, chemical companies had emphasized technological solutions to EH&S issues. At the time, the emphasis was undoubtedly on "end-of-pipe" solutions, that is, control technologies to treat waste and pollutants after they are generated in order to comply with the regulatory requirements. However, even in these early years, some in the chemical industry had started to think beyond end-of-pipe control technologies to pollution prevention.

Pollution prevention is a technological approach focused on preventing pollution at the source (as opposed to end of pipe), by modifying the design of the product or processes. A pioneer in this area is 3M Corporation, which launched an aptly named "Pollution Prevention Pays" (or 3P) program in 1975. The program aims to remove pollutants at the source through product reformulation, process modification, equipment redesign, and recycling and reuse of waste materials. In addition, as the program's name implies, pollution prevention certainly pays. By 2010, the program has saved the company more than US$1.2 billion, while only accounting for savings in the first year of each pollution prevention measure; that is, the actual savings through time from the reductions in raw material requirements, energy consumption, and pollution control expenses are likely much higher.

These early pollution prevention efforts led to the industry's long-term focus on eco-efficiency, that is, generation of more economic value while reducing natural resource consumption and environmental impacts. Until the early 2000s, efforts to develop sustainability metrics in the chemical industry remained focused on defining measures of eco-efficiency (Schwarz et al., 2002; Institution of Chemical Engineers, IChemE, 2002; Saling et al., 2002; Tanzil and Beloff, 2006). In general, these metrics measure environmental burden (e.g., energy use in primary fuel equivalents, global warming potential, and toxicity potential) associated with each unit produced or economic value generated by a chemical operation. The goal was to reduce these metrics through the adoption of eco-efficient technologies or through better housekeeping (e.g., preventing leaks). These and similar eco-efficiency metrics have been widely used both in technology development and for corporate management.

In the mean time, various other sustainable design and technological development concepts have also emerged. They include, most notably,

- life-cycle assessment (LCA) and life-cycle design, which extend the eco-efficiency concept beyond the gates of the chemical plant to incorporate impacts from other stages of the product or material life cycle, including resource extraction, transportation, product use, and disposal (Keoleian and Menerey, 1993; Saling et al., 2002);
- green chemistry, which focuses on the design of chemical products and processes that reduce or eliminate the use and generation of hazardous substances as well as other environmental impacts and hazards (Anastas and Warner, 1998;

- green engineering, which focuses on the design of chemical processes that maximize economic objective while minimizing pollution and risk to human health and the environment (Allen and Shonnard, 2002; Nguyen and Abraham, 2003);
- industrial ecology, which seeks to optimize use of material and energy resources by studying the interactions between industrial entities, for example, identifying waste streams that can be used as raw material or energy source by other industrial plants (Graedel and Allenby, 2003); and
- inherently safer chemical processes, which seek to reduce or eliminate process hazards through material substitution, alternative reaction routes, and process intensification and simplification (Hendershot, 2004).

These design and technology development concepts have been adopted by the chemical industry to various extents. The "Twelve Principles of Green Chemistry" (Table 2.1) is arguably the most recognized set of principles. The USEPA's Presidential Green Chemistry Challenge Awards evaluate new chemical products and processes on criteria based on these principles. For example, Dow Chemical Company and BASF were among the recipients of the Awards in 2010 for their joint development and commercialization of a new environmentally friendly synthetic pathway for the production of propylene oxide, a high-volume building-block chemical. The advantages of this technology compared to other synthetic pathways were demonstrated using BASF's life-cycle-based eco-efficiency analysis tool. Other recent award recipients include Procter and Gamble (P&G) and Cook Composites and Polymer Company in 2010 for a new high-performance low-volatile organic compound (VOC) paint; and the insecticide producer Clarke in 2011 for the development and commercialization of an environmentally safe larvicide that is produced through a solventless process. As we recall the controversies involving chemical products and operations in the 1960s through the 1980s discussed earlier, the chemical industry today has come a long way with these innovations.

Table 2.1 Twelve principles of green chemistry.

1	Prevent waste
2	Maximize the integration of all process materials into the finished product
3	Use and generate substances with little or no toxicity
4	Design chemical products with less toxicity while preserving the desired functions
5	Minimize auxiliary substances (e.g., solvents, separation agents)
6	Minimize energy inputs
7	Prefer renewable feedstock over nonrenewable ones
8	Avoid unnecessary derivations and minimize synthesis steps
9	Prefer selective catalytic reagents over stoichiometric reagents
10	Design products for post-use decomposition and no persistence in the environment
11	Use in-process monitoring and control to prevent formation of hazardous substances
12	Use inherently safer chemistry that minimizes the potential for accidents

Source: Anastas and Warner (1998).

2.2.3
Corporate Sustainability Strategies

Unlike in many other industries, the management of sustainability issues in the chemical industry was typically never relegated to one corner of a company's EH&S office. With the high-profile public pressure, many of the sustainability issues have long been front and center to the senior management and executives of chemical companies. This is true especially for larger, research-driven chemical companies, where technology development is an important piece of the sustainability puzzle.

Large US chemical manufacturers were among the first in formulating public corporate-level response to sustainability. In 1989, DuPont announced its first sustainability goals. This came after the company was faced with tremendous public pressure in the 1970s as the world's largest producer of chlorofluorocarbons (CFCs), the compound blamed for the destruction of the earth's protective ozone layer. DuPont decided to lead the industry's turnaround in phasing out CFCs and developing alternatives, which earned DuPont the position to work on the issue with different stakeholder groups (Tanzil et al., 2005). By 1989, following the Bhopal incident, Edgar Woolard, DuPont CEO at the time, decided that the company had to align itself to where society wants it to be. He began a series of public conversations on "corporate environmentalism" and publicly committed the company to a number of sustainability goals – including 90% reduction in the emissions of air carcinogens and 70% reduction in air toxins. A target for greenhouse gas (GHG) reduction was added in 1993. These were among the first public environmental improvement targets in industries and helped earn DuPont the reputation as a pioneer in sustainability. In 1997, Chad Holliday, the new CEO, further revolutionized the company with its "sustainable growth" program – redirecting the company's growth strategy to high-value, high-technology areas that involve less waste and emissions. Integration of sustainability into the business also became a key focus, with sustainability measures integrated into the company decision processes and individual performance assessment metrics.

In 1992, Dow Chemical Company, which is the largest chemical manufacturer in the United States, established a Sustainability External Advisory Council (SEAC). The SEAC involves representatives from NGOs, academia, businesses, and government to advise the company's leadership on EH&S, and other sustainability issues. At the time of its formation, the SEAC was considered the first of its kind in industry and provided a stakeholder engagement venue at the executive level, complementary to the CAPs at the grass-root operational level. The SEAC advised Dow in the development of its 2005 EH&S Goals, which were announced in 1996 and contain a set of aggressive and specific public goals to improve the company's EH&S management and performance. These aggressive public goals led to a company-wide effort to achieve them. Dow regularly updated the public on its progress on the 2005 EH&S Goals through its annual EH&S report and other communication channels. In the end, in 2005, Dow came close to meeting its aggressive targets in most areas, fell quite short in some (mainly in supply chain safety), and performed better than targets in some others (mainly in reducing emissions). Nevertheless, Dow was able to establish a reputation as a sustainability leader through its transparency and continual public update on its effort to meet its aggressive targets.

These are examples from two large US chemical manufacturers. These two companies have since established new sets of public goals, which are discussed below. Most large chemical companies around the world today have also established public sustainability goals. These public goals force transparency and a coordinated effort from the whole company to meet them.

2.3
An Evolving Framework

The chemical industry's response to sustainability changes with the evolving global framework on sustainability. Much of these changes were driven by the evolution in the range of sustainability issues and the risks and opportunities associated with them, as well as by the level of public and corporate awareness on these issues.

2.3.1
New Issues and Regulations

While the public pressure of the 1960s, 1970s, and 1980s were much more focused on pollution and the safety of chemical operations, the range of sustainability issues today are a lot broader. They encompass workplace and social issues as well as other environmental issues. Foremost among these is the issue of climate change. Despite the recent legislative problems with climate change regulations in the United States and other countries, the issue is receiving increasing public attention. There is a rising public demand for corporate response to the issue of climate change from the chemical industry as well as from other industries, which are often customers of the chemical industry. The high and volatile costs of energy also make energy and GHG an increasingly important issue to the chemical industry.

Changes in regulations also contribute to the evolving sustainability framework. Recent environmental directives from the European Union forced more life-cycle thinking and greater transparency from the industry. These include

- the Restriction of Hazardous Substances (RoHS) Directive, which restricts the use of lead and other hazardous materials in electronic equipment;
- the Water Framework Directive, which aims at improving the aquatic environment, including a requirement for the cessation or phasing out of discharges, emissions, and losses of a set of high-priority chemicals within 20 years;
- the Registration, Evaluation, Authorization, and Restriction of Chemicals (REACH) Directive, which places greater responsibility on the industry to protect the health and safety of the public, including the requirement to provide information on the risk and safety of chemical products.

These European directives, especially REACH, has far-reaching impacts in the industry as they affect not only European companies but also other chemical companies that market their products in Europe.

2.3.2
Sustainability as an Opportunity

Furthermore, global sustainability issues also provide new market opportunities to the chemical industry. For example, the new sustainability goals of both DuPont and Dow no longer focused only on reducing impacts, but also on increasing societal benefits from their business. DuPont's 2015 Sustainability Goals, announced in 2006, include not only targets for continual reduction of the company's environmental footprint, but also a new set of market-facing goals, including goals to double revenue from products that improves energy efficiency, reduce GHG emissions, or protect safety for its customers, as well as increasing revenue from products made from nondepletable resources.

Similarly, Dow 2015 Sustainability Goals, also announced in 2006, include goals to enhance the benefits of its products. It includes the goal to increase the percentage of products that exhibit the advantages of sustainable chemistry, as well as "actively working toward, and committed to achieving, at least three breakthroughs by 2015 that will significantly help solve world challenges," such as energy and climate, access to clean water in the developing world, food, housing, and health.

European chemical manufacturers are also increasingly taking more of a product life-cycle perspective, including enhancing the sustainability benefits of their products. In 2008, BASF published a corporate carbon footprint that included emissions from their operations as well as from other life-cycle stages, including resource extraction, customer use, and disposal. It has a sustainability management goal of "create business opportunities, and minimize risk," recognizing the opportunities to develop products that support energy efficiency, renewable energy development, and other sustainability goals, as well as internal efforts and external services to reduce environmental and social risks to itself and its customers.

2.3.3
Recent Industry Trends

Along with the increase in public awareness on sustainability issues, the breadth of the industry's response to sustainability has also increased both in range of issues being managed and the number of companies proactively managing them.

An annual benchmarking study by the Institute for Sustainability at the American Institute of Chemical Engineers (AIChE) illustrates some of the recent changes in the industry. Since 2007, the Institute for Sustainability has produced an annual AIChE Sustainability Index™, which assesses the sustainability of the chemical industry along seven factors: strategic commitment, environmental performance, safety performance, product stewardship, social responsibility, innovation, and value chain management (Cobb *et al.*, 2007, 2009). Currently, the AIChE Sustainability Index™ assessment is focused on 10 large multinational chemical companies operating in the United States. Recent analysis of the last few years of data revealed the following trends:

- An increasing number of companies are publishing sustainability reports. Although not all of the large chemical companies have public sustainability goals, they all have formal internal programs to manage sustainability issues.
- Notable performance improvements were observed among the 10 large chemical companies in energy efficiency and process safety. These may reflect the increasing role of energy cost as a driver for sustainability and the Responsible Care's new process safety program requirements.
- Almost all of the 10 large chemical companies have implemented comprehensive product stewardship and risk communication programs in recent years. These can be attributed to the implementation of the Responsible Care product stewardship requirements as well as the implementation of REACH reporting system in the companies.
- An increasing number of companies have implemented environmental and social criteria for their suppliers, as well as audit programs, as part of their supplier management programs.

Not all chemical companies have developed mature and comprehensive sustainability programs. Larger companies tend to lead, perhaps because they have greater internal resources and because they are more exposed to reputational risks due to their sizes. Nevertheless, a survey of chemical industry executives by Accenture (2010), as well as various other recent surveys, shows an increasing strategic emphasis on the management of sustainability issues in the chemical and other industries. The Accenture survey also revealed the increasing role of customer demand for sustainability in influencing the chemical companies' sustainability programs. The public pressure on the chemical industry appears to have somewhat morphed into greater collaboration, as the chemical industry is increasingly involved in partnerships with NGOs, academia, governments, and supply chain partners.

The public reputation of the chemical industry, too, appears to have improved. The declining public favorability ratings for the chemical industry in both the United States and Europe have bottomed out in the 1990s (Boswell, 2001; Milmo, 2001). More importantly, those studies indicate that the industry's favorability ratings are significantly higher among communities near chemical plants and among people who are more familiar with the chemical industry. In 2004, for the first time, the favorability rating for the chemical industry in Europe exceeds its unfavorability rating. Since then, the chemical industry's favorability rating in Europe has stayed just below 50% CEFIC (The European Chemical Industry Council, (2011). To be sure, the chemical industry still has many challenges to overcome. However, in terms of how the industry is viewed by the public, the industry today is a far cry from where it was a few decades ago.

2.4
Conclusions: the Sustainability Drivers

Protecting the chemical industry's reputation and social license to operate has been the initial force driving the industry's environmental and sustainability efforts. From the Responsible Care program to the various corporate sustainability initiatives, the industry's initial responses were largely shaped by high-profile incidents and controversies involving chemical products and processes, and the resulting negative public opinion.

Throughout the large few decades, however, the chemical industry's response has been increasingly proactive and wide ranging, reflecting the increasing awareness, and evolution of sustainability issues among the general public. Protecting reputation and social license to operate certainly remains a key driver for sustainability in the industry. It requires careful consideration and management of process safety as well as the safety, environmental, and social impacts of chemical products and processes throughout the cradle-to-grave life cycle. As history taught us, the recent gains in the industry's reputation and public goodwill can still evaporate with one or two high-profile incidents or controversies.

Aside from the management of reputational risks, other drivers for sustainability in the chemical industry have also emerged. They include

- cost reduction, starting from the early pollution prevention efforts to today's increased emphasis on resource efficiency due to the high cost and price volatility of energy, raw material, and water resources;
- innovation, which is increasingly driven by the customers' demand for safe, low-emission, and resource-efficient products;
- new markets in products that address societal concern on climate change, clean water availability, and other sustainability issues;
- partnership opportunities with communities, NGOs, governments, and the supply chain.

While the reputational risks affect larger entities more than the smaller companies, the cost reduction, innovation, new market, and, to some extent, partnership drivers present opportunities to all entities in the chemical industry and its supply chain. Thus, as sustainability evolves from a risk issue to an issue involving both risks and opportunities, one can expect that the industry's response to sustainability will continue to increase in both depth and breadth.

References

Accenture (2010) Sustainability Strategies for High Performance in the Chemicals Industry, http://www.accenture.com/SiteCollectionDocuments/PDF/Accenture_Chemicals_POV_Sustanbility_v1.pdf.

Allen, D.T. and Shonnard, D.R. (eds) (2002) *Green Engineering: Environmentally Conscious Design of Chemical Processes*, Prentice Hall, Upper Saddle River, NJ.

Anastas, P.T. and Warner, J.C. (1998) *Green Chemistry: Theory and Practice*, Oxford University Press, New York.

Bélanger, J.M. (2005) Responsible care in Canada: the evolution of an ethic and a commitment. *Chem. Int.*, **27** (2), 4–9, http://www.iupac.org/publications/ci/2005/2702/1_belanger.html. (accessed December 2012).

Boswell, C. (2001) Public perception and the chemical industry: the US perspective. *Chem. Market Rep.*, **260** (3), 13–15.

Carson, R. (1962) *Silent Spring*, Houghton Mifflin Company.

CEFIC (The European Chemical Industry Council) (2011) Facts and Figures 2011: The European Chemical Industry in a Worldwide Perspective, http://www.cefic.org/Facts-and-Figures/.

City of Mississauga (undated) Mississauga Train Derailment (1979), http://www.mississauga.ca/portal/home?paf_gear_id=9700018&itemId=5500001.

Cobb, C., Schuster, D., Beloff, B., and Tanzil, D. (2007) Benchmarking sustainability. *Chem. Eng. Prog.*, **103** (6), 38–42.

Cobb, C., Schuster, D., Beloff, B., and Tanzil, D. (2009) The AIChE sustainability index: the factors in detail. *Chem. Eng. Prog.*, **105** (1), 60–63.

Desai, U. (2002) *Environmental Politics and Policy in Industrialized Countries*, MIT Press.

Hoffman, A.J. (1999) Institutional evolution and change: environmentalism and the U.S. Chemical Industry. *Acad. Manage. J.*, **42** (4), 351–371.

Hook, G.E.R. (1996) Editorial: responsible care and credibility. *Environ. Health Perspect.*, **104** (11), 1138–1139.

Geiser, K. (2005) Limits of risk management and the new chemicals policies. In B. Beloff, M. Lines and D. Tanzil (eds). *Transforming Sustainability Strategy into Action: The Chemical Industry*. John Wiley & Sons, Inc.

Gottlieb, R. (1993) *Forcing the Spring: The Transformation of The American Environmental Movement*, Island Press.

Graedel, T.E. and Allenby, B.R. (2003) *Industrial Ecology*, Prentice Hall, Englewood Cliffs, NJ.

Hendershot, D.C. (2004) A new spin on safety. *Chem. Process.*, **67** (5), 16–23.

ICCA (International Council of Chemical Associations) (2004) Responsible Care® Global Charter.

Institution of Chemical Engineers (IChemE) (2002) The Sustainability Metrics: Sustainable Development Progress Metrics Recommended for Use in the Process Industries, Institution of Chemical Engineers, Rugby.

Keoleian, G.A. and Menerey, D. (1993) Life Cycle Design Guidance Manual: Environmental Requirements and the Product System, EPA 600/R-92/226, Risk Reduction Engineering Laboratory, Office of Research and Development, US Environmental Protection Agency, Cincinnati, OH.

Lapierre, D. and Moro, J. (2001) *It was Five Past Midnight in Bhopal*, Full Circle Publishing, New Delhi.

Milmo, S. (2001) Public perception and the chemical industry: the european perspective. *Chem. Market Rep.*, **260** (3), 11–12.

Natural Resources Defense Council (1997) The Story of Silent Spring: How a Courageous Woman Took on the Chemical Industry and Raised Important Questions about Humankind's Impact on Nature, http://www.nrdc.org/health/pesticides/hcarson.asp.

Nguyen, N. and Abraham, M.A. (2003) 'Green engineering: defining the principles' — results from the Sandestin conference. *Environ. Prog.*, **22** (4), 233–236.

Saling, P., Kicherer, A., Dittrich-Krämer, B., Wittlinger, R., Zombik, W., Schmidt, I. et al. (2002) Eco-efficiency analysis by BASF: the method. *Int. J. Life-Cycle Assess.*, **7** (4), 203–218.

Schwarz, J., Beaver, E., and Beloff, B. (2002) Use sustainability metrics to guide decision-making. *Chem. Eng. Prog.*, **98** (7), 58–63.

SustainAbility (2004) External Stakeholder Survey: Final Report for the Global Strategic Review of Responsible Care®.

Tanzil, D. and Beloff, B.R. (2006) Assessing impacts: overview on sustainability indicators and metrics. *Environ. Qual. Manage.*, **15** (4), 41–56.

Tanzil, D., Rittenhouse, D. and Beloff, B. (2005) DuPont: growing sustainably. In B. Beloff, M. Lines and D. Tanzil (eds).

Transforming Sustainability Strategy into Action: The Chemical Industry. John Wiley & Sons, Inc.

Yosie, T.F (2005). Taking performance to a new level: the responsible Care® Global Charter. In B. Beloff, M. Lines and D. Tanzil (eds). *Transforming Sustainability Strategy into Action: The Chemical Industry*. John Wiley & Sons, Inc.

3
From Industrial to Sustainable Chemistry, a Policy Perspective

Karl Vrancken and Frank Nevens

3.1
Introduction

With the surge of environmental awareness, industries (including chemistry) have been facing an increasing amount of environmental legislation. The development of this legislation has been following scientific insights and public concerns. New developments have been based on scientific analysis and modeling, as well as on public awareness campaigns (Popp, Hafner, and Johnstone, 2011). Science has been shown to play a pivotal role in assessing environmental risks and problems and in proposing relevant and effective solutions for them (Sundqvist, Letell, and Lidskog, 2002). The interactions among scientists, nongovernmental organizations (representing the citizens), industry, and policy makers have been a solid basis for developing environmental legislation. Furthermore the progressive development of environmental policy has been guided by, but has also guided in itself, the development of new technological solutions. This is exemplified by the Sevilla Process on Integrated Pollution Prevention and Control (IPPC), which also had a strong influence on the chemical industry. With the shifting focus from pollution control to integrated prevention and resource efficiency, a strong role remains to be played by all stakeholders involved. The policy approach also proves to be shifting from "government" to "governance," which is characterized by an increasing involvement of the private sector and civil society in environmental policy making (Cocklin, 2009). The widening of environmental problems – from local water quality to global climate change, from local soil pollution to worldwide resource efficiency – calls for a more systemic, inter- and transdisciplinary approach. A framework for such an approach is provided by "transition management" (Rotmans *et al.*, 2000; Loorbach, 2007).

Management Principles of Sustainable Industrial Chemistry: Theories, Concepts and Industrial Examples for Achieving Sustainable Chemical Products and Processes from a Non-Technological Viewpoint,
First Edition. Edited by Genserik L.L. Reniers, Kenneth Sörensen, and Karl Vrancken.
© 2013 Wiley-VCH Verlag GmbH & Co. KGaA. Published 2013 by Wiley-VCH Verlag GmbH & Co. KGaA.

3.2
Integrated Pollution Prevention and Control

3.2.1
Environmental Policy for Industrial Emissions

In the 1970–1980s, the effects of water and air pollution became clearly observable and tangible, with for example, the loss of biological life in rivers and the effects of acid rain on the quality of forests. Problems of that kind showed that environmental impacts of (industrial) activities did not halt at national borders and, in many cases, produced long-range effects. The overall awareness of these effects resulted in multiple science–policy collaborations, stakeholder discussions, and transnational initiatives, such as the Convention on Long-Range Transboundary Air Pollution (CLRTAP) (Sundqvist, Letell, and Lidskog, 2002). On a European level, the European Commission initiated specific policy measures, eventually resulting in the Framework Directives for Water and Waste, which established a general framework for minimizing loss of environmental quality (European Commission, 2000, 2008a). On a local or national level, policy initiatives were largely compartmented. From an industry perspective, this resulted in separate legislation and permitting for the fields of water, air, waste, and soil. Such fragmented approaches were given up with the introduction of the IPCC Directive (IPPC Directive, 96/61/EC).[1] This Directive specified its aim to "protect the environment as a whole" by making sure that minimizing effects in one environmental compartment would not result in damage to another. A specific result of this Directive was the introduction of a single integrated permit for certain categories of industrial installations, indicating the conditions for all types of environmental impact. The Directive further specified that reduction of environmental effects should not only be based on end-of-pipe solutions, but also on preventive, process-integrated measures. After a recasting process, the IPPC Directive merged with five other directives into the Industrial Emissions Directive (IED, 2010/75/EC) (European Commission, 2010).

The IPPC (European Commission, 2008b) and IED Directives both oblige operators and regulators to take an integrated, overall view of the potential of the installation to consume and pollute. The overall aim of such an integrated approach is to improve the design, construction, management/control, and even decommissioning of industrial processes so as to ensure a high level of protection for the environment as a whole. The central general principle of the IED Directive (given in Article 3) states that operators should take all appropriate preventative measures against pollution, particularly by application of the best available techniques (BAT) enabling them to improve their environmental performance (EIPPCB, 2010).

---

1) The IPPC Directive initially had the code 96/61/EC, but was later recodified and received a new code 2008/1/EC, after a revision process, when the IPPC Directive merged with five other directives into the Industrial Emission Directive (IED, 2010/75/EC).

3.2.2
Best Available Techniques and BREFs

The term "best available techniques" is (as defined in Article 3(10) of the IED) is "the most effective and advanced stage in the development of activities and their methods of operation which indicate the practical suitability of particular techniques for providing the basis for emission limit values and other permit conditions designed to prevent and, where that is not practicable, to reduce emissions and the impact on the environment as a whole." Article 3(10) goes on to clarify this definition as follows:

- "Techniques" indicate both the technology used and the way in which the installation is designed, built, maintained, operated, and decommissioned.
- "Available" techniques mean that the techniques are developed on a scale that allows implementation in the relevant industrial sector, under economically and technically viable conditions, taking into account the costs and advantages that are reasonably accessible to the operator.
- "Best" means most effective in achieving a high level of protection of the environment as a whole.

BAT thus corresponds to technologies and organizational measures with minimum environmental impact and acceptable cost (Dijkmans, 2000). The concept of BAT was operationalized through a stakeholder process that established an extensive exchange of information between Member States, the European Commission, nongovernmental organizations, and the industries concerned. This process is also known as the "Sevilla Process," (Schoenberger, 2009) referring to the place where most of the meetings take place, and where the central EU coordination desk resides (the European Integrated Pollution Prevention and Control Bureau, EIPPCB). A major outcome of the information exchange is the Best Available Technique Reference Documents (BREFs) (EIPPCB, 2012).

The concept of BAT is based on the idea of identifying the front-runner initiatives with regard to environmental protection in specific industrial fields (process technology, effluent treatment, and operational practices) and to check whether their practice can be generalized over the sector concerned (i.e., to find out whether, and under what conditions, the techniques are available). This "check" is an interactive process based on existing information, with involvement of and discussion with all relevant stakeholders. These stakeholders, from industry (sector organizations, suppliers), NGOs (European Environment Bureau), Member States (permit writing authorities), and the European Commission are gathered in the Technical Working Group (TWG). Once a technique is considered BAT, it becomes a benchmark for permit requirements in all EU Member States, which increases the urge for its broader transfer. A consequence of the practice of BAT determination is a tension in the stakeholder process whether to consider "front-runner" techniques or merely look at techniques that already are "common practice"; the ultimate level of performance achieved by BAT will depend on the way the stakeholder process is set up, managed, and concluded. This aspect has sparked some discussions on whether BAT can be seen as a driver for innovation (Ganzleben, 2000; Gislev, 2000). The extent to which this potential bias is actually the case differs between sectors and is strongly prone to interpretation. Nevertheless, it is clear that the IPPC has been a driver for enhanced environmental performance. It has raised the level of environmental protection in industry and resulted in a common high level of technology introduction in European. Projects on the introduction of IPPC in "new" EU countries showed that this regulatory framework results in a step-change introduction of environmental technologies in the industry.

The BAT approach does not remain limited to the European territory and to large-scale industrial plants. In Belgium, BAT documents have been made for over 40 sectors, including typicals small and medium-sized entreprise (SME) sectors such as car repair shops, fuel stations, and swimming pools (Vrancken and Huybrechts, 2011; Vrancken, Vercaemst, and Dijkmans, 2004). BREFs are also used as reference documents outside Europe, for instance by the OECD, the International Finance Corporation (member of the World Bank), and UNIDO (United Nations Industrial Development Organization) (Schoenberger, 2009). The concept is also being introduced as a basis for industrial environmental policy in the Southern Mediterranean region (IAT, 2011).

3.2.3
Integrated Pollution Prevention and Control in the Chemical Sector

The BREFs give a good insight into the activities of a sector, its key environmental issues and the techniques to minimize the impact on the environment. BREFs have been written for all major industrial sectors. Because of the complexity of the chemical industries, eight BREFs have been written to cover the whole sector. Seven of them are "vertical" (covering specific activities), one is "horizontal" (treating common themes for the whole chemical sector) (EIPPCB, 2012). The following is an overview of these:

- Large Volume Organic Chemicals (LVOC)
- Chlor-Alkali Manufacturing Industry (CAK)
- Large Volume Inorganic Chemicals – Ammonia, Acids, and Fertilizers Industries (LVIC-AAF)
- Large Volume Inorganic Chemicals – Solids and Others Industry (LVIC-S)
- Production of Speciality Inorganic Chemicals (SIC)
- Manufacture of Organic Fine Chemicals (OFC)
- Production of Polymers (POL)
- Common Waste Water and Waste Gas Treatment/Management Systems in the Chemical Sector (CWW).

The BREFs give a general introduction to the size and activities of the sector under study and define key environmental issues. As the division of the chemicals sector into seven subsectors is somewhat arbitrary, there are no clear-cut demarcations between them. The description of applied techniques and processes and the selection of BAT are therefore specified for a range of cornerstone products and additional illustrative products. This scope is selected by the TWG.

3.3
From IED to Voluntary Systems

The BAT approach has provided a strong incentive for minimizing the environmental impact of industrial installations. The new IED confirms and strengthens the use and value of BAT in the environmental permitting policy. In the years to come, the BREFs will keep their strong role in the setting of standards for environmental protection (Vrancken and Huybrechts, 2011). On the other hand, it must be admitted that there is a solely technological focus. The definition of BAT clearly mentions "techniques" rather than "technologies." This means that management principles and operational measures are also included. Nevertheless, the overall focus lies in the implementation of technological measures. In 2000, Cunningham had already pointed out that BAT discussions may have a tendency to preserve the overall status quo and may lead to a technology "lock in." It was suggested that companies may be stimulated to more innovative responses by application of other tools available within the Directive (Cunningham, 2000). These include prevention measures, voluntary schemes, and management systems. More recently, it has been recognized that for some polluters and some types of pollution problems, the IED-type command-and-control regulation is still necessary, but that at the same time the approach starts to run into the law of diminishing returns (Zarker, 2008).

In recent years, voluntary approaches to pollution prevention programs have gained a global relevance. With the BAT scheme, they share the element of being based on an agreement between government and industry. As opposed to the BAT approach, the agreement often rests in a nonformal commitment from industry. They frequently include collaborative arrangements between individual businesses, industry associations, and regulatory agencies or central governments. The spectrum of voluntary approaches varies from industry-initiated and developed guidelines for pollution prevention, over codes of good practice or cleaner production principles, up to international organizational standards (e.g., ISO) (Chittock and Hughey, 2011).

On the basis of experience from Japan, Canada, the United States, Australia, and the UK, nine key features of such programs have been identified (Table 3.1).

The voluntary arrangements add an alternative approach to the command-and-control regulation, while still building on stakeholder involvement and reaching an agreement between the regulator and the industry. They provide the possibility to shift focus from technological (often end-of-pipe) solutions to prevention- and responsibility-based industrial operations, with an appeal for empowerment of the industrial operator or sector organization. As such, they provide a strong additional feature in the policy framework and form a good complement to the regulatory framework.

Table 3.1 "Best Practice" design features for voluntary pollution prevention programs.

Design feature	Overall justification
Adequate and consistent funding	Sufficient funds need to be committed to run the program for its lifetime, for credibility, maintaining industry relationships, and removal of a perceived industry barrier
Collaborative relationship with industry	Provides credibility and trust with key "players" at the implementation phase and allows focus on "at risk" industry sectors
Single sector program focus	Allows appropriate initiatives and inclusion of specific sector information and resources providing added value to the industry involved
Setting credible goals	Credible targets need to be established from current practices and agreed upon, to allow participants and authorities to evaluate progress
Info-regulation and resources available	Shows commitment to industry, provides a conduit for information dissemination (codes of practice and guidelines), allowing technological or competitive advantage to be maintained
Threat of credible enforcement	Provides a backstop for the program, maintains credibility, and motivates some participants to achieve. A transitional leniency component for some participants is required
Regular and credible monitoring	Essential for tracking performance improvements, retaining credibility, and level playing field for participants. Potential to invite a third party to maintain consistency
Visible participant benefits	Provides an incentive to prospective sites and industry sectors to participate and a marketable feature for the instigating authority
Transparent provision of program results	Provides validity for the program and all parties involved to interested or associated stakeholders.

Taken from Chittock and Hughey (2011).

From an integrative and prevention-based perspective, voluntary systems provide a step forward toward a more sustainability-based regulatory approach. Still, this route is focused on the industrial plant as a central unit or the sector organization as a main representative body. In the following paragraph, we explain why there is need for a next step.

3.4
Sustainability Challenges for Industry

3.4.1
Introduction

Environmental policy in the twenty-first century faces new challenges that go beyond air and water quality. Current key issues have a global rather than a local dimension. Several concurrent crises have either sprung up or accelerated during the last decade: climate change, biodiversity loss, fuel prices, food availability and prices, water scarcity, and, of late, the financial crisis and the economy on the edge of recession (UNEP, 2011). UNEP calls for the transition to a "green economy" that "results in improved human well-being and social equity, while significantly reducing environmental risks and ecological scarcities."

The European Commission calls for concerted action on resource efficiency in its flagship initiative. "Continuing our current patterns of resource use is not an option. In response to these changes, increasing resource efficiency will be key to securing growth and jobs for Europe. It will bring major economic opportunities, improve productivity, drive down costs, and boost competitiveness. It is necessary to develop new products and services and find new ways to reduce inputs, minimize waste, improve management of resource stocks, change consumption patterns, optimize production processes, management and business methods, and improve logistics. This will help stimulate technological innovation, boost employment in the fast developing 'green technology' sector, sustain EU trade, including by opening up new export markets, and benefit consumers through more sustainable products" (European Commission, 2011).

The transition to a green economy (e.g., by building a resource-efficient industrial system) goes beyond the borders and capacities of the individual installation/actor. It is a responsibility that surpasses the level of the individual company and goes beyond measures for process optimization as presented in the previous paragraphs.

As Peter Senge states in *The Necessary Revolution*, 'what we need is a firm switch from "business/solutions as usual" tactics to really transformative strategies that are essential for creating a flourishing and sustainable world' (Senge *et al.*, 2008). Therefore, we should implement revolutionary – not just incremental – changes in the way we live, work, move, think, learn .... Several authors call for remaking the way we make things (McDonough and Braungart, 2002), rethinking the way we think about things (De Bono, 2009). Our complex society deals with persistent problems that are deeply rooted in our structures and institutions and for which no tailor-made solutions are readily available. The necessary radical changes toward sustainable consumption and production imply a strong orientation toward system thinking and system innovation (Tukker *et al.*, 2008).

Additionally, more than being reactively triggered by a sense of urgency (push), effective solutions for sustainable development result, to a great extent, from explicit and proactive, creative visions on the future (pull) (Davoudi, 2000).

3.4.2
Policy Drivers for Sustainable Chemistry

The far-reaching challenges described above necessitate an alternative approach to policy making: an approach building on an open and participatory concept, coupling regulatory tools with incentives and stakeholder involvement. This approach is called "governance," to distinguish it from the more traditional "government" paradigm. Governance stands for the linking of actors, interests, and perspectives through improving multiactor decision making (Koppenjan and Klijn, 2004).

A workshop of the American Chemical Society (ACS) and American Institute of Chemical Engineers (AIChE) identified the following five qualities of governmental management strategies to encourage a transition to sustainability (Satterfield et al., 2009):

1) regulations that are outcome-oriented to drive innovation and sufficiently flexible to accommodate it;
2) government management strategies that are broad enough to have industry-wide impacts and ensure a level playing field, yet flexible enough to be tailored to specific industrial operations;
3) a regulatory framework that integrates local, regional, national, and international concerns to avoid conflicts in jurisdiction and allows issues to be addressed on their appropriate geographical level;
4) careful, ongoing evaluation to ensure that regulations are delivering the anticipated benefits and are not negated by unanticipated outcomes;
5) government regulations that identify and prioritize sustainability challenges by consulting a range of stakeholders.

The BAT concept and the voluntary approaches each hold all of these characteristics. Widening of these government approaches to governance is supported by the introduction of a wider concept: transition management.

3.4.3
Transition Concept

Transitions are processes of radical, structural change of society and the sociotechnical systems composing it. They encompass fundamental turns in the established structures, cultures, and modes of action. As a consequence, transitions are long-term processes (transition practice thinks in "generations"), trying to tackle complexity and insecurity in societal systems (Rotmans, 2003; Loorbach, 2007; Grin, 2010; Nevens et al., 2012). Transitions build on a multilevel perspective (Geels, 2002) that distinguishes landscape, regime, and niches.

- **Landscape**: At the landscape level, "gradients of force" are in play – these are dominant trends and evolutions from which it is difficult to deviate and which are rigid in the sense that it is difficult to change them on an individual base (e.g., globalization, climate change, aging populations). Nevertheless, these prevailing evolutions, and trends exert external pressure on the systems in place.
- **Regime**: A "regime" refers to the dominant culture, structure, and practice embodied in physical and immaterial infrastructures (e.g., roads, power grids, routines, actor networks, regulations, government and policy). Regimes are the backbone of the stability of societal systems and have a characteristic rigidity that very often prevents innovations from altering the standing structures fundamentally.

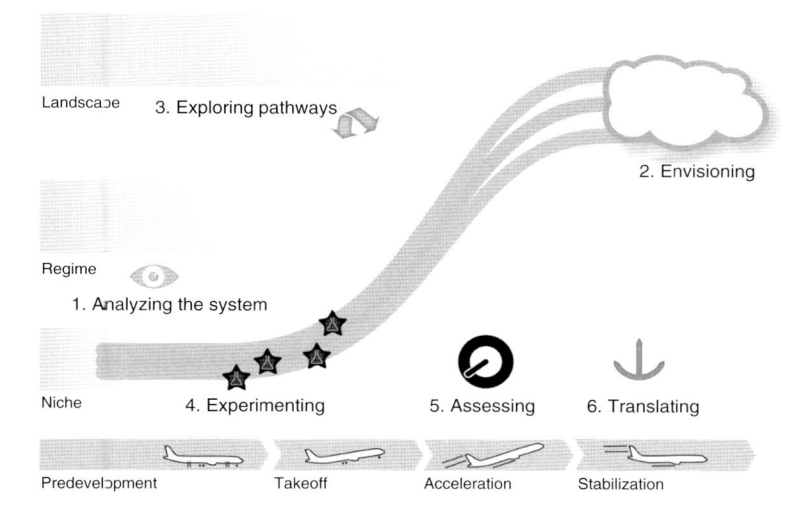

Figure 3.1 Transition management approach (Nevens et al., 2012).

- **Niche**: Niches are protected, and often little-visible, small-scale segments in society. In "niches," novelties are created and tested. These novelties can be (combinations of) new technologies, new rules and legislation, new concepts, new organizations, and so on. Niches contain incubators for transition experiments and proofs of concept of radical innovations.

Theoretically, transitions are characterized by a combination of mutually reinforcing steps/activities. This sequence fits into a concept of "logical frameworks" and envisages a (temporal) logic between different actions that combine into a consistent process of change. In the VITO transition approach, six major actions are distinguished: analyzing the system, envisioning the future, exploring pathways, experimenting, assessing the process, and translating. This is depicted in Figure 3.1.

1) "**Analyzing the system**": The first step in changing a system is getting to know it. The elements of a profound system description and outline, quantitatively as well as qualitatively, are actually determining the relevant players and their interrelations, the key system functions, the institutions and regulations, flows and barriers, and so on.
2) "**Envisioning the future**": A change trajectory toward a more sustainable society or system is mainly initiated by an appealing and inspiring vision. A vision entails a clear image/narrative of desirable systems based on shared principles of sustainable development. Inspiring visions replace "having to" by "wanting to," "reactive" by "proactive" and "creative."
3) "**Exploring pathways**": Starting from an inspiring and clear vision, different strategies to reach the desired system can be outlined. This "backcasting" exercise results in a number of strategic pathways that should be followed in order to co-establish the desired system configuration. Models/scenarios

can assess and underpin the effectiveness and feasibility of pathways and the alignment of envisaged actions.

4) **"Experimenting"**: Transition experiments are real-life developments of drastically alternative ways of working and/or thinking, fitting into envisaged new system approaches. Experiments develop in "niches," where they can develop with a certain degree of protection from ruling regimes and their institutions, legislation, power, routines, and so on, and show their effectiveness and feasibility. Successful experiments link the future vision with concrete action potential and hence can be major triggers to enable takeoff and acceleration of transition.

5) **"Assessing"**: In the course of the different trajectories toward the vision, instruments can be designed for an effective follow-up of actions that are undertaken. Products, processes, technologies, organizations, and so on, can all be the subject of different types of monitoring and assessment, examining their compliance with the diverse sustainability criteria of the new systems. Assessment-based knowledge of effectiveness, speed, and "distance to transition" can help to steer ongoing transition activities. Therefore, monitoring instruments should basically not be designed to "measure" but to trigger action and to enhance system change in a desired direction.

6) **"Translating"**: This action expresses that in order to actually initiate system change, experiences from the different typical transition activities have to be incorporated and multiplied in the actions of the relevant system stakeholders. In this manner, the lessons learned from experiments, backcasting, or scenario exercises and envisioning result in an effective dynamic process of change. In this context, the term, "stakeholders" includes governments, industry, consumers, researchers, entrepreneurs, and the more.

Transition management (Loorbach and Rotmans, 2010) was developed as a mode of governance for sustainable development. It involves the setup of a stakeholder process (the transition arena) that gets involved in the process as depicted above. Typically, the six steps will not be consecutive, but will rather develop in a congruent and interconnected way.

The application of the transition concept in the chemical industry is explored further in Chapter 4.

3.5
Conclusion

Industrial environmental policy has seen a shift in focus over the past two decades from command-and-control government to stakeholder-based governance. This change has been inspired by the widening scope of environmental problems. The initial focus on local emissions and nuisance evolved to an international regulation and integrated approaches to pollution prevention and control. The involvement of the industry, as also authorities and NGOs, has been a key factor in the success of the European permit system based on BAT. The introduction of voluntary systems called upon and allowed for a further intrinsic motivation of the industrial sectors. The new challenges for industrial policy reach far beyond the installation level. Resource efficiency and sustainability issues require alternative governance approaches to motivate and drive industrial sectors, such as the chemical industry, toward a transition to sustainability.

References

Chittock, D.G. and Hughey, K.F.D. (2011) A review of international practice in the design of voluntary pollution prevention programs. *J. Cleaner Prod.*, **19**, 542–551.

Cocklin. C. (2009) in *International Encyclopedia of Human Geography* (eds R. Kitchin and N. Thrift), Elsevier, pp. 540–545.

Cunningham, D. (2000) IPPC, BAT, and voluntary agreements. *J. Hazard. Mater.*, **78**, 105–121.

Davoudi, S. (2000) Sustainability: a new vision for the British planning system. *Planr. Perspect.*, **15**, 123–137.

De Bono, E., 2009. *Think!: Before it's Too Late: Twenty Three Reasons Why World Thinking is So Poor*. Vermillion, Ireland, 264 pp. ISBN: 0091924096.

Dijkmans, R. (2000) Methodology for selection of best available techniques (BAT) at the sector level. *J. Cleaner Prod.*, **8**, 11–21.

EIPPCB (2010) Standard texts for BREFs IEF 22-4-3. *European IPPC Bureau*, European Commission.

EIPPCB (2012) European IPPC Bureau, http://eippcb.jrc.es/reference, (accessed 2012).

European Commission (2000) Directive 2000/60/EC, establishing a framework for community action in the field of water policy. *Off. J. Eur. Comm.*, **L327/1**.

European Commission (2008a) Directive 2008/1/EC, concerning integrated pollution prevention and control. *Off. J. Eur. Comm.*, **L24/8**.

European Commission (2008b) Directive 2008/98/EC, on waste and repealing certain Directives. *Off. J. Eur. Comm.*, **L312/3**.

European Commission (2010) Directive 2010/75/EC, on industrial emissions (integrated pollution prevention and control). *Off. J. Eur. Comm.*, **L334/17**.

European Commission (2011) A Resource Efficient Europe – Flagship Initiative under the Europe 2020 Strategy, COM (2011)21, Brussels.

Ganzleben, C. (2000) IPPC a Driver for Innovation in the Pulp and Paper Industry, *citeseerx.ist.psu.edu/index*, (accessed 2012).

Geels, F.W. (2002) Technological transitions as evolutionary reconfiguration processes: a multi-level perspective and a case-study. *Res. Pol.*, **31**, 1257–1274.

Gislev, M. (2000) European innovation and information exchange about BAT. European Conference on "The Sevilla Process: A Driver for Environmental Performance in Industry", Stuttgart, April 6 and 7, 2000.

Grin, J., Rotmans, J., and Schot, J. (2010) *Transitions to Sustainable Development: New Directions in the Study of Long Term Transformative Change*, Routledge, New York.

IAT (2011) BAT4MED-Project, *www.bat4med.org* (accessed January 2012).

Koppenjan, J. and Klijn, E.H. (2004) *Managing Uncertainties in Networks: A Network Approach to Problem-Solving and Decision-Making*, Routledge.

Loorbach, D. (2007) *Transition Management*, International Books, Utrecht.

Loorbach, D. and Rotmans, J. (2010) The practice of transition management: examples and lessons from four distinct cases. *Futures*, **42** (3), 237–246.

McDonough W. and Braungart M. (2002). *Cradle to Cradle. Remaking the way we Make Things*. North Point Press, New York, 193 pp. ISBN: 0865475873.

Nevens, F., De Weerdt, Y., Vrancken, K., and Vercaemst, P. (2012) Transition in Research Research in Transition, Vision on Transition Series #1, VITO, Mol.

Popp, D., Hafner, T., and N.J. (2011) Environmental policy vs. Public pressure: innovation and diffusion of alternative bleaching technologies in the pulp industry. *Res. Pol.*, **40**, 1253–1268.

Rotmans, J. (2003) Transitiemanagement, Sleutel Voor een Duurzame Samenleving, Koninklijke Van Gorcum, Assen, Nederland.

Rotmans, J., Kemp, R., Van Asselt, M., Geels, F., Verbong, G., and Molendijk, K. (2000) Transities en Transitiemanagement, de Casus van een Energiearme Emissievoorziening. ICIS/MERIT, Maastricht, The Netherlands, 123 p.

Satterfield, B., Kolb, C.E., Peoples, R., Adams, J.L., Schuster, D.S., Ramsey, H.C., Stechel, E., Wood-Black, F., Garant, R.J., and Abraham, M.A. (2009) Overcoming nontechnical barriers to the implementation of sustainable solutions in industry. *Environ. Sci. Technol.*, **43**, 4221–4226.

Schoenberger, H. (2009) Integrated pollution prevention and control in large industrial installations on the basis of best available techniques – the sevilla process. *J. Cleaner Prod.*, **17**, 1526–1529.

Senge, P., Smith, B., Kruschwitz, N., Laur, J., and Schley, S. (2008) *The Necessary Revolution. How Individuals and Organizations are Working together to Create a Sustainable World*, Nicholas Brealey Publishing, London, p. 416.

Sundqvist, G., Letell, M., and Lidskog, R. (2002) Science and policy in air pollution abatement strategies. *Environ. Sci. Policy*, **5**, 147–156.

Tukker A., et al. (2008). *System Innovation for Sustainability. Perspectives on Radical Changes to Sustainable Consumption and Production*. Greeleaf Publishing, Sheffield, 470 pp. ISBN: 978 1903063037.

UNEP (2011) Towards a Green Economy, Pathways to Sustainable Development and Poverty Eradication, A Synthesis for Policy Makers, www.unep.org/greeneconomy, (accessed 2012).

Vrancken, K. and Huybrechts, D. (2011) Evolution of BAT, as seen by the BAT-centre, in *De VlaamseOmgevingsvergunning* (eds I. Larmuseau and P. Bernaerteds) Uitg., Van den Broele, Belgium (in Dutch).

Vrancken, K., Vercaemst, P., and Dijkmans, R. (2004) Implementation of the IPPC-concept in the Flemish region of Belgium, through Vito's BAT-centre. *Planeta Mag.*, **XII** (2), 11–12.

Zarker, K.A. and Kerr, R.L. (2008) Pollution prevention through performance-based initiatives and regulation in the united states. *J. Cleaner Prod.*, **16**, 673–685.

4
Sustainable Industrial Chemistry from a Nontechnological Viewpoint

Genserik L.L. Reniers, Kenneth Sörensen, and Karl Vrancken

4.1
Introduction

According to the well-known Brundtland definition, *sustainable development* is "the development that fulfills the needs of the present generation without compromising the ability of future generations to meet their own needs" (United Nations, 1987). Such a definition obviously implies the need for a long-term vision within every industrial sector, thus also (and maybe especially) within the chemical industry, which depends largely on the availability of natural resources. But an essential question to be able to apply the sustainable development concept within the chemical industry is to have a solid and unambiguous understanding of the meaning and implications of what constitutes "sustainable chemistry." We therefore define *sustainable chemistry* as "the contribution of chemistry (synthesis and separation of atoms, molecules, chemicals, materials, and their mixtures) to sustainable development (i.e., perform future activities in a more sustainable manner but also tackle historic problems) through all its activities." The last words of this definition are particularly interesting: "through all its activities." Let us discuss the meaning of this important part of the definition of sustainable chemistry.

The difference between "technological" and "nontechnological" deserves to be better explained at this point. Technology implies conversion techniques, separation technologies, microtechnology, catalysis, membrane technology, process intensification, and so on. Technology is the obvious path of innovation that leads toward what is called *"greener chemistry."* In current industrial practice, to achieve sustainable industrial chemical processes and products, companies, research centers, and academia tend to focus mainly on technological solutions such as cleantech, green technology, improved process intensification, new catalysts, new membranes, ecofining, and so on. Thus, sustainable chemistry is currently translated into making typically engineering matters more sustainable. Engineering matters indicate, for example, harvesting of biomass, winning of petroleum, (petro)chemical processes (e.g., making them more sustainable via separation processes, process intensification, etc.), optimization of (bio)chemical products,

Management Principles of Sustainable Industrial Chemistry: Theories, Concepts and Industrial Examples for Achieving Sustainable Chemical Products and Processes from a Non-Technological Viewpoint,
First Edition. Edited by Genserik L.L. Reniers, Kenneth Sörensen, and Karl Vrancken.
© 2013 Wiley-VCH Verlag GmbH & Co. KGaA. Published 2013 by Wiley-VCH Verlag GmbH & Co. KGaA.

use of chemical products, waste handling and recycling, and so on. Such research represents an absolute condition to increase sustainability in the chemical industry.

Nonetheless, *nontechnological innovation* is equally necessary to enhance sustainable chemistry. Managers are discovering that the indicators that gauge sustainability can also be indicators of efficacy – that is, how well a company is *run*. From the management of corporate liabilities to new market ventures, a sustainable business strategy can improve all segments of corporate activity. It can be argued that management with focus on sustainability, is a good proxy for gauging overall management capabilities at strategic, tactic, and operational levels. It is obvious that enhancing sustainability in a nontechnological way, helps spark (nontechnological as well as technological) innovations that eventually improve chemical process efficiencies, tap new markets, streamline productions and materials use, reduce pollution, and lead to many other benefits. But what can be understood by the word *"nontechnological"*?

The nontechnological side of chemical sustainability implies management and business models, OR techniques, economic models, transition models, measuring models, and so on. Hence, it includes all activities, approaches, concepts, techniques, and so on that are not directly technology-driven. It should be very clear that such nontechnological solutions are also essential to achieve adequate success in sustainable chemistry. For example, cluster management, sustainable supply chain management, chemical leasing, integrated management systems and sustainable business models, societal expectations, and so on are all important aspects of truly sustainable industrial chemistry. Actually, nontechnological approaches can be classified into three perspectives: (i) clustering: make industrial plants, sites, and clusters more sustainable from a collaborative perspective, (ii) logistics: transportation of chemicals, and (iii) optimization of services (e.g., SHESQ).

It is thus obvious that a chemical enterprise has several possible ways of installing organizational improvements to achieve its aim of optimizing chemical process or product sustainability. The overall perspectives can be translated into four concrete managerial topics: (i) managing intraorganizational sustainability; (ii) managing horizontal interorganizational sustainability; (iii) managing vertical interorganizational sustainability; and (iv) sustainable chemistry in a societal context. Figure 4.1 displays the different concrete managerial topics.

As a first managerial topic, intraorganizational sustainability management can be achieved through several possible approaches. It is obvious that corporate social responsibility has to be embedded within any organization committed to improve sustainability. Moreover, especially in the chemical industry, management should measure sustainability of chemical processes and products and thus assessment methods and tools should be available. Furthermore, by integrating a company's safety, health, environment, security, and quality management systems, chemical processes and products may be efficiently approached from a holistic people, planet, and profit viewpoint.

The second managerial topic is concerned with horizontal interorganizational management and concerns cross-plant management between corporations situated on the same level of the supply chain. This innovative perspective for enhancing

4.1 Introduction

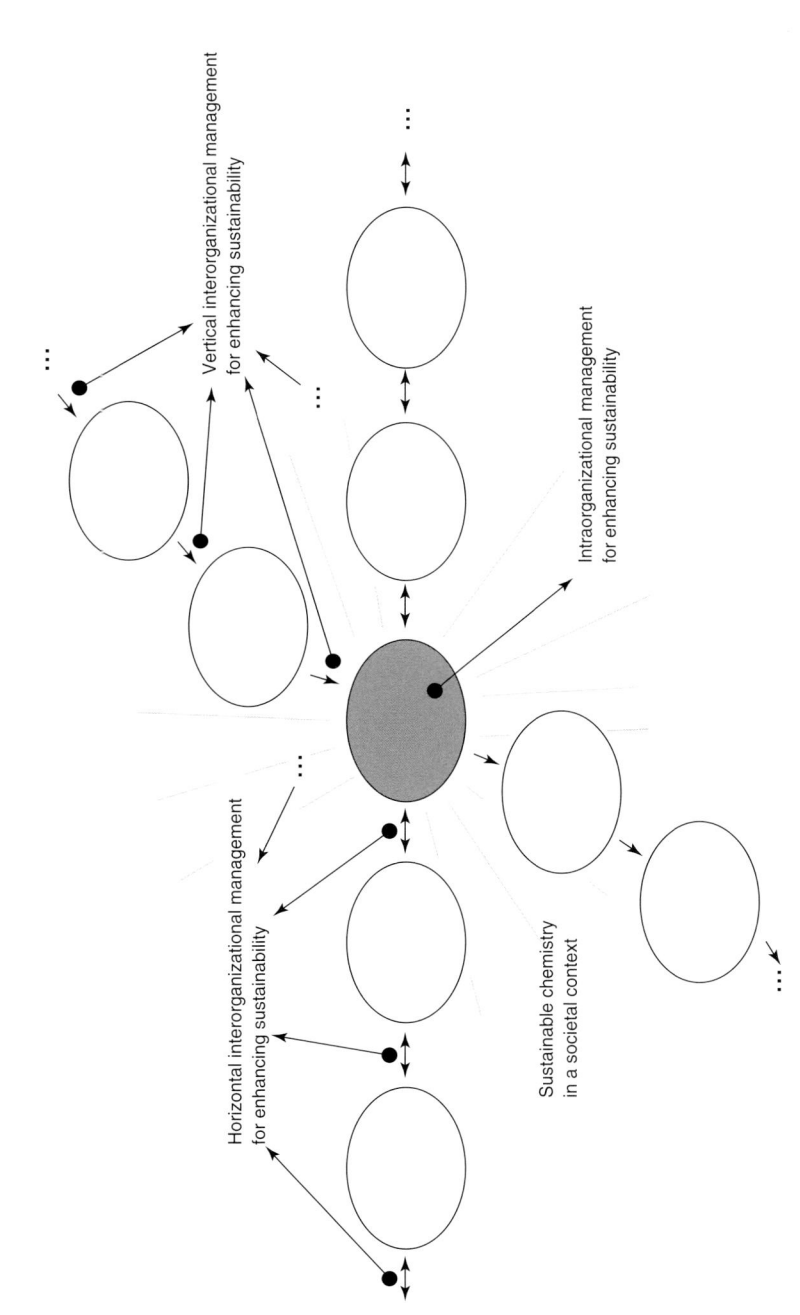

Figure 4.1 Different managerial topics for achieving sustainable chemistry in a nontechnological way.

sustainability in a nontechnological way is called *multiplant management or cluster management*. Cluster management can lead to innovation in chemical processes and products, as well as improved safety and security, amongst others.

Vertical interorganizational management represents the third managerial topic and envisions supply chain management issues. A supply chain policy trying to optimize sustainability within a chemical industrial area is aimed at realizing a more efficient chemical logistics chain.

The fourth managerial topic deals with sustainable chemistry in a societal context. Several subtopics are important in this regard: meeting with societal expectations, experience with transition in industrial context, and stakeholder interactions in support of chemical industry transition.

The transition from a "traditional" chemical plant toward a truly "sustainable" corporation may only be achieved through the use of adequate business models directed at cooperation and multiplant management. The four managerial perspectives for elaborating these business models and leading toward the "sustainable company" vision are discussed in greater depth in the following sections.

4.2
Intraorganizational Management for Enhancing Sustainability

Sustainability can be integrated in the organization on a management level, or in the processes and products. Corporate social responsibility (CSR) programs aim at supporting sustainability at a management level. The industrial activity is embedded in a societal context, resulting in a responsibility to control and report possible impacts on all involved stakeholders. The CSR approach has found widespread implementation in the chemical industry.

Intraorganizational sustainability assessment is performed either at a process or at a product level. A wide variety of tools and methods have been developed over the years. Each of these has specific characteristics with regard to system boundaries, data needs, applicability, width of impact analysis, necessary expertise, cost, time, and level of detail of the assessment. It has not been possible to come up with "the one" general sustainability assessment method, as the assessment as such is inherently integrated and dependent on the user need. Furthermore, current methods increasingly allow integration of ecological and economical aspects, but a true integration of social aspects remains to be established.

Operational risks may be classified into four distinct management domains: safety and health, security, environment, and quality. Management systems and norms are available to organizations in each of the four domains (e.g., ISO 9001:2000 (quality), OSHAS 18001:2007 (safety and health), and EMAS2, ISO 14000:2004 (environment)). The traditional approach of employing a different management system in every domain for dealing with operational risks in a company has resulted in parallel, and thus entirely separate, management systems. Every domain has its own history, specificities, and insights, for example, company know-how (quality), government and other stakeholders' expectations (safety and environment), and societal evolutions (security). As such, it is understandable that separate obligations, models, and instruments have emerged in the different domains: domain-specific legislation, management models, risk analysis techniques, education and training sessions, distinct functions within companies (e.g., the prevention advisor, the environmental coordinator, the security liaison officer), and so on. For every management domain, different performance indicators are used as well. Nevertheless, public as well as private bodies recognize the need for an integrative approach to deal with the four domains, in addition to a specialist approach, and framed within the sustainability concept of organizations. Karapetrovic and Jonker (2003) indicate that integrating standardized management systems leads to synergetic effects and significant savings in business operations. According to Smith (2002), an integrated management system has the additional advantage that improvements are implemented simultaneously within all four domains. This way, an integrated management system works in a proactive manner. The embedding of sustainability concepts into the system results in a continuous improvement in all business operations.

4.3
Horizontal Interorganizational Management for Enhancing Sustainability

Although collaborative arrangements within many industries are well known and often successful and appreciated, further optimization of these arrangements is at times possible. By augmenting collaborative agreements and relationships and by linking up with other firms on the same level of the market, a company may enjoy options otherwise unavailable to it, such as access to certain infrastructure, innovation possibilities, better access to markets, pooling or swapping of technologies and production volumes, access to specialized competencies, lower risk of R&D, enjoying larger economies of scale, benefiting from economies of scope, more know-how on SHESQ (People: SHS; Planet: E; Profit: Q), increasing energy efficiency within the cluster, decreasing waste streams within the cluster, and so on. (De Man *et al.*, 2000; Contractor and Lorange, 2002). This way, collaborative arrangements often lead to innovation and more sustainable solutions and situations. For example, Seuring and Müller (2008) discuss the cross-integration of integrated chain management and supply chain management from a conceptual viewpoint and describe five case studies proving that collaborative arrangements are the only way to improve the competitiveness of the supply chain while reducing environmental burden.

The development of industrial eco-parks or industrial ecology is another example of how companies may improve their sustainable behavior through collaboration. Sterr and Ott (2004) indicate that larger regional industrial areas may be more suitable for creating sustainable industrial ecosystems. At the same time, these authors also point out that the larger the industrial region, the more difficult it is to establish the necessary trust and coordination among the actors for setting up successful collaborative agreements.

Therefore, a framework to allow and enable companies to take joint strategic decisions (e.g., investments) should be shaped on the basis of a multiplant or cluster vision. Multiplant opportunities with respect to innovativeness and sustainability should be identified and mapped.

An innovative cluster-level policy and the resulting innovative policies on individual companies' levels should lead to strategic decisions being implemented on tactic and operational levels, aiming at making chemical products and processes more sustainable.

Cluster management may also lead to the transition of managing safety, health, environment, security, and quality within a group of chemical plants. By taking people (safety and security), planet (environment), and profit (quality) simultaneously into account in a cross-plant SHESQ management system/concept, improvement steps for individual plants may be achieved. Ethical aspects may also be introduced into this cross-plant management system and, by extension, to the individual plants. This way, a chemical industrial area may increase its efficiency, as well as greatly improving its internal and external multicorporate image. Such an approach may also lead to an important competitive advantage for the individual companies participating in the cluster initiative.

4.4
Vertical Interorganizational Management for Enhancing Sustainability

Another crucial factor in the development of sustainable chemistry is managing logistic processes within and between (chemical) companies in an optimal way. Obviously, technical innovations such as process and product improvement and the replacement of fossil fuels with renewable raw materials can deliver an important contribution to sustainability. However, the benefits of these innovations can be easily undone by an inefficient organization of the supply chain, both within a company and between different companies. To put it bluntly, a production process improvement that reduces CO_2 emissions by 20% can easily be undone if twice as many kilometers must be completed to deliver this product to the customers. Notwithstanding the obvious advantages of logistic optimization, examples of major inefficiencies in the organization of logistics processes between companies (for example, large quantities of identical products are transported in both directions between chemical clusters) or within firms (e.g., large stocks resulting from a lack of demand forecasting) are all too common.

Logistics and supply chain management are not equivalent to physical distribution, but also include the sound management of the information necessary for an effective control of the supply chain. Where performance measures of supply chains in the past have been predominantly cost based, a broad consensus has formed in the post-Kyoto period that logistics operations should proceed in a more sustainable manner (European Commission, 2007; Lambert and Cooper, 2000; Lamming and Hampson, 1996). Consequently, Christopher (1999) indicates that the optimization of the various components of the supply chain increasingly takes into account criteria related to sustainability and the environment. Simultaneously, various social criteria such as working time restrictions are also taken into account. Cost and sustainability are both important, but conflicting criteria are present and a trade-off must be made between them.

Logistics in the chemical industry also differs significantly from logistics in other sectors and therefore requires a specific approach. This has an impact on many aspects of the supply chain such as inventory management, production management, and distribution management. In the chemical industry transported products are often hazardous, which imposes important restrictions (permits, regulations) on the organization of the supply chain and possibly affects its flexibility. Another difference is that chemicals, unlike most other types of freight, can be transported by pipeline. This transport mode has some very specific properties, such as a lack of flexibility on the one hand, but a very small variable cost on the other. The fact that often large quantities of homogeneous products have to be carried makes the chemical industry an important subject for a clearing system. In such a system, a competing supplier can perform the physical production and delivery of a generic product to a customer clients if this is advantageous (for example, because the competitor is located much closer to the customer). The competitor will then invoice the supplier for the services provided, or perform similar services and incur similar costs. In this way, unnecessary transportation and logistics operations are avoided. Evidently the success of such a system requires a high degree of openness and trust between the various chemical companies. Similarly, a shift toward multimodal transportation and the corresponding increase in sustainability can only be done through collaborative agreements, as the capacity to fill entire trains with chemical products can only be delivered by consortia of chemical companies.

Chemical leasing is a trend that should also be viewed in terms of the march toward sustainability. In this approach, the supplier sells the *functionality* of a chemical product, rather than the product itself. Backed by the United Nations Industrial Development Organization, chemical leasing provides ample opportunities for both the producer and the service provider to lower overconsumption of products and thus increase efficiency.

4.5
Sustainable Chemistry in a Societal Context

As sustainability typically joins aspects of people, profit, and planet, it involves a need to discuss and determine values and visions. Introducing sustainability therefore necessarily has a societal component. It forces, or stimulates, the industrial actors to cross the installation limits and consider their role in the societal context. It bases itself on interaction with stakeholders. These stakeholders are suppliers, customers, other installations in the same cluster, consumers, policy makers, neighbors, and the broader public.

A shift of the chemical industry toward different types of processes and products will have an effect on all stakeholders. It will involve a shift in supply of raw materials, logistics, secondary processing, applications, and consumer behavior. Such changes in the different layers of the sociotechnical system are called *transitions* (Rotmans, 2003). Transitions are multilevel and multi-actor processes that result in system changes. In order for a transition to result in the realization of a (typically) long-term ambition, a multidisciplinary approach is needed with broad involvement of different stakeholders. There is a need for innovation and initiative in industry, a need for creation of a suitable policy framework by the government, a need to supply a scientific and technological basis for knowledge workers. The innovation should not only be in the ideas and techniques but also in the applied methodologies. In the sustainability context this is seen as a need for open innovation, sharing of ideas and concepts, introduction of new business models, and a shift in policy from government to governance. All these concepts require an intensified exchange of information and an increased interaction between stakeholders.

In Belgium and in the Netherlands, initiatives have been taken at regional and local levels to stimulate the transition toward sustainable chemistry. This has resulted in new developments of biochemical clusters in the port regions in the Rhine-Scheldt Delta. Local conditions and historical backgrounds result in a different evolution of these clusters. The Flemish Initiative on Sustainable Chemistry (FISCH) is a broad sectoral initiative to stimulate the transition from petro-based to bio-based chemistry in Flanders (Belgium). It fits in the broader SusChem-initiative at the European level.

4.6
Conclusions

Leveraging the search for truly sustainable chemical products and processes from a managerial perspective and disseminating the nontechnological ideas and concepts to captains of industry and to leaders of the public sector, as well as to company management (within all organizational levels and from all different departments and disciplines), is not an easy task. Although it is obvious that realizing sustainable chemistry strongly depends on the levels at which innovative technological breakthroughs are possible and develop, several managerial topics should also be considered as essential "nontechnological" ways to stimulate innovation and sustainable technology.

Integrating management systems, for example, leads to continuous improvement of the sustainable chemistry concept within corporations, leading to competitive advantages. Adequate cross-company policies lead to innovative and more effective business solutions within industrial parks, for example, increased energy efficiencies, less waste streams, and improved safety and security. Optimizing the chemical supply chain leads among others to more efficient economic and ecological product streams. Meeting with societal expectations results in sharing a common goal and long-term vision with the stakeholders and leads to a permit to operate in the long run.

Sustainability is a concept rather than a clear definition. It is a human concept based on values and principles and therefore it is in constant evolution. Trying to achieve sustainable industrial chemistry is aiming at a moving target. A broad range of nontechnological aspects help to increase the focus.

References

Christopher, M. (1999) *Logistics and Supply Chain Management: Strategies for Reducing Cost and Improving Service*, International Journal of Logistics Research and Applications 2nd edn, Vol. 2, Pearson Education, p. 103.

Contractor, F.J. and Lorange, P. (2002) Cooperative strategies in international business, in *Joint Ventures and Technology Partnerships between Firms*, Pergamon, Amsterdam.

De Man, A., van der Zee, H., Geurts, D., Kuijt M., and Vincent, N. (2000) *Competing for Partners*, Pearson Education, Zeist, p. 211.

European Commission (2007) Freight Transport Logistics Action Plan. Communication of the Commission of the European Communities.

Karapetrovic, S. and Jonker, J. (2003) Integration of standardized management systems: searching for a recipe and ingredients. *Total Qual. Manag.*, **14** (4), 451–459.

Lambert, D.M. and Cooper, M.C. (2000) Issues in supply chain management. *Ind. Market. Manag.*, **29**, 65–83.

Lamming, R. and Hampson, J. (1996) The environment as a supply chain management issue. *Br. J. Manag.*, **7**, S45–S62.

Seuring, S. and Müller, M. (2008) From a literature review to a conceptual framework for sustainable supply chain management. *J. Cleaner Prod.*, **16**, 1699–1710.

Smith, D. (2002) *IMS: The Framework*, British Standards Institute, London.

Sterr, T. and Ott, T. (2004) The industrial region as a promising unit for eco-industrial development–reflections, practical experience and establishment of innovative instruments to support industrial ecology. *J. Cleaner Prod.*, **12**, 947–965.

United Nations (1987) Report of the world commission on environment and development. General Assembly Resolution 42/187, December 11, 1987.

Rotmans, J. (2003) Transitiemanagement: sleutel naar een duurzame samenleving, Van Gorcum Uitgeverij, Assen.

Part II
Managing Intra-Organizational Sustainability

5
Building Corporate Social Responsibility – Developing a Sustainability Management System Framework

Stefan Maas, Genserik L.L. Reniers, and Marijke De Prins

5.1
Introduction

A study of some theoretical approaches to corporate social responsibility (CSR) forms an ideal preamble for the specific construction of a sustainability management system. The divergent character of CSR is characterized by the multiple motives that are responsible for the choice of a sustainable organizational profile. Referring to Cochius and Moratis (2010), internal desire, external pressure, and financial added value can be identified as essential reasons for choosing to improve sustainability within organizations. Besides, history has cultivated a large list of different notions with respect to the CSR concept. Two interpretations deserve special attention, because of their significant impact in both public and private contexts. First, Carroll (1991) introduced the sustainable responsibilities of organizations by means of a pyramidal approach. He described economic responsibility as the base of the pyramid. A second layer comprises legal responsibility, followed by ethical and philanthropic responsibility as third and fourth layers. Second, Elkington created the so-called "triple bottom line" and postulated that organizations should simultaneously pay attention to financial, social, and ecological prosperity, referred to as the well-known *"Profit, People, and Planet"* (Elkington, 1994). Furthermore, the diffuse character of CSR is manifested by the distinction that must be made between those companies systematically embedding the CSR idea into their daily practices and those using CSR only as a marketing tool. Using CSR only as a marketing instrument leads to an imbalance between the internal approach of CSR and the external communication about sustainability efforts. Such practice can be referred to as *"greenwashing."*

The target group of this chapter consists of those organizations willing to implement management systems for installing and continuously improving corporate sustainability in order to legitimatize their external statements with corresponding internal approach.

Considering this target group, we may proceed with a definition and some general characteristics of an operational management system with respect to issues such as planet, people, or profit. To realize continuous improvements concerning these issues, the importance of consecutive planning, implementing, checking, and reviewing activities in the generally known Deming cycle is highlighted. In contrast with traditional management systems, some special conditions with respect to sustainability management systems need to be mentioned. The double role of stakeholders as they function as instigators as well as receptors of a sustainability policy (De Colle, 2006), and the importance of such a policy through the whole supply chain are essential topics that especially have to be taken into account (Jørgensen, 2008).

As basic information, the standards and guidelines currently available and forming the essential source of inspiration to implement sustainability management systems, should be listed. It is obvious that organizations can model their CSR policy on the basis of a large number of standards and guidelines. Those standards, often linked with sustainability-related topics such as quality, safety, and environment are justly endorsed as cornerstones of organizational sustainable behavior. However, some shortcomings impede the implementation of an optimal, systematic, and holistic sustainability policy within organizations, when simply using these contemporary standards. A first problem is linked with the ability to obtain certifications by making use of a specific standard. Such certifications are often seen as a distortion of the CSR concept and a barrier to the idea of realizing continuous improvements, since CSR should be viewed as an inherent feature of good business instead of just a willingness to obtain a certification (UNIZO, 2008). Moreover, during the last years, an abundance of CSR-related standards has been published. Most of these only cover a small domain within the total context of CSR. Standards such as ISO 9001, ISO 14001, and OHSAS 18001, for example, respectively only focus on quality, environment, and safety, and consequently are of too small a capacity to provide a holistic corporate sustainability standard. Thus, organizations that want to implement an integrated sustainability management system can actually attain this goal by making use of several specific standards. In addition to these specific standards, holistic sustainability standards do exist, but they are usually restricted to a national application. The usages of Sigma Guidelines mainly in the United Kingdom and Q-RES Guidelines, mainly in Italy, prove this statement (De Prins, 2009). This phenomenon, in combination with the current global efforts to acquire ever more certifications, leads to a proliferation of CSR-related standards and certifications. Organizations become uncertain which standards to use and which concepts are essential to implement in order to become a sustainable organization. A classification of the current CSR-related standards on the basis of their specificity and worldwide application, as visualized in Figure 5.1, shows the current absence of a holistic and widespread sustainability standard.

The recent publication of guideline ISO 26000 in November 2010 may form a possible cornerstone in this regard. The guideline is not intended for certification purposes and aspires to provide an international platform for every organization wanting to create its own CSR profile. Although ISO 26000 admits that each CSR policy depends on specific organizational and environmental factors, the guideline has been drafted to provide organizations with some general concepts that are essentially applicable before they can implement CSR standards (ISO, 2010).

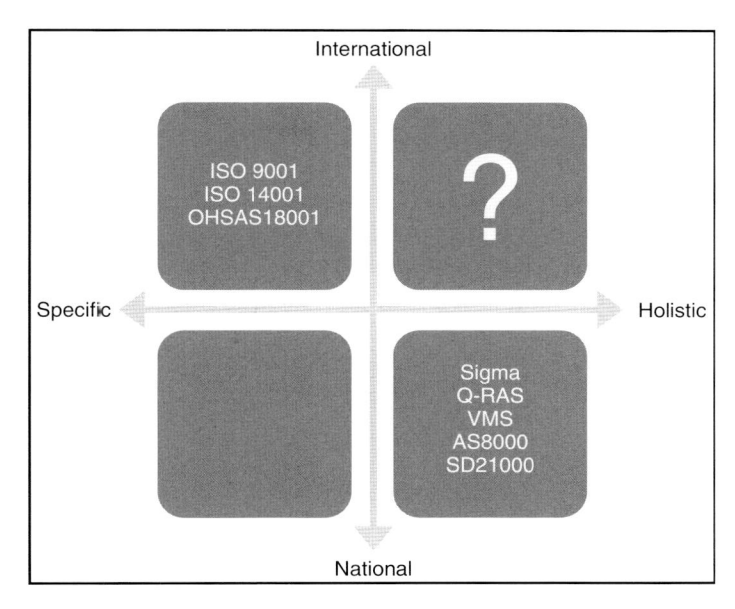

Figure 5.1 Classification of CSR-related guidelines.

In view of the previously mentioned theories and concepts, the central research question of this chapter is, "How can an organization systematically embed its sustainability aspirations and what are the different points of interest that need to be dealt with?"

5.2
Development of a CSR Management System Framework

Through its seventh chapter, ISO 26000 proclaims a clear structural breakdown as a result of which the integration process of CSR can be appointed as a threefold challenge. The first topic is related with the need to raise the essential awareness and to build the necessary competencies for social responsibilities inside any organization. The determination of a clear direction and the selection of a finite amount of strategic CSR pillars is the second important instruction, before this strategy can be translated into an operational planning, execution, and monitoring action. The third challenge is to identify specific procedures, instructions, and actions that embody the chosen strategy.

In combination with this threefold challenge, CSR should really be seen as a product of people. Sustainable results more than ever depend on the efforts made by management, stakeholders, and employees. In order to make continuous sustainable progress, commitment of these three clusters of people is indispensable.

These lines of thought are the basic ingredients of a CSR management system framework, which is visualized in Figure 5.2. In agreement with traditional

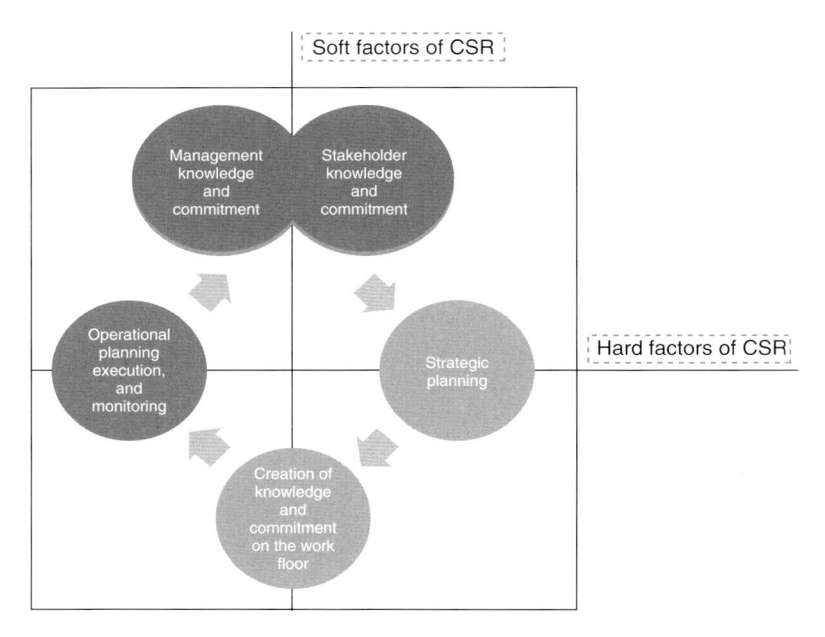

Figure 5.2 Illustration of a conceptual sustainable management system framework on the base of ISO 26000.

management systems, the idea of continuous improvement is depicted by the circular outline of the figure.

Figure 5.2 shows five inherent and characterizing components of a genuine CSR policy, which are connected with each other. The combined knowledge and commitment from central management and relevant stakeholders forms the foundation of a strategic and genuine CSR policy. This should be evident for company employees. Only after this fulfillment is realized, does an organization become able to concretize its strategy into operational actions and procedures. An accurate evaluation of these actions and their results should optimally result in more knowledge and a higher level of commitment on the part of management and stakeholders. From this point of view, the upper sphere of knowledge and commitment can be interpreted as a reservoir that is being filled permanently with the essential fuel to undertake a new and more detailed run through the CSR management system. CSR should, in this context, be interpreted as a gradual and endless process.

Human influence makes a huge contribution in the creation of the conceptual and paradigmatic sustainability cycle. Attitudes of people function as influencing factors within the model. As a result, and in contrast with traditional management systems, the system created within this study pays attention to rational as well as emotional or human aspects of entrepreneurship. The two parameters forming the axes of the model are denominated in Figure 5.2 as the soft and the hard factors of CSR. The development of an ever-extending level of knowledge and commitment among management, stakeholders, and employees belongs to the soft factor, as these challenges are focused on an enlargement of the human contribution and emotional importance within a general CSR policy. The construction of strategic planning for CSR and the operational execution are part of the hard factor. This denomination refers to the quite impersonal character of these two challenges.

In its attempt to establish a sustainable policy and to prevent possible failures, each organization should consciously pay attention to the design of each sphere within the illustrated framework and try to actively fill the eventual gaps in each of them. With this in view, the following subsections discuss a sequence of possible measures to be strived for, and these are described for each sphere.

5.2.1
Management Knowledge and Commitment (Soft Factor)

A higher level of understanding and commitment from central management toward the interest of CSR often starts with the participation in either private or public CSR projects. This external stimulus and expertise is often vital to build up a structured CSR policy within an organization. "Organizations sometimes join associations or peer organizations to establish or promote socially responsible behavior within their area or activity or within their respective communities" (ISO, 2010). An adapted mission or statement of principles is another way to concretize the management understanding and commitment. Such a sustainable mission should spread sustainable thoughts and approaches through the entire organization. From this point of view, a mission is really authenticated as a primary source of inspiration for employees, especially with due observance of this quotation: "Companies whose employees understand the mission enjoy 29% greater return than other firms." (Jackson, 2005). In addition, organizational management is encouraged to address its first sustainability efforts to the most enthusiastic employees with respect to an eventual CSR implementation. Management should make use of this enthusiasm by gathering some enthusiastic CSR ambassadors in a so-called CSR Team, as the liaison between management and the workforce. Such a modification of the internal organizational structure should facilitate a smoother transfer of motivation and information between these two groups of people.

5.2.2
Stakeholder Knowledge and Commitment (Soft Factor)

In order to realize a structured CSR policy, organizational management should take care that (at least the most important) stakeholders are involved during the exercise of defining the sustainable direction of the organization. ISO 26000 explicitly names those stakeholders as cofounders of the organizational CSR profile. "Although an organization itself may believe it understands its social responsibility, it should nevertheless consider involving stakeholders in the identification process to broaden the perspective on the core subjects and issues" (ISO, 2010). Stakeholders should be identified and categorized. On the basis of this categorization, the most renowned and relevant stakeholders should be drawn into the governance structure of the organization. This idea of engaging the most relevant stakeholders is being concretized ever more in organizations by installing and appreciating stakeholder panels.

5.2.3
Strategic Planning – the Choice of Sustainable Strategic Pillars (Hard Factor)

The interaction between attitudes of management and stakeholders should result in the choice of a limited number of strategic CSR pillars, which need to form the foundations of the CSR policy. Such pillars lead in turn to a CSR frame that determines the limits of the CSR policy and enables embracing all the sustainability efforts of the company.

In this context, ISO 26000 provides aid by enumerating a list of seven core subjects and many more core issues related to CSR. An organization needs to select those topics that are relevant and significant for the organization. "Once an organization has identified the broad range of issues relevant to its decisions and activities, it should look carefully at the issues identified and develop a set of criteria for deciding which issues have the greatest significance and are most important to the organization" (ISO, 2010). Moreover, among these relevant and significant issues, the organization should make a distinction between issues easy to influence and topics not easy to influence through its supply chain. This classification lays the foundation of a domino effect of future objectives, allowing the treatment of highly influential issues, and at the same time allowing the organization to raise its voice with regard to contemporary noninfluential but significant issues. "It is also important to recognize that the process of integrating social responsibility throughout an organization does not occur all at once or at the same pace for all core subjects and issues. It may be helpful to develop a plan for addressing some social responsibility issues in the short term and some over a longer period of time" (ISO, 2010). Consultation and negotiation with some major suppliers or clients are possible ways in which to increase this influence; and once prominent partners are convinced of the importance of sustainability efforts, they will most likely intensify instead of weaken these efforts, making it possible to fulfill objectives in the long term.

5.2.4
Knowledge and Commitment from the Workforce (Soft Factor)

Once the prospective CSR policy is written, it has to be communicated to all the internal employees, as internal understanding of the content together with internal commitment with respect to CSR can be interpreted as crucial factors toward an increased sustainability performance. "Communication is critical to many different functions in social responsibility including [...] helping to engage and motivate employees and others to support the organization's activities in social responsibility" (ISO, 2010). On the one hand, this process should occur from a top-down approach. During the recruitment process itself, management can decide to use the sustainability mentality of potential employees as a distinguishing parameter. Subsequently, after commencement of employment, the employee should continue to be informed regularly by central management or by the CSR Team about the sustainability course being followed by means of frequent information sessions. On the other hand, a simultaneous bottom-up approach is also required. Employees must feel that they are in possession of a key that opens the gate toward an operational fulfillment of the chosen strategy. The design of some motivation programs or suggestion systems, which enable employees to convey their ideas, for example, on a central intranet, is a concrete example of such a bottom-up approach.

5.2.5
Operational Planning, Execution, and Monitoring (Hard Factor)

Once employees are informed and convinced of the importance of a structured CSR policy, management and the CSR Team can proceed to develop and implement an operational planning of specific activities and instructions. Within this operational section, two issues are essential within the context of a systematic CSR policy.

The first element is the elaboration of a regular and proactive action plan so that execution of sustainability ideas may not evolve into a random process. Such an action plan contains desired actions and instructions in the course of a specific time span. An organization that just finds itself in the beginning phase of a structured CSR policy should principally look to central investments and quick wins in these action plans, as these quick wins generate a high amount of swift profits, which, in an optimal scenario, raise the internal motivation toward sustainability efforts. However, as the CSR policy becomes more mature, it becomes more important that these central attempts are substituted by local efforts on the work floor. A CSR policy evolves in this context from a centrally oriented one to a distributed one, where sustainability improvements depend even more on modified actions on the work floor instead of quick wins by central management. Every department is being confronted with ever more specific and modified instructions for each strategic issue. Once it has reached this phase, CSR can be described as a black box, which, even though it is fed with essential input by the management team and by the CSR Team, basically depends on local attempts. In the same optimal scenario as before, every department should evolve toward forming a self-guided team that conceptualizes and drafts action plans drafted in a decentralized way.

A second prerequisite is the application of consistent key performance indicators, which can be considered as an ever-returning issue within a long-term CSR policy and which enables organizations to make their efforts measurable. "There are many different methods that can be used to monitor performance on social responsibility, including reviews at appropriate intervals, benchmarking, and obtaining feedback from stakeholders. [...] One of the more common methods is measurement against indicators" (ISO, 2010). An eventual underestimation of the importance of indicators within an organization implicates the impossibility of examining whether realized actions have led to a significant progress. Above all, a major challenge is to classify the chosen indicators in a structured way for each strategic pillar as the proof of a well-reasoned selection process. The operational information and scores on indicators, which become available after the completion of the integral conceptual management system, can be compared with the essential fuel to drive the internal motor of that same sustainability management system. The scores realized on indicators should be seen as the material that drives an organization forward to new achievements and sharper objectives. In this context, CSR should really be seen as a gradual process, which can never be ratified as being finished.

The completion of the integral conceptual management system may lead to the formal announcement of obtained results and desired actions for the future, based on information that can be filtered out of the circle. "An organization should, at appropriate intervals, report about its performance on social responsibility to the stakeholders affected. [. . .] Reporting to stakeholders can be done in many different ways, including meetings with stakeholders, letters describing the organization's activities related to social responsibility for a defined period, website information, and periodic social responsibility reports" (ISO, 2010). Management and the most prominent stakeholders can choose to publish this kind of information as a documented proof of significant efforts and realized results.

5.3
Conclusions

This chapter suggests a conceptual sustainability management system framework, which is mainly based on the recently published umbrella guideline ISO 26000. Five inherent features of CSR are linked with each other within a coherent and systematic setting: (i) management knowledge and commitment, (ii) stakeholder knowledge and commitment, (iii) strategic planning, (iv) knowledge and commitment on the work floor, and (v) operational execution and monitoring. Furthermore, each of these essential building blocks, required to obtain and sustain CSR within any organizational context, is explained and discussed. The authors are convinced that building CSR into the DNA of any organization will become common practice in the near future. The framework proposed in this chapter can serve this goal adequately.

References

Carroll, A. (1991) The pyramid of corporate social responsibility: toward the moral management of organizational stakeholders, Bus. Horiz., 4, 39 – 48.

Cochius, T. and Moratis, L. (2010) ISO 26000. Handleiding voor MVO, Van Gorcum, Assen, 261 p.

De Colle, S. (2006) CSR and Management Systems, LIUC University of Castellanza, 21 p.

De Prins, M. (2009) Managementinstrumenten voor MVO, Centrum voor Duurzaam Ondernemen, Hogeschool-Universiteit Brussel, 6 p.

References

Elkington, J. (1994), *Cannibals with Forks; The Triple Bottom Line of 21st Century Business*, Capstone Publishing, Ltd, Oxford, pp. 69–96.

ISO (2010) International Standard ISO 26000. *Guidance on Social Responsibility*, International Standards Organization, Genève, p. 106.

Jackson, L.A. (2005) You can't do it alone: building an effective team brings bottom-line results, *Black Enterprise*, **35** (12), 63.

Jørgensen, T.H. (2008), Towards more sustainable management systems: through life cycle management and integration, *J. Cleaner Prod.*, **16**, 1071–1080.

UNIZO 2008 Duurzaam Ondernemen, Drijfveer of Label? April 21, 2008, *http://www.unizo.be/mvo/* (accessed 9 May 2011).

6
Sustainability Assessment Methods and Tools

Steven De Meester, Geert Van der Vorst, Herman Van Langenhove, and Jo Dewulf

6.1
Introduction

Sustainable development, as defined by Brundtland (World Commission on Environment and Development, WCED, 1987), is a very valuable concept. It has the merit that society started thinking of the earth's capacity and the social system on a longer term. Likewise, the chemical industry realized that an unequal and polluted world, with depleted resources would not be a healthy climate for business. As such, sustainable development was introduced in many different companies, and became an important concept at the management level. However, although the Brundtlandt definition is valuable as a concept, it is not very tangible in practice. The "triple bottom line" (Elkington, 1994) introduced by Elkington is more tailored for practical implementation in industry. This concept states that sustainable development should benefit the three Ps: People, Planet, and Profit, changing the perception that environmental and social considerations are not compatible with economic benefits.

While this definition gives a good overarching direction, it does not give a concrete answer on how to get there. It is for this purpose that assessment is necessary where macroscale improvement is induced by sustainable development at the microscale (Huppes and Ishikawa, 2009). So when management decides in its strategy to develop a company in a sustainable way, the only way to see whether improvement has been achieved is by assessment of the impact of its processes, products, and services. It is only when the results of assessments are available, that the most objective sustainable decisions can be made in management or policy strategies.

Owing to the holistic nature of sustainable development, however, the necessary assessment is not a straightforward task. Much work has already been done in social, environmental, and economic research groups, but while this information is valuable, it is very extended and fragmented. This has resulted in the development of different combinations of principles, depending bottom-up on the available methodologies and top-down on the chosen concept (Figure 6.1). In this chapter, an overview is presented of the methodologies and tools that are available, how

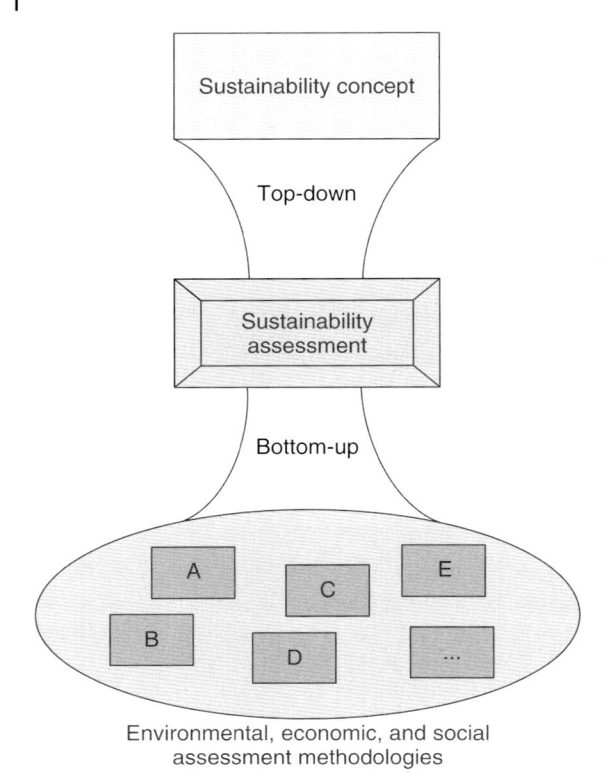

Figure 6.1 The top-down/bottom-up structure of sustainability assessment research.

exactly they work, and which fit best in the concept; that is, the goal and scope of a sustainability assessment of products, processes, and services. Determining the latter, that is, constructing an assessment framework in the light of the available means, is the first and essential phase of assessments.

6.2
Sustainability Assessment Framework

In order to obtain a sound assessment, a framework should be constructed where several conceptual and methodological choices have to be made. The starting point is determining a "function" within the studied system that has to be assessed and to which the results can be linked. This can range from a process, product, or service to a company, sector, or a region. In this chapter, focus is on the microscale, as macroscale improvement is only the result of improvements at the microscale. As soon as this is identified, a system boundary can be defined to decide which part of the production chain can be included and which can be excluded or cut off (International Organization for Standardization, ISO, 2006). This means that the study previous phases of the life cycle focuses on a specific part of the life cycle, which can be a process within a company (process level); a process chain including all supporting utilities at the company level ("gate-to-gate" boundary); a cradle-to-gate system, including the full production chain; or a cradle-to-grave study, which encompasses the use phase and end-of-life strategy in addition to the previous. Within the selected list of unit operations, a foreground and background system can be selected, where the first is defined as "those processes of the system that are regarding their selection or mode of operation directly affected by decisions analyzed in the study" or as the "case-specific processes," while the second is defined as "those processes that are operated as part of the system but that are not under direct control or decisive influence of the producer of the good" or as the "market-average processes" (European Commission–Joint Research Centre, JRC–Institute for Environment and Sustainability, IES, 2010).

Most economic assessments focus solely on the monetary interactions between the foreground and background systems, and thus require a decision regarding the costs and benefits that are of interest to the practitioner. The system boundary in environmental and social assessment is often more complex, as interactions occur between the technosphere on the one hand and the ecosphere and social system on the other. These can be quantified by introducing "elementary flows" in case of the ecosphere and "elementary interactions" or positive and negative "pressure" in the case of the social system (Figure 6.2). Some economic assessments however also try to grasp these interactions by monetizing the "external" costs.

The amount and nature of the unit operations and interactions included in the study depend mainly on the concept of the study and the available time and means. As it is impossible to give one "absolute" advice, a simple and sound suggestion is to start thinking with involved stakeholders, "if we had no budget or time limitations, what would we include?" and then to include what is possible in the scope and within the means available for the study (UNEP SETAC Life Cycle Initiative, 2009). To avoid confusion, each study should include a system diagram, where the system boundary is well defined with respect to the three different spheres. Subsequently, a detailed and reliable quantitative or (semi) qualitative inventory should be collected within the system boundary. Doing this is the most challenging, labor- and time-intensive phase (Finnveden et al., 2009), but it is the key to useful results.

The collected data can be site specific and thus collected within a real production process chain or modeled with software. Alternatively, to save time, generic/average data can be used. For the latter, which is often applied for background processes, two sources exist: first, databases containing previous studies such as ecoinvent (The Ecoinvent Centre, 2011), the European Reference Life Cycle Database (ELCD) (European Commission, JRC, IES, 2011); and second, a hybrid assessment can be chosen, which couples microscale assessments to meso or macroscale input–output (IO) databases. This approach uses lower quality IO data to fill data gaps in a more detailed assessment (Suh et al., 2004) and where the gaps are "purchased" from the IO database (Peters and Hertwich, 2006). In the best case, the IO tables are disaggregated to a more detailed level, such as a sector or a product group, where the best fitting and most specific IO is used or created (Suh, 2009).

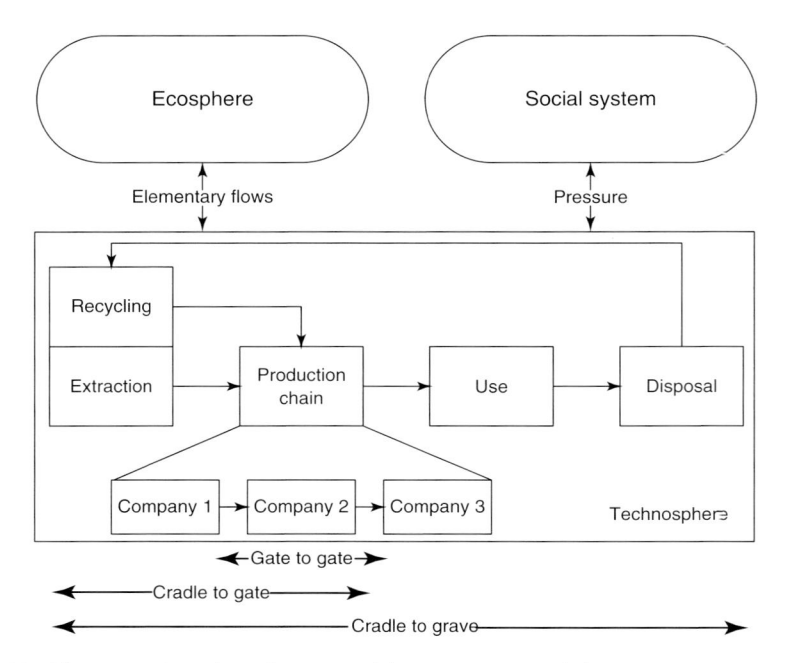

Figure 6.2 The system boundary of a sustainability assessment includes certain unit operations within the technosphere and determines interactions with the ecosphere and social system. Based on European Commission–Joint Research Centre, JRC–Institute for Environment and Sustainability, IES (2010).

Depending on the scope of the study, the assessment can be executed in a backward-looking way, that is, attributional, or in a forward-looking way, that is, consequential (or previously marginal, prospective, etc.) (Curran, Mann, and Norris, 2005; Thomassen et al., 2008; Sandén and Kalström, 2007). The latter approach is very promising, as it is able to grasp the larger picture of the impact of decisions in the foreground system, on the background system, the ecosphere, and the social system. In this way, issues such as partial equilibrium modeling, including market mechanisms in the supply/demand system (Ekvall, 2002), experience curves, including the lower price and higher efficiency of large-scale production of new technologies (Zamagni et al., 2008), and rebound effects, including changes in consumption (Thiesen et al., 2008) can be captured. Furthermore, this can be coupled to the hybrid approach to model the impact of a change in a micro system on the meso and macro levels, which allows a better understanding of the sustainability of the total system. However, this research area needs further development, as many IO databases are still based on attributional data (Finnveden et al., 2009) and furthermore, it might lead to an unachievable data collection.

As a final step, the resulting data inventory should be allocated to the "function" of the studied system, allowing to express the final impact per unit of function. While it is suggested by International Organization for Standardization, ISO (2006) to first try to avoid allocation through system expansion or by division in subsystems,

Lundie, Ciroth, and Huppes (2007) conclude that this is not always possible and state that, in most cases, a choice is made between a physical parameter such as mass, energy, and exergy, or an allocation based on economical value. Weidema however, states that system expansion is a sound strategy to avoid allocation in consequential assessments (Weidema, 2003).

From all the points set out above (functional unit, system boundaries, data inventory, scale, etc.), it can be concluded that defining a good sustainability framework for every sustainability assessment of chemicals or chemical processes is a very important step, which defines the impact assessment and final interpretation of every study and its results. It is therefore important for every type of evaluation to have a good background and a good communication strategy on those choices that are made in these first steps in order to understand and interpret the final results. This can only improve the credibility of the assessment and the executor.

6.3
Impact Indicators and Assessment Methodologies

Assessing the impact of a product system is a very complex task, as it tries to grasp the different aspects of the broad concept of sustainability in the 3 Ps: People, Planet, and Profit. All economic interactions, the impacts of elementary flows to the ecosystem, and pressure to the social system have to be included and characterized, with a list of indicators for the different impacts as the final outcome. It is therefore very important to understand how these indicators are constructed and applied in assessment methodologies. Furthermore, for the purpose of decision making and external communication, the different indicators are often aggregated, which simplifies decision making on the one hand, but often results in a loss of relevant information on the other.

Two strategies can be applied to construct indicators, as is demonstrated in Figure 6.3 (Jolliet *et al.*, 2004):

- **Assessing impact at the midpoint level**. Indicators are chosen at an intermediate position of the cause–effect impact pathway. It is stated that this point should be taken where it is judged that further modeling includes too much uncertainty (European Commission–Joint Research Centre, JRC–Institute for Environment and Sustainability, IES, 2010).
- **Assessing impact at the endpoint level**. This strategy tries to include the effect of midpoint indicators on endpoint indicators, which are better understandable from the concept of sustainability.

In the case of endpoint modeling, further aggregation is often not necessary as the number of endpoints is limited. In the case of aggregation of midpoint indicators, a choice can be made between the concept of strong and weak sustainability, where weak sustainability has a viewpoint that one impact category can be compensated by another, whereas strong sustainability does not accept this substitutability (Cabeza Gutés, 1996). Following the definition of strong sustainability, each

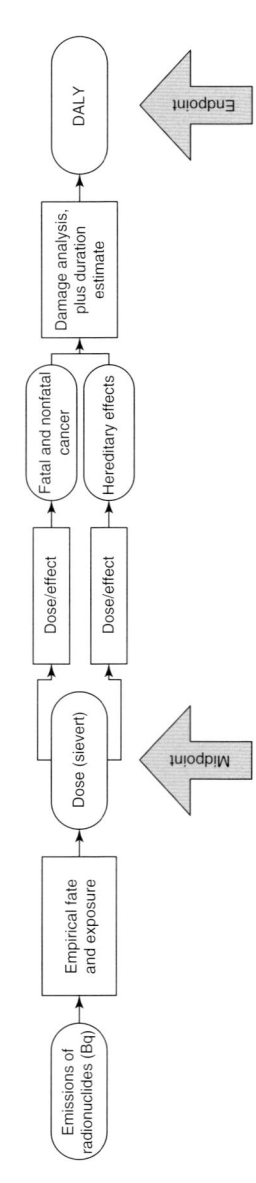

Figure 6.3 Midpoint and endpoint indicators in the cause effect chain (DALY = disability-adjusted life years).

indicator has a value that cannot be compensated and therefore results are left as they are, without aggregation. In the case of weak sustainability, a composite indicator (CI) (Gasparatos, El-Haram, and Horner, 2008) can be constructed through normalization and weighting.

Normalization, that is, putting the result versus reference information, can be internal or external. Internal normalization is especially useful for comparison of different options and takes case-specific data by using, for example, a division by the maximum value (Xu *et al.*, 2006; Krajnc and Glavic, 2005; Diaz-Balteiro and Romero, 2004). External normalization is more often used and is aimed at understanding the relative magnitude of each indicator in, for example, a certain region or population (e.g., per capita (European Commission–Joint Research Centre, JRC–Institute for Environment and Sustainability, IES, 2010) or sector (International Organization for Standardization, ISO, 2006; van Oers and Huppes, 2001)). The following weighting step quantifies the relative significance of each indicator within the goal and scope of the assessment. Weighting is an arbitrary and controversial step and therefore often the simplest methodology is chosen, that is, assigning equal weights (Singh *et al.*, 2009). When this approach is not suitable, three options are possible (Soares, Toffoletto, and Deschênes, 2006; Seppälä and Hämäläinen, 2001):

- Using a distance-to-target approach. This could also be seen as an additional normalization step, and does not give any information on the relative importance of impact categories to each other (Soares, Toffoletto, and Deschênes, 2006).
- Using a monetization approach by using, for example, the willingness-to-pay (WTP) approach.
- Using a panel approach with involved stakeholders is often preferred, so more systematic approaches to assign weights are used, mainly borrowed from the long-existing multicriteria decision analysis (MCDA) discipline where alternatives can be ranked on the basis of expert judgment (Linkov *et al.*, 2004). A promising tool to support this analysis is the analytic hierarchy process (AHP) (Saaty, 1990) consisting out of four steps (Figure 6.4):
  - Structuring the problem in a goal (assigning weights), in criteria for ranking the indicators, and finally in the different options that can be chosen
  - Comparing the criteria pairwise and give scores from 1 (equal importance) to 9 (one is extremely more important compared to the other).
  - Pairwise comparison of the options within each criterion by giving scores from 1 to 9.
  - Combining scores of options and criteria.

In this context, several assessment methodologies have been developed, each with its specific starting assessment framework and list of midpoint and endpoint indicators. In the following, first, the indicators will be discussed and secondly, an overview of different methodologies using these indicators will be presented, both for the three dimensions of sustainability.

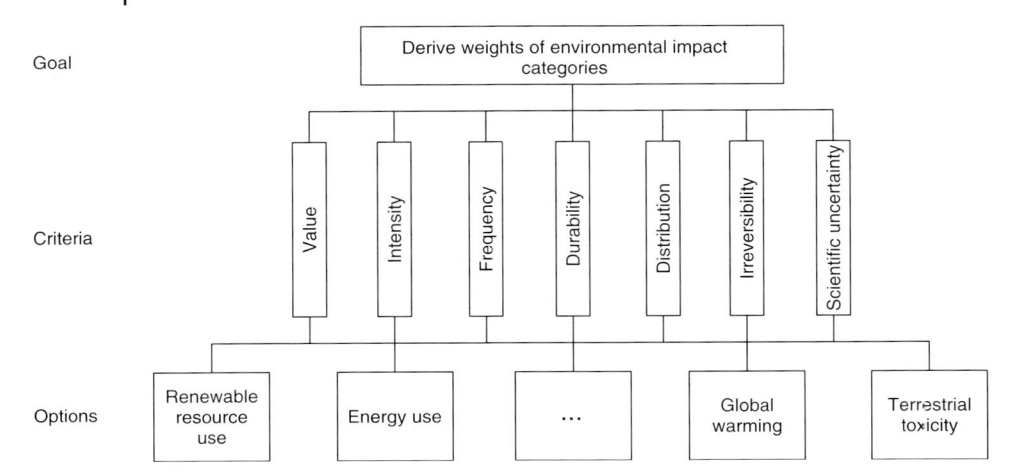

Figure 6.4 The analytic hierarchy process (AHP) used in the weighting step of sustainability assessment.

6.3.1
Environmental Impact Assessment

The concept of sustainable development actually grew mainly from environmental concerns. This, together with the large complexity of the environment, has led to a large list of indicators with many resulting impact assessment methodologies, and therefore this is the most elaborate research area discussed in this chapter. Impact categories and methodologies related to emissions are discussed first. Resource use is treated separately, as this is a very relevant and important parameter in modern industry. In addition to this, specific technology indicators focusing on internal performance can be constructed.

6.3.1.1 Emission Impact Indicators

6.3.1.1.1 Midpoint Indicators The overall formula to calculate a midpoint indicator is given by

$$\text{Category Indicator} = \sum_s \text{Characterisation Factor (s)} \times \text{Inventory data (s)}$$

The inventory data of substance s is given per functional unit, while the characterization factor expresses the contribution of substance s to an impact category (in a unit of the equivalence factor) per unit of inventory data (Pennington et al., 2004), as such giving an estimation of the relative importance of that inventory data in a given impact category (Pennington, 2001). An overview of the most common midpoint indicators is given in (Table 6.1).

6.3.1.1.2 Endpoint Indicators Endpoint or damage categories are at the level of ultimate societal concern and thus easier to link with the sustainability

6.3 Impact Indicators and Assessment Methodologies

Table 6.1 An overview of the most common midpoint impact categories, together with their characterization and equivalence factors. An overview of the most common midpoint indicators is given in Table 6.1

Impact category	Impact specification	Most commonly used characterization factor(s) or formula to calculate the indicator	Most commonly used equivalence unit(s) for the characterization factor(s). Expressed per unit of inventory
Ozone depletion	Accounts for the depletion of the protective ozone in the earth's stratosphere mainly due to emissions of halogens (Singh et al., 2007).	Ozone depletion potential (ODP)	CFC-11-eq
Climate change	Refers to the change of the climate and temperature due to anthropogenic emissions disturbing the adsorption capacity of the atmosphere (Pennington et al., 2004).	Global warming potential (GWP)	CO_2-eq
Photochemical ozone creation potential (POCP)	Refers to excessive concentrations of ozone and its intermediate reaction products. It is influenced by different volatile chemicals such as NO_x, OH-reactive hydrocarbons, and CO (Pennington et al., 2004).	Smog potential (SP)	Ethylene-eq (Jolliet et al., 2003; Goedkoop and Spriensma, 2001; Itsubo and Inaba, 2003; Guinée et al., 2002) Volatile organic compounds (VOC)-eq (Goedkoop et al., 2009; Potting and Hauschild, 2005) NO_x-eq (Singh et al., 2007; Heijungs et al., 2002)
Acidification (can be subdivided into terrestrial and aquatic acidification)	Acidification gives an indication about the increase in the hydrogen ion concentration in water and soil systems due to deposition of inorganic substances (Goedkoop et al., 2009).	Acidification potential (AP)	SO_2-eq (Goedkoop et al., 2009) H^+ moles-eq (Singh et al., 2007) m^2 unprotected ecosystem (Potting and Hauschild, 2005)

(continued overleaf)

Table 6.1 (continued)

Impact category	Impact specification	Most commonly used characterization factor(s) or formula to calculate the indicator	Most commonly used equivalence unit(s) for the characterization factor(s). Expressed per unit of inventory
Eutrophication (in many cases subdivided into terrestrial and aquatic eutrophication)	Aquatic eutrophication is the result of anthropogenic nutrients (specially N and P) enriching and disturbing the natural nutrient balance in aquatic environments (Pennington et al., 2004) through giving rise to biomass (e.g., algae) growth (Singh et al., 2007). Terrestrial eutrophication occurs when a soil is enriched with the otherwise restricting nutrient, nitrogen, and nitrogen-adapted species thus get a competitive advantage (Pennington et al., 2004). Phosphorus is of less importance in the terrestrial environment, because it is seldom a restrictive nutrient (Potting and Hauschild, 2005).	Eutrophication potential (EP)	N-eq (Singh et al., 2007)NO_3^--eq P-eq/PO_4^{3-}-eq (Jolliet et al., 2003) Or combination of previous SO_2-eq (Jolliet et al., 2003) N-eq (Toffoletto et al., 2007) m^2 unprotected ecosystem (Potting and Hauschild, 2005)
Species and organism dispersal plus gene dispersal	The dispersal of invasive species and organisms due to anthropogenic actions or processes can result in substantial change of the natural animal and plant populations of an invaded region. dispersal of invasive genes from genetically modified organisms can cause harm to a region's natural composition (Jolliet et al., 2004). This indicator is still under development.	—	—

Noise	A noise indicator is often not calculated in LCA studies, because it is stated that the existing noise of processes is taken for granted (Potting and Hauschild, 2005) or is very local and difficult to interpret in relation to other impact categories.	$NN_d = P_d \times T_{proc} \times NNF_{LP}$ NN_d = noise nuisance at distance d from point source P_d = number of persons at distance d T_{proc} = duration of noisy process (h) NNF_{LP} = noise nuisance factor	person hour (Potting and Hauschild, 2005)
Odor	Odor is a subjective nuisance, but above a certain level, some odors from emissions are "experienced" as stench by everyone (Guinée et al., 2002). Despite the fact that odor is mentioned several times as a possible impact category (Toffoletto et al., 2007), it is not often calculated.	It is calculated as the reciprocal of the odor threshold value (1/OTV). Guinée et al. (2002)	m^3 (Guinée et al., 2002)
Ionizing radiation	This impact category is related to the release of radioactive material to the environment (Goedkoop et al., 2009).	Ionizing radiation potential (IRP)	kBq (Becquerel) U-235 air-eq (Goedkoop et al., 2009) Bq-eq carbon 14 in air (Jolliet et al., 2003)
Soil salination	Soil salination refers to the increasing salt concentrations in the soil. It can be calculated for irrigation practices based on the sodium adsorption ratio (SAR) and electrical conductivity (EC) (Feitz and Lundie, 2002).	Salination potential (SP) $SP = EF_i \times [Na_J] \times V_i$ with EF_i the site-specific equivalence factor based on EC and SAR, Na_j the sodium concentration, and V_i the total volume of irrigation water	Na^+-eq (Feitz and Lundie, 2002)

(continued overleaf)

Table 6.1 (continued)

Impact category	Impact specification	Most commonly used characterization factor(s) or formula to calculate the indicator	Most commonly used equivalence unit(s) for the characterization factor(s). Expressed per unit of inventory
Human toxicity	Human toxicity is an adverse effect on humans as a result of exposure to a chemical (Pennington et al., 2004). Further distinction can be made between carcinogenic and noncarcinogenic or even respiratory impacts for indoor air pollution (Singh et al., 2007). Midpoint indicators for this category are mainly based on reference substances, but can alternatively be more qualitative with respect to the numbers of persons exposed.	Human toxicity potential (HTP) Particulate matter formation potential (PMFP)	1,4-Dichlorobenzene-eq (Goedkoop et al., 2009) Chloroethylene-eq (Jolliet et al., 2003) person µg m^{-3} (Potting and Hauschild, 2005) *Carcinogenic:* Benzene-eq (Singh et al., 2007) *Noncarcinogenic:* Toluene-eq (Singh et al., 2007) *Respiratory:* PM 2.5-eq (Singh et al., 2007) PM 10-eq (Goedkoop et al., 2009)
Ecotoxicity	Ecotoxicity is an adverse effect on organisms and/or the functioning of the ecosystem as a result of exposure to a chemical released in the environment (Pennington et al., 2004). Apart from reference substances, the volume of water and soil that is exposed can also be seen as a midpoint indicator. In this case, spatial diversity can be introduced by a site factor (SF) Potting and Hauschild (2005).	Ecological toxicity potential (ETP)	2,4-Dichloro-phenoxyacetic acid-eq (Singh et al., 2007) 1,4-DCB-eq (Guinée et al., 2002) Triethylene glycol-eq (Jolliet et al., 2003; Toffoletto et al., 2007) m^3 (Potting and Hauschild, 2005)

concept. The most known endpoint model is presented in the Eco-indicator 99 impact assessment methodology (Goedkoop and Spriensma, 2001) and uses two damage categories for emissions, following the so-called areas of protection (AoP) proposed by Udo de Haes et al. (1999):

- Damage to human health
- Damage to ecosystem quality or diversity.

Calculating the impact of a certain flow on one of these endpoints is not straightforward as it requires complicated cause–effect models. Current environmental endpoint modeling has learnt much from the longer existing risk assessment (RA) approach, which starts from hazard identification and is followed by a release and exposure assessment and hazard characterization phase (dose/response) to obtain a final risk characterization and interpretation. Similarly, endpoint modeling typically uses four steps (Goedkoop and Spriensma, 2001):

- Fate analysis linking an emission to a temporary concentration. This requires extensive data and good models. A harmonized approach is presented in "The Tool" (Wegmann et al., 2009). Three parameters are calculated in this model, which can also be used as midpoint indicators: overall persistence, characteristic travel distance, and transfer efficiency.
- Exposure analysis, linking this temporary concentration to a dose (used for human health, not for ecosystem quality)
- Effect analysis, linking the concentration or dose to the effects. Typically, dose–response curves are constructed, allowing comparisons between the predicted environmental concentration (PEC) and the predicted no-effect concentration (PNEC) based on acute (EC_{50}/LC_{50}) or preferably chronic (no observed effect concentration NOEC) data (Van Leeuwen and Hermens, 1995). While this should actually be determined for all species, often a more generic species sensitivity distribution (SSD) approach is used, focusing on a full community or ecosystem, based on hazardous concentrations (HCs) (Larsen and Hauschild, 2007).
- Damage analysis, linking the effects to a damage indicator. The most-used endpoint indicators are
  - disability-adjusted life years (DALYs) for human health, representing years of "healthy" life lost (WHO, 2011).
  - potentially affected fraction (PAF) or potentially disappeared fraction (PDF) of the species, for ecosystem quality. The first gives an indication of the fraction of species affected above their NOEC and the second adding what happens beyond this NOEC (Pennington, 2001).

Figure 6.5 is an example of the determination of the endpoint indicator from an SSD curve, plotting the NOEC of a set of species. A benchmark can then be put on the concentration where the NOEC value is exceeded for 5% or 50% of the species, $HC5_{NOEC}$ and $HC50_{NOEC}$, respectively (Smit, 2007).

These curves are typically sigmoid, implying that there is a nonlinear relationship between the dose/concentration and the impact. Accounting for this is not possible because of the limitations of the current life-cycle assessment (LCA) methodology.

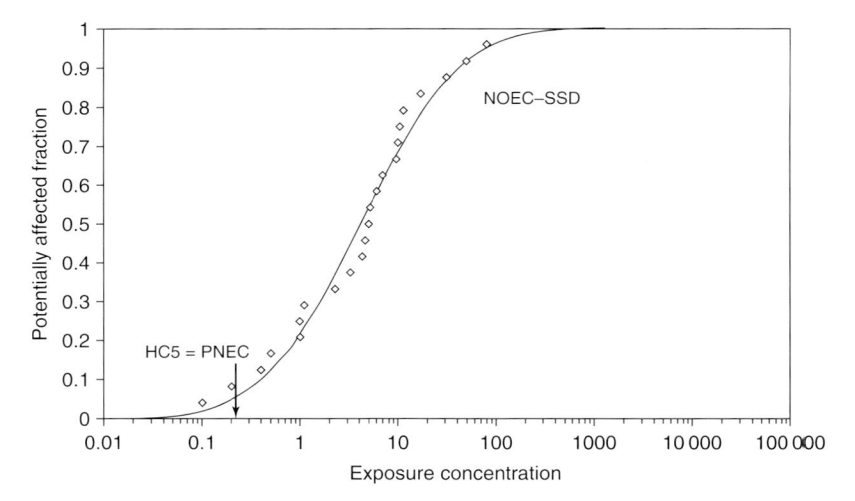

Figure 6.5 A species sensitivity distribution based on NOEC values, where the exposure concentration is plotted versus the potentially affected fraction (PAF) of the species. HC5 is the point where 5% of the species is potentially affected (Smit, 2007).

However, linearity can be assumed when using marginal models, as the changes of the emission levels are often relatively small compared to the total level (Heijungs et al., 2002). In addition to this, some impact curves are assumed to be of the nonthreshold linear type, which means that marginal and average approaches are equal (Heijungs et al., 2002), implying that linearity can also be assumed at concentrations less than the PNEC or HC5 (Huijbregts et al., 2000). The assumption of linearity by using an average or marginal approach has led to sound life-cycle impact assessment (LCIA) endpoint methodologies such as ReCiPe 2008 (Goedkoop et al., 2009), or the consensus model for toxicity USEtox™ (Rosenbaum et al., 2008), the latter based on previous work in the OMNIITOX project (Molander et al., 2004).

6.3.1.2 Resource Impact Indicators

The assessment of environmental impacts has been dominated for a long time by the characterization of emissions, but now, in the beginning of the twenty-first century, there are numerous indications that technology assessment should go beyond this and account for the emerging depletion of resources as well. The environment (ecosphere) can indeed be considered as a reservoir for energy, material, and land needs (from the technosphere). From this point of view, chemical industry and other economic activities can only exist as long as they exploit these (Dewulf et al., 2008). In addition to this, resources cost money and are thus directly coupled to economic profit. Because of this relevance, several methodologies were developed to account for the use of resources by processes and products. Elaborating the work of Steen (2006), resources can be subdivided into five midpoint types:

- Mass and energy
- Land use

- Exergy consumption
- Emergy consumption
- Relation to use of deposits.

In addition to this, one endpoint indicator is constructed, focusing on future consequences of resource extractions.

6.3.1.2.1 Midpoint Indicators

Mass and Energy Mass and energy are very tangible concepts, which are easily applicable owing to their omnipresent use in industrial applications. For material use, the material input per service (MIPS) unit indicator quantifies (in kilograms) the amount of resources needed to manufacture a product or service (Ritthoff, Rohn, and Liedtke, 2002). All materials taken away from the ecosphere and technosphere, depending on the system boundary of the study, are counted and subdivided into five different input categories:

- Abiotic or nonrenewable resources
- Biotic or renewable resources
- Earth movements in agriculture and silviculture (consumption/erosion and alteration through farming and forestry)
- Water
- Air.

When applying a similar approach at the level of national economies, this is called the Total Material Requirement (TMR) (Ritthoff, Rohn, and Liedtke, 2002). In recent years, water use is becoming a more relevant indicator, especially in regions with low water availability. Recent developments focus not only on surface- and groundwater use, but also on the available levels (Milà i Canals et al., 2009). Furthermore, the quality of the water that is discharged back into the environment in comparison to that of the input water can be included (Goedkoop et al., 2009).

For energy use, the concept of cumulative energy demand (CED) or cumulative energy requirement analysis (CERA) is used as a measure for the primary energy demand of a product or service, that is, the energy content (the upper heating value in joules) of energy carriers that have not yet been subjected to any conversions is quantified (Wrisberg and Udo de Haes, 2002) while taking losses due to transformation and transport fully into account (Klöpffer, 1997). Generally, eight categories are used:

- Nonrenewable resources:
    - Fossil
    - Nuclear
    - Primary forest.
- Renewable resources
    - Biomass
    - Wind
    - Solar

- Geothermal
- Water (Hydro-energy).

Land Use   In a world where biomass is gaining importance as an alternative for fossils, the inclusion of land use in assessments is essential. Two starting points can be taken:

- From an inventory point of view, that is, occupied area multiplied with the time, in m^2a, (Krajnic and Glavic, 2005), not considering.
- From an impact point of view, that is, including transformation and resulting effects such as impact on biodiversity, biotic production, ecological soil quality, and so on (Milà i Canals et al., 2007). Despite the difficulty of finding a starting point for transformation impacts, inasmuch as used land today is usually already transformed (Schmidt, 2008), assessing these types of impact is necessary. This can also be seen in the amount of work done on the assessment of biodiversity, where Delbaere (2002) has identified over 600 impact indicators. Examples focus on threatened vascular plant species (Schmidt, 2008), red-listed species (Kyläkorpi et al., 2005), threatened and endangered species (Singh et al., 2007), global species diversity (Jeanneret et al., 2006), and so on.

Exergy Consumption and Entropy Production   While energy is based on the first law, exergy is based on the second law of thermodynamics, which states that all processes and activities generate entropy. Exergy thus quantifies (in joules) the quality of all types of mass and energy (Dewulf et al., 2000) and the amount of useful work that can be obtained from a system or resource when it is brought to equilibrium with the chosen surroundings or "dead state" through reversible processes (Dewulf et al., 2008). As such, materials are delivered by the ecosphere and their exergy content is degraded in the technosphere.

Similarly to CED, the cumulative exergy consumption (CExC) can be defined as the sum of the exergy contained in all resources entering the supply chain of the selected product system. This resource indicator can be subdivided into 10 categories (fossil, nuclear, wind, solar, water, primary forest, biomass, water resources, metals, and minerals), including both flows and stocks (Boesch et al., 2007).

This approach can be applied in LCA, as an exergetic life-cycle assessment (ELCA) (Dewulf et al., 2008), which is elaborated by Dewulf et al. (2007) in the CEENE (cumulative exergy extracted from the natural environment) methodology, where wind, solar, primary forest, and biomass equivalents are elaborated as "renewable resources" and "land occupation and transformation."

As the concept of exergy is a measure of "usefulness," it has often been coupled to economics in two directions. Extended exergy accounting (EEA), gives an exergetic equivalent to a monetary cost. By doing so, other production costs such as labor, capital, and environmental remediation activities can be added to the exergy of the resources (fuel) needed. The conversion can be made by using a case- and time-dependent equivalence coefficient equal to the total influx of exergy in a given society in a certain year divided by corresponding monetary circulation (Sciubba and Ulgiati, 2005). Thermoeconomics does the opposite by giving monetary values to exergy streams by writing monetary balances on components or subsystems of a system (Dewulf *et al.*, 2008).

Emergy The starting principle of emergy is solar (equivalent) energy/exergy that creates, helps develop, and maintains all biophysical processes on earth. The emergy concept draws up a balance of all solar energy/exergy flows that are necessary and are thus "embodied" in the final product (in solar emJoules (sej)) (Bastianoni *et al.*, 2007). Apart from the constant input of solar energy on earth, geothermal, and tidal energy are the two other constant forms of energy that can be rescaled to solar equivalent (Odum, 1988; Hau and Bakshi, 2004). Emergy is calculated based on the transformity concept, which is defined as "the solar emergy required to make 1 J of a service or product" (Odum, 1996). This approach is interesting for ecologists, as it includes the contribution of ecological processes to human welfare (Hau and Bakshi, 2004), and is furthermore comprehensive in communication. However, putting the system boundary at the sun might be of less interest in industrial applications.

Relation to Use of Deposits Methods based on deposits use an R factor, which is a function of natural reserves (R_i in kg) of the resource i combined with its rate of extraction (DR_i in kilograms per year). An example is the abiotic depletion potential (ADP), which is used in the CML 2002 (Guinée *et al.*, 2002) method, where the ADP_i is derived for each extraction of element i and is seen relative to the depletion of antimony as a reference (DR_{ref} and R_{ref} in Sb-eq).

$$ADP_i = \frac{\frac{DR_i}{(R_i)^2}}{\frac{DR_{ref}}{(R_{ref})^2}}$$

The biotic depletion potential can be constructed in the same way, with another reference, for example, the reserve of African elephants (Guinée *et al.*, 2002).

6.3.1.2.2 Endpoint Indicators

Future Consequences of Resource Extractions As it is often difficult to estimate the total reserves of a certain resource, an alternative approach can be chosen, by accounting for the consequences of future extractions. This can be elaborated in an environmental context by quantifying the energy needed to extract the resource as megajoules per kilogram (Singh *et al.*, 2007; Goedkoop and Spriensma, 2001) or economically as money per kilogram extracted (Goedkoop *et al.*, 2009).

6.3.1.3 Technology Indicators

Technological indicators are frequently used in chemical industry and give specific information about the performance of the studied system within a specific chosen system boundary, often at the level of the gate-to-gate boundary. Examples for these are

- Yield, dividing the mass of product by the mass of raw material needed
- Waste indicators, dividing the amount of waste by the amount of product
- Recyclability, being the quotient of the amount of recycled material and the sum of all materials used
- Renewability, quantifying the share of renewable materials in the total amount of raw materials
- Energy efficiency, giving an indication how much energy is used for the product, and how much is directed to waste.

6.3.1.4 Assessment Methodologies

The methodological principles explained in the previous sections are the basis of many different assessment methodologies, with different names, depending mainly on choices concerning the system boundary and the chosen impact indicators.

6.3.1.4.1 Life-Cycle Methodologies

An LCA takes the cradle-to-gate or the full cradle-to-grave production chain into account, following the principles of the ISO 14040/44 (International Organization for Standardization, ISO, 2006) and International Reference Life Cycle Data System (ILCD) (European Commission–Joint Research Centre, JRC–Institute for Environment and Sustainability, IES, 2010) guidelines. Many methodologies have been proposed to execute an LCA, such as the Eco-indicator 99 (Goedkoop and Spriensma, 2001), Recipe 2008 (Goedkoop et al., 2009), USES-LCA (Huijbregts et al., 2000), LIME (Itsubo and Inaba, 2003), IMPACT 2002+ (Jolliet et al., 2003), EDIP2003 (Potting and Hauschild, 2005), LUCAS (Toffoletto et al., 2007), CML 2002 (Guinée et al., 2002), Carbon Footprint (European Commission, JRC, IES, 2007), and so on. In general, these methodologies have a very similar structure, and differ mainly in the exact definition of the impact assessment stage. A very comprehensive LCA methodology is the ecological footprint, originally developed by Wackernagel and Rees (Wackernagel and Rees, 1996) for the assessment of land use of populations and economies, but modified by Huijbregts et al. (2008) for inclusion in LCA as a combination of direct land use, land use for CO_2 sequestration, and land use for nuclear energy use. This methodology has been extended in the Sustainable Process Index (SPI) (Narodoslawsky and Krotscheck, 1995, 2000; Dewulf and Van Langenhove, 2006). Similarly, a water footprint can be constructed for nations (Chapagain et al., 2006), or for use in LCA, accounting for blue water, that is, ground and surface water, green water, that is, rainwater stored as soil moisture, and gray water, that is, water polluted during production (The Water Footprint Network, 2011). Unlike these more "default" LCA methodologies, the ECEC/ECOLCA methodology uses a hybrid approach with exergy or emergy as impact indicator (Ukidwe and Bakshi, 2007; Zhang et al., 2008a).

6.3.1.4.2 Gate-to-Gate Methodologies

If the assessment focuses on the gate-to-gate boundary of a production chain, the green degree (Zhang et al., 2008a) methodology based on, and thus very similar to, the WAR algorithm (Young, Scharp, and Cabezas, 2000) can be used. A flow sheet of this methodology is

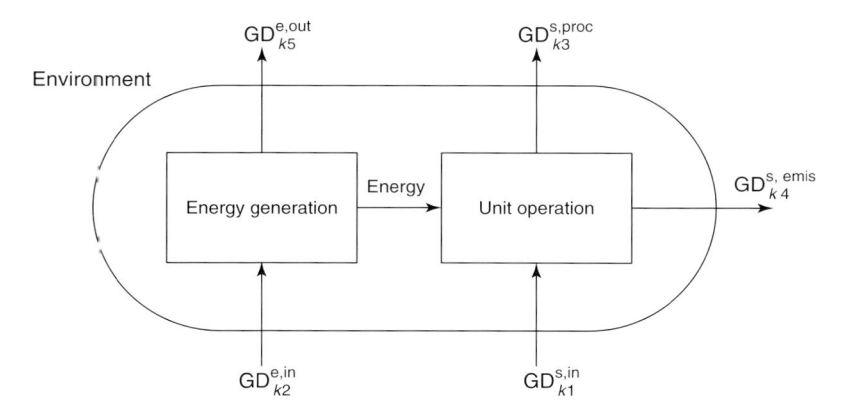

6.3 Impact Indicators and Assessment Methodologies

Figure 6.6 The process flow sheet of the green degree methodology.

presented in Figure 6.6. The green degree of substances (GD_i^{su}) is calculated by

$$GD_i^{su} = -\sum_j^9 (100\alpha_{i,j}\varphi_{ij}^N)$$

where

$$\varphi_{ij}^N = \frac{\varphi_{i,j}}{\varphi_j^{max}}, \varphi_j^{max} = \max(\varphi_{i,j})$$

$\alpha_{i,j}$ is the weighting factor of substance i for impact category j

$\varphi_{i,j}$ is the environmental impact potential of substance i for the nine midpoint impact categories j

$\varphi_{i,j}^N$ is the relative impact potential obtained by normalizing $\varphi_{i,j}$ by φ_j^{max}, which is the maximum value for category j among all the substances reported. The consequence is that for each impact category the indicator is normalized from 0 to 1. For mixtures, the green degree values of the substances are linearly added to each other.

The green degree value of the production of a unit (ΔGD^u) expresses the change that is caused by the material and energy conversions taking place in the unit and causing additional environmental impact. It is calculated by following equation

$$\Delta GD^u = \sum_{k3} GD_{k3}^{s,proc} + \sum_{k4} GD_{k4}^{s,emis} + \sum_{k5} GD_{k5}^{e,out} - \sum_{k1} GD_{k1}^{s,in} - \sum_{k2} GD_{k2}^{e,in}$$

With $GD_{k1}^{s,in}$ the green degree value of the input stream of the unit operation. k1 indicates the different input streams, such as raw materials, solvents, or catalysts. $GD_{k2}^{e,in}$ indicates the green degree value of an energy source fed into the energy generation system, such as natural gas, coal, or oil. $GD_{k3}^{s,proc}$ represents the green degree value of the process stream exiting the unit, possibly to another unit, and $GD_{k4}^{s,emis}$ is the green degree value of the emission or discharge stream from that unit directly to the environment. $GD_{k5}^{e,out}$ is the green degree value of an emission or discharge stream from the energy generation system into the environment. When assessing the production of a unit, $\Delta GD^u > 0$ indicates that the unit operation is benign to the environment and $\Delta GD^u < 0$ indicates that the unit operation adds pollution to the environment.

A semiquantitative methodology for process assessment in the early phase of development is presented by Biwer and Heinzle (2004), where substances get A/B/C scores in 14 impact categories, which can further be aggregated to two environmental factors.

6.3.1.4.3 Methodologies Using Technology Indicators

Several methodologies focus on performance of a system by using technological indicators:

- Fijal (2007) proposes a methodology including a raw material unit index, an energy unit index, a waste generation unit index, a production unit index, and a packaging unit index.
- The Green Chemistry metrics (Anastas and Warner, 1998) have been extensively discussed by Lapkin and Constable (2008) and have been elaborated in the EATOS (Environmental Assessment Tool for Organic Synthesis) methodology (Eissen and Metzger, 2002). These indicators focus on mass and energy efficiency, emissions, the amount of redox reactions quantified as a hypsicity indicator, and on economic cost associated with reactions.
- Dewulf and Van Langenhove (2005) combine exergy with principles of industrial ecology, to account for efficiency, reuse of materials, recoverability of waste materials, renewability, and toxicity.
- Lou *et al.* (2004) use emergy to assess efficiency and emissions of a production plant.

6.3.1.4.4 Shortcut Tool Kits

If the means for the assessment are limited, several shortcut tool kits can be used to have a first indication of the sustainability of the functional unit, which is, in this case, often a single process or product, not considering for supporting processes or supply chains:

- The Green Alternatives Wizard uses a database to replace hazardous chemicals on the basis of their properties (Massachusetts Institute of Technology, 2006).
- A solvent selection guide (Curzons, Constable, and Cunningham, 1999) accounts for environmental health and safety aspects of different solvents.
- The EcoScale (Van Aken, Strekowski, and Patiny, 2006) and the iSUSTAIN™ (Sopheon Corporation, 2011) are semiquantitative methodologies based on the 12 principles of green chemistry, using a system of scores or penalties for process parameters and technological properties of organic synthesis routes.
- The PBT profiler assesses persistence, bioaccumulation, and aquatic toxicity based on chemical structure (USEPA, 2011).

6.3.2
Economic Impact Assessment

Economic performance is the main driver of the industrial system. In most situations, it will even be *the* decisive factor whether a product or process will be chosen or not. Furthermore, the assessment of economics is rather straightforward in comparison to environmental and social assessment, as only one type of flow, namely, the monetary one, is assessed. As such, the discussion of midpoint and endpoint indicators and aggregation is not necessary.

A very important determining factor for the result of the assessment, however, is selecting a relevant cost breakdown structure to decide which direct and indirect costs and benefits should be included. Here, two main approaches are distinguished:

- **Using a company/user perspective**. Kawauchi and Rausand (1999) define three cost categories for the oil and chemical process industry:

    $$LCC = LAC + LOC + LLC$$

    with LAC, the life acquisition cost (e.g., equipment purchase, installation cost, commissioning cost, insurance spares, reinvestment cost, design cost, and administration cost), LOC, the life ownership cost (e.g., man-hour cost, spare parts consumption cost, logistics support cost, energy consumption cost, and insurance cost), and LLC, the life loss cost (e.g., cost of deferred production, hazard cost, warranty cost, loss of image, and prestige cost). Utne (2009) goes further and defines costs as

    $$LCC = CAPEX + OPEX + RISKEX + ENVEX + DISPEX$$

    where CAPEX + OPEX are the capital (production) and operational costs, respectively; RISKEX are the costs caused by accidents and fatalities; ENVEX are the environmental expenditures, mainly based on eco-taxes for climate change and acidification; and DISPEX are the costs for disposal.
- **Using a societal perspective**. This approach focuses, in addition to the previous cost categories, on the indirect or "external" costs. These externalities arise when the social or economic activities of one group of persons have an impact on another group and when that impact is not fully accounted, or compensated for, by the first group (ExternE, 2011). Efforts to quantify these costs, mainly by using the WTP and willingness to accept (WTA) principles, are made by the Environmental Priority Strategy (EPS) (Steen, 1999), the ExternE projects (ExternE, 2011), and the NEEDS project (NEEDS, 2011); the latter two available in the software tool EcoSense (Stuttgart University, 2011). Including these external costs, however, might not be an appropriate approach in sustainability assessments, since double counting issues with the environmental and social assessment might occur. Furthermore, within economic assessment, several eco-taxes already include some external costs (Pearce and Turner, 1990; Hanley, Shogren, and White, 1997).

Determining the "value" of money is another point influencing economic assessments. First, the value of the benefits should be considered as indirect, revenue-enhancing effects that can be generated next to the direct profit (Wynstra and Hurkens, 2005). Second, the value of money changes with time due to market mechanisms. The time dependency of the value of money can be included by discounting, based on changes in inflation, cost of capital, investment opportunities, and personal consumption preferences (Gluch and Baumann, 2004). This is mostly captured with the net present value principle, calculated by

$$\text{NPV} = \sum_{t=0}^{t=n} \frac{C_t}{(1+r)^t}$$

with n the number of years of analysis, r the discount or interest rate, and C_t the estimated costs in year t.

6.3.2.1 Economic Impact Indicators

The result of the cost breakdown should result in a final indicator, which aggregates the different values obtained during the assessment, and at the same time covers the discounting issue. In this context, Huppes *et al.* (2004) suggest seven different indicators:

- The NPV
- The average yearly cost (AYC), which is the sum of all yearly costs divided by the functional running time. This is called the steady-state cost (SSC) when the functional time is infinite.
- The annuity factor (A), which accounts for the regularly paid constant amount over the functional lifetime (fn). An eternal annuity factor has an infinite functional lifetime.

$$A = \text{NPV} \times \frac{r}{1-(1+r)^{-\text{fn}}}$$

- internal rate of return (IRR)The is the discount rate (r) which makes the NPV equal to 0, and thus makes the present value of benefits equal to the present value of costs.
- Profit is the benefits minus the costs.
- Payback time is the measure of the time required to return the initial investment and can thus be calculated by dividing the initial investment costs by the yearly net benefits.
- The benefit–cost ratio (BCR) is the ratio of the present value of the future benefits and the present value of the future costs, discounted at the same rate.

6.3.2.2 Assessment Methodologies

Owing to the relevance of money, economic assessment has widespread use. Therefore, many different names have been assigned to this type of study, mostly depending on the sector where it is applied. Originating from the building sector, whole-life costs (WLCs) and life-cycle costing (LCC) are used, where WLC is seen broader than LCC, by including all indirect, tangible, and intangible social, environmental, and business costs, and the benefits from consumption and production (International Organization for Standardization, 2008). The Center for Waste Reduction Technologies (CWRTs) introduced total cost assessment (TCA) to include environmental and health risks and costs (American Institute of Chemical Engineers, AIChE Center for Waste Reduction Technologies, CWRT, 2000). Similarly, total, full, or true cost accounting or full cost pricing, total cost of ownership, cost effectiveness analysis (CEA), or cost–benefit analysis can be used. However, it can be questioned what the difference between these methodologies really is. It is indeed stated that LCC is the theory behind all economic assessment methodologies, but depending on the perspective and application, choices are being made concerning the costs to be included (which and whose) and the time frame of the assessment (Huppes et al., 2004).

6.3.3
Social Impact Assessment

The third dimension of sustainable development differs quite significantly from environmental and economic assessments, as there are no "flows" going in and out a system, complicating the boundary, inventory, and impact assessment. At the same time, social issues are often hardly distinguishable from their economic context. This is why some sources such as Kruse et al. (2009) rather mention socioeconomic assessment. Aspects such as fair trade have been implemented for a longer time, but methodological developments concerning a quantified social impact assessment, compatible with environmental and economic impact assessments in industry, are still in a state of infancy and research has not advanced significantly in the last decades (Hunkeler and Rebitzer, 2005). This is why the UNEP SETAC Life Cycle Initiative started a working group and published its report "Guidelines for Social Life Cycle Assessment of Products" in 2009 (UNEP SETAC Life Cycle Initiative, 2009). The merit of this report is not only solving some of the obstacles but also boosting the debate and research.

In comparison to environmental assessment, the natural link between the physical input/output and an impact such as the change in environment is absent (Dreyer, Hauschild, and Schierbeck, 2006). Collecting social data is therefore not straightforward, and could result in a never-ending data collection (Nazarkina and Le Bocq, 2006). It is stated that the causal link of social issues is situated at a company/stakeholder level, and not at process level (Dreyer, Hauschild, and Schierbeck, 2006; Nazarkina and Le Bocq, 2006; Spillemaeckers et al., 2004). Furthermore, social data are difficult to quantify, and sometimes a more qualitative or semiquantitative approach might be preferred. In this context, Swarr (2009) quotes Einstein, "Not everything that can be counted counts, and not everything that counts can be counted". However, this is not necessarily a problem as Jørgensen et al. (2009) questioned several large Danish companies and concluded that a full list of quantitative life-cycle indicators for the social aspects is not essential. Industry just wants to take well-informed decisions, so it might be sufficient to simply define a process to assess the social impact in a reliable way (Swarr, 2009), rather than to construct a complete quantitative methodology.

Table 6.2 The structure of the social assessment methodology developed by the UNEP SETAC Life Cycle Initiative (2009) with stakeholder categories and impact indicators at the midpoint and endpoint levels.

Stakeholder categories	Impact categories	
	Midpoint	Endpoint
Workers/ employees	Freedom of association and collective bargaining, child labor, fair salary, working hours, forced labor, equal opportunities/discrimination, health and safety, and social benefits/social security	Human rights Working conditions Health and safety Cultural heritage Governance Socioeconomic repercussions
Consumer	Health and safety, feedback mechanism, consumer privacy, transparency, and end-of-life responsibility	
Society (national and global)	Public commitments to sustainability issues, contribution to economic development, prevention, and mitigation of armed conflicts, technology development, and corruption	
Local community	Access to material and immaterial resources, delocalization and migration, cultural heritage, safe and healthy living conditions, respect of indigenous rights, community engagement, local employment, and secure living conditions	
Value chain actors	Fair competition, promoting social responsibility, supplier relationships, and respect of IP rights	

6.3.3.1 Social Impact Indicators

T the UNEP SETAC LCI selected a list of midpoint and endpoint impact categories (Table 6.2) based on the definition "social impacts are consequences of positive or negative pressures on social endpoints (i.e., well-being of stakeholders)". These categories should then be assessed by indicators that are appropriate and fitting in the study. However, the midpoint–endpoint modeling might become very difficult. Further work on this topic is thus required, optionally using work done by other assessment frameworks such as Global Reporting Initiative (GRI) indicators, IChemE Sustainability metrics, and so on. A suggestion for a generic methodology could be to include an obligatory and a mandatory list of indicators (Dreyer, Hauschild, and Schierbeck, 2006; Kruse *et al.*, 2009).

As social impacts are site specific, Dreyer, Hauschild, and Schierbeck (2006) consider unlimited generic data unacceptable, regardless of the high data demand. Therefore, it is suggested that site-specific data from the most relevant processes should be gathered. For example, impacts that are not under the assessment executor's influence could then be handled with generic data. Alternatively, a hot-spot approach is suggested for identifying unit operations located in a region where problems risks or opportunities occur, where, in these hot spots, site-specific data are assessed (UNEP SETAC Life Cycle Initiative, 2009; Hauschild, Dreyer, and Jørgensen, 2008). Furthermore, a hybrid approach can be used, where macro data, for example, social data coupled to Gross National Product/Gross Domestic Product (GNP/GDP), can be coupled to micro assessments (Norris, 2006; Hutchins and Sutherland, 2008; UN, 2005).

6.3.3.2 Assessment Methodologies

Parallel to the construction of an impact assessment structure, several authors have tried to develop quantitative methodologies. The following are among the others:

- A societal LCA has been developed by Hunkeler (2006), which transforms the indicator "working hours" to the ability to acquire four regionalized societal necessities: housing, health care, education, and necessities.
- A working environment (WE) LCA can be used to grasp the direct impact, as lost work days (LWDs), of health and safety in the working environment (Schmidt et al., 2004; Kim and Hur, 2009).
- The Gabi methodology suggests a qualified working time (QWT) approach based on working time and qualification profile. This can be extended with a health and safety aspect and humanity of working conditions (Back et al., 2009; Makishi, Barthel, and Fischer, 2006).
- Weidema (2006) focuses on one endpoint indicator: the quality-adjusted life years (QALYs) concept, which is the elaborated version of the DALY indicator, including health-related quality of life. Six damage categories are addressed: life and longevity, health, autonomy, safety security and tranquillity, equal opportunities, and participation of influence. It is furthermore stated that this indicator can be linked to an overall endpoint indicator of human well-being.
- Labuschagne and Brent (2006) construct a social impact indicator (SII) based on 18 impact indicators in 4 resource groups (internal human resources, external population, stakeholder participation, and macro-social performance).

6.3.4 Multidimensional Assessment

According to Elkington's three dimensions of sustainability, a win-win-win situation should be obtained. Therefore, the assessment should focus on the people-, planet-, and profit-related issues similarly. Originally, most research was aimed at the concept of eco-efficiency and dematerialization or decoupling economic growth from environmental impact (World Business Council for Sustainable Development, WBCSD, 2005), but nowadays, the inclusion of social issues is high at the agenda.

It is generally assumed that the life-cycle approach will take an important role in sustainability assessment because of its holistic viewpoint (Klöpffer, 2003, 2005), which seems essential to grasp the broad concept of sustainable development. Indeed, a good evolution in one part of the life cycle can still have negative overall consequences. As such, Klöpffer (2008) proposes two possibilities for life-cycle sustainability assessment (LCSA):

LCSA = LCA + LCC + SLCA

LCSA = new LCA

The second option would mean that a completely new methodology should be constructed. However, up to now, the first option is most chosen, and, in most cases, LCSA is rather the streamlining and combination of knowledge in the different research areas.

The most well-known example is the SEE balance tool (Saling, 2010) developed by BASF, building on their previously developed eco-efficiency methodology (Saling et al., 2002; Kicherer et al., 2007; BASF, 2011). The environmental impact assessment consists of 11 midpoint categories: global warming potential (GWP), ozone depletion potential (ODP), acidification potential (AP), POCP, solid wastes, water emissions, energy consumption, raw material consumption, land use, toxicity potential, and a risk (of accidents) potential. This is combined with an economic LCC value and social indicators in five stakeholder groups: employees, consumer, local and national community, future generations, and international community. Following this, an external normalization is executed on the basis of national statistics such as the GDP or impact per GDP, while weights are determined by so-called relevance and societal factors. Doing so allows this methodology to have a communicative visualization, the so-called SEE-cube.

Other methodologies using the same strategy are the BEES (Building for Environmental and Economic Sustainability) tool (Lippiatt, 1999) using MCDA principles (Lippiatt and Boyles, 2001) and the Green Productivity methodology (Hur, Kim, and Yamamoto, 2004) accounting for eco-efficiency, while Öko-Institut has developed a PROduct Sustainability Assessment (PROSA) framework for sustainability assessment (Öko-Institut, 2007). The latter proposes a "usual" LCSA, such as the SEE balance, but gives more attention to a broader analysis of the product system in the market to allow a better scoping of the study. Furthermore, it has the unique feature that it also includes a fourth dimension of sustainability, that is, utility, which is based on three perspectives:

- Practical utility from the perspective of the users/consumers (durability, performance, reliability, etc.)
- Symbolic utility accounting for the perception of stakeholders (prestige, enjoyment, etc.)
- Societal utility or "public value," which focuses on the essential contribution of the product or service to societal objectives (basic needs, education, poverty reduction, etc.).

A simpler and user-friendly LCSA is presented in the Life-cycle INdeX (LINX) (Khan, Sadiq, and Veitch, 2004), which accounts for four impact categories: environment and resources (11 indicators), cost (3 indicators), technology (4 indicators), and sociopolitical factors (3 indicators). Each of these parameters is assessed with the use of a specific monograph where the value of the calculated parameter is directly linked to a penalty score, which eases the aggregation and interpretation step.

Other methodologies assess sustainability or eco-efficiency on a non-life-cycle basis. Examples of these are Sustain-Pro (Carvalho, Gani, and Matos, 2008), which uses the WAR algorithm and indicators from the IChemE metrics list, with added safety indices and seven own-developed indicators, and the Quotes for environmentally WEighted RecyclabiliTY and eco-efficiency (QWERTY/EE) methodology (Huisman, Boks, and Stevels, 2003), which was originally designed to assess the eco-efficiency of waste treatment and recycling options for electronic products.

6.3.5
Interpretation

The interpretation phase of the results of a sustainability assessment is of essential importance. It requires knowledge of the assessment framework with the choices and assumptions that are made in the study and the used indicators and possible aggregation steps. Only in this case, well-founded conclusions can be made at the end of the study. Determining the uncertainty of the final result can be a helping tool for the interpretation. According to Huijbregts (2011), three types of uncertainty can be determined, where only the last one is often quantified in assessment studies:

- Decision rule uncertainty includes the methodological choices that are made.
- Model uncertainty is caused by limitations in knowledge or techniques to quantify the impacts (e.g., endpoint cause–effect models, characterization factors, etc.)
- Statistical uncertainty is mainly related to uncertainty in the data inventory and is the uncertainty that is currently most analyzed by using a semiquantitative pedigree matrix base on scores in five criteria: reliability, completeness, temporal differences, geographical differences, and further technological differences (Luncie, Ciroth, and Huppes, 2007).

6.4
Conclusions

Assessment is essential for sustainable development in industrial progress. It is a necessary control tool to check whether certain chemical processes, products, and services and thus industrial companies, truly enhance sustainable development rather than just talking about it. It helps decision makers in making the right "sustainable" choices and will help companies to remain or become competitive.

As the concept of sustainable development originates mainly from environmental concern, many efforts have been made to assess environmental impacts. The relationships between the functional unit and the technosphere, as also their relations with the ecosphere are relatively well characterized. Furthermore, owing to the relevance of money, economic assessment is quite well established. The social dimension is elaborated; however, often it is, unlike the environmental and economic dimension, executed at a more macro level, which causes difficulties when linking social indicators to a company or product system.

It is clear for the assessment community that developing new methodologies is not the (only) way to go. Instead a harmonization and, in the best case, a specific standardization for the chemical industry of the knowledge in the three dimensions should be pursued. Therefore, next to research that reduces model uncertainty and includes more consequential or scenario-related issues, research aimed at lowering decision rule uncertainty by searching for streamlined goal setting, scoping, inventory collection, and aggregation of results is needed. Only when this is done, sustainability assessment can evolve further to the center of the decision-making process. This does not mean that performing sustainability assessments should be relegated to the background; on the contrary, it should be encouraged. Not only does it give a good understanding of the studied product system, it is also able to find bottlenecks in production and possible solutions that enhance sustainable development of the chemical industry and thus of society as a whole. However, the assessment and the following interpretation should both be executed with proper knowledge.

References

American Institute of Chemical Engineers (AIChE) Center for Waste Reduction Technologies (CWRT) (2000) Total Cost Assessment Methodology: Internal Managerial Decision Making Tool, AIChE.

Anastas, P. and Warner, J. (1998) *Green Chemistry: Theory and Practice*, Oxford University Press, New York.

Back, T., Bos, U., Colodel, C., and Lindner, J. (2009) Retrieved from http://www.lcacenter.org/LCA9/presentations/295.pdf. (accessed 2011).

BASF (2011) Retrieved from http://www.nutrition.basf.com/pdf/luprosil.pdf. (accessed 2011).

Bastianoni, S., Facchini, A., Susani, L., and Tiezzi, E. (2007) Emergy as a function of exergy. *Energy*, 32, 1158–1162.

Biwer, A. and Heinzle, E. (2004) Environmental assessment in early process development. *J. Chem. Technol. Biotechnol.*, 79, 579–609.

Boesch, M.E., Hellweg, S., Huijbregts, M.A., and Frischknecht, R. (2007) Applying cumulative exergy demand (CExD) indicators to the ecoinvent database. *Int. J. LCA*, 12, 181–190.

Cabeza Gutés, M. (1996) The concept of weak sustainability. *Ecol. Econ.*, 17, 147–156.

Carvalho, A., Gani, R., and Matos, H. (2008) Design of sustainable chemical processes: Systematic retrofit analysis generation and evaluation of alternatives. *Process. Saf. Environ. Prot.*, 86, 328–346.

Chapagain, A., Hoekstra, A., Savenije, H., and Gautam, R. (2006) The water footprint of cotton consumption: an assessment of the impact of worldwide consumption of cotton products on the water resources in the cotton producing countries. *Ecol. Econ.*, 60, 186–203.

Curran, M., Mann, M., and Norris, G. (2005) The international workshop on electricity data for life cycle inventories. *J. Clean. Prod.*, 13, 853–862.

Curzons, A., Constable, D., and Cunningham, V. (1999) Solvent selection guide: a guide to the integration of environmental, health and safety criteria into the selection of solvents. *Clean Products Processes*, 1, 82–90.

Delbaere, B. (2002) European Centre on Nature Conservation on Behalf of the European Topic Centre on Nature Protection and Biodiversity. European Community Biodiversity Clearing-House Mechanism. Retrieved from http://biodiversity-chm.eea.europa.eu/information/indicator/

F1034860208/F1034948411/F1036740916. (accessed 2011).

Dewulf, J., Bösch, M., De Meester, B., Van der Vorst, G., Van Langenhove, H., Hellweg, S. et al. (2007) Cumulative exergy extraction from the natural environment (CEENE): a comprehensive life cycle impact assessment method for resource accounting. *Environ. Sci. Technol.*, **41**, 8477–8483.

Dewulf. J., Langenhove, H.V., Muys, B., Bruers, S., Bakshi, B.R., Grubb, G. et al. (2008) Exergy: its potential and limitations in environmental science and technology. *Environ. Sci. Technol.*, **42**, 2221–2232.

Dewulf, J. and Van Langenhove, H. (2005) Integrating industrial ecology principles into a set of environmental sustainability indicators for technology assessment. *Resour. Conserv. Recycl.*, **43**, 419–432.

Dewulf, J. and Van Langenhove, H. (2006) *Renewables-Based Technology*, John Wiley & Sons Ltd, Chichester.

Dewulf, J., Van Langenhove, H., Mulder, J., van den Berg, M.M., van der Kooi, H.J., and Arons, J.D. (2000) Illustrations towards quantifying the sustainability of technology. *Green Chem.*, **2**, 108–114.

Diaz-Balteiro, L. and Romero, C. (2004) In search of a natural systems sustainability index. *Ecol. Econ.*, **49**, 401–405.

Dreyer, L., Hauschild, M., and Schierbeck, J. (2006) A framework for social life cycle impact assessment. *Int. J. LCA*, **11**, 88–97.

Eissen, M. and Metzger, J. (2002) Environmental performance metrics for daily Use in synthetic chemistry. *Chem. Eur. J.*, **8**, 3580–3585.

Ekvall, T. (2002) Cleaner production tools: LCA and beyond. *J. Clean. Prod.*, **10**, 403–406.

Elkington, J. (1994) Towards the sustainable corporation: Win-win-win business strategies for sustainable development. *Calif. Manag. Rev.*, **36**, 90–100.

European Commission, JRC, IES (2007) Retrieved from http://lca.jrc.ec.europa.eu/Carbon_footprint.pdf. (accessed 2011).

European Commission–Joint Research Centre (JRC)–Institute for Environment and Sustainability (IES) (2010) International Reference Life Cycle Data System (ILCD) Handbook–General guide for Life Cycle Assessment–Detailed guidance, 1st edn, Publications Office of the European Union, Luxembourg, March 2010.

European Commission, JRC, IES (2011) Retrieved from http://lca.jrc.ec.europa.eu/lcainfohub/datasetArea.vm. (accessed 2011).

ExternE (2011) Retrieved from http://www.externe.info/. (accessed 2011).

Feitz, A. and Lundie, S. (2002) Soil salinisation: a local life cycle assessment impact category. *Int. J. LCA*, **7**, 244–249.

Fijal, T. (2007) An environmental assessment method for cleaner production technologies. *J. Clean. Prod.*, **15**, 914–919.

Finnveden, G., Hauschild, M., Guinée, J., Heijungs, R., Hellweg, S., Koehler, A. et al. (2009) Recent developments in life cycle assessment. *J. Environ. Manag.*, **91**, 1–21.

Gasparatos, A., El-Haram, M., and Horner, M. (2008) A critical review of reductionist approaches for assessing the progress towards sustainability. *Environ. Impact Assess. Rev.*, **28**, 286–311.

Gluch, P. and Baumann, H. (2004) The life cycle costing (LCC) approach: a conceptual discussion of its usefulness for environmental decision-making. *Build. Environ.*, **39**, 571–580.

Goedkoop, M., Heijungs, R., Huijbregts, M., De Schryver, A., Struijs, J., and Van Zelm, R. (2009) Recipe 2008: A Life Cycle Impact Assessment Method Which Comprises Harmonised Category Indicators at the Midpoint and the Endpoint Level, 1st edn, Report I: Characterisation. VROM.

Goedkoop, M. and Spriensma, R. (2001) The Eco-indicator 99. A Damage Orientated Method for Life Cycle Impact Assessment. Methodology Report, 3rd edn., Pré Consultants, Amersfoort.

Guinée, J., Gorrée, M., Heijungs, R., Huppes, G., Kleijn, R., de Koning, A. et al. (2002) *Handbook on Life Cycle Assessment: Operational Guide to the ISO Standards*, Series: Eco-Efficiency in Industry and Science, Kluwer Academic Publishers, Dordrecht.

Hanley, N., Shogren, J., and White, B. (1997) *Environmental Economics: in Theory and Practice*, MacMillan Press Ltd, London.

Hau, J. and Bakshi, B. (2004) Promise and problems of emergy analysis. *Ecol. Model.*, **178**, 215–225.

Hauschild, M., Dreyer, L., and Jorgensen, A. (2008) Assessing social impacts in a life cycle perspective - lessons learned. *CIRP Ann. Manuf. Technol.*, **57**, 21–24.

Heijungs, R., Goedkoop, M., Struijs, J., Effting, S., Sevenster, M., and Huppes, G. (2002) Retrieved from media.leidenuniv.nl/legacy/recipe%20phase%201.pdf. (accessed 2011).

Huijbregts, M. (2011) A Proposal for Uncertainty Classification and Application in PROSUITE. Retrieved from http://www.prosuite.org/c/document_library/get_file?uuid=ce93fd1b-5c04-4cb0-bd65-98b740fe5b27&groupId=12772. (accessed 2011).

Huijbregts, M., Hellweg, S., Frischknecht, R., Hungerbühler, K., and Hendriks, A. (2008) Ecological footprint accounting in the life cycle assessment of products. *Ecol. Econ.*, **64**, 798–807.

Huijbregts, M., Thissen, U., Guinée, J., Jager, T., Kalf, D., van de Meent, D. et al. (2000) Priority assessment of toxic substances in life cycle assessment. Part I: calculation of toxicity potentials for 181 substances with the nested multi-media fate, exposure and effects model USES-LCA. *Chemosphere*, **41**, 541–573.

Huisman, J., Boks, C., and Stevels, A. (2003) Quotes for environmentally weighted recyclability (QWERTY): concept of describing product recyclability in terms of environmental value. *Int. J. Prod. Res.*, **41** (16), 3649–3665.

Hunkeler, D. (2006) Societal LCA methodology and case study. *Int. J. LCA*, **11** (6), 371–382.

Hunkeler, D. and Rebitzer, G. (2005) The future of life cycle assessment. *Int. J. LCA*, **10**, 305–308.

Huppes, G. and Ishikawa, M. (2009) Eco-efficiency guiding micro-level actions towards sustainability: Ten basic steps for analysis. *Ecol. Econ.*, **68**, 1687–1700.

Huppes, G., van Rooijen, M., Kleijn, R., Heijungs, R., de Koning, A., and van Oers, L. (2004) Life Cycle Costing and the Environment. Retrieved from http://www.rivm.nl/milieuportaal/images/Report%20LCC%20April%20%202004%20final.pdf. (accessed 2011).

Hur, T., Kim, I., and Yamamoto, R. (2004) Measurement of green productivity and its improvement. *J. Clean. Prod.*, **12**, 673–683.

Hutchins, M. and Sutherland, J. (2008) An exploration of measures of social sustainability and their application to supply chain decisions. *J. Clean. Prod.*, **16**, 1688–1698.

International Organization for Standardization (ISO) (2006) ISO 14040/44. *Environmental Management–Lifecycle Assessment–Requirements and Guidelines*, International Organization for Standardization, Geneva.

International Organization for Standardization (2008) ISO 15686–1. *Buildings and Constructed Assets. Service Life Planning Part 5: Life Cycle Costing*, International Organization for Standardization, Geneva.

Itsubo, N. and Inaba, A. (2003) A New LCA method: LIME has been completed. *Int. J. LCA*, **8**, 305.

Jeanneret, P., Baumgartner, D., Freiermuth, R., and Gaillard, G. (2006) Life Cycle Impact Assessment Method for the Impact of Agricultural Activities on Biodiversity, Agroscope FAL Reckenholz, Switzerland.

Jolliet, O., Margni, M., Charles, R., Humbert, S., Payet, J., Rebitzer, G. et al. (2003) Impact 2002+: a New life cycle impact assessment methodology. *Int. J. LCA*, **8**, 324–330.

Jolliet, O., Müller-Wenk, R., Bare, J., Brent, A., Goedkoop, M., Heijungs, R. et al. (2004) The LCIA midpoint-damage framework of the UNEP/SETAC life cycle initiative. *Int. J. LCA*, **9**, 394–404.

Jørgensen, A., Hauschild, M., Jørgensen, M., and Wangel, A. (2009) Relevance and feasibility of social life cycle assessment from a company perspective. *Int. J. LCA*, **14**, 204–214.

Kawauchi, Y. and Rausand, M. (1999) Life Cycle Cost (LCC) Analysis in Oil and Chemical Process Industries, Trondheim, RAMS Group, NTNU.

Khan, F., Sadiq, R., and Veitch, B. (2004) Life cycle iNdeX (LInX): a new indexing procedure for process and product design and decision-making. *J. Clean. Prod.*, **12**, 59–76.

Kicherer, A., Schaltegger, S., Tschochohei, H., and Ferreira Pozo, B. (2007) Eco-efficiency, combining life cycle assessment and life cycle costs via normalization. *Int. J. LCA*, **12**, 537–543.

Kim, I. and Hur, T. (2009) Integration of working environment into life cycle assessment framework. *Int. J. LCA*, **14**, 290–301.

Klöpffer, W. (1997) In defense of the cumulative energy demand. *Int. J. LCA*, **2**, 61.

Klöpffer, W. (2003) Life-cycle based methods for sustainable product development. *Int. J. LCA*, **8**, 157–159.

Klöpffer, W. (2005) Life cycle assessment as part of sustainability assessment for chemicals. *Environ. Sci. Pollut. Bull.*, **12** (3), 173–177.

Klöpffer, W. (2008) Life cycle sustainability assessment of products. *Int. J. LCA*, **13** (2), 89–95.

Krajnc, D. and Glavic, P. (2005) How to compare companies on relevant dimensions of sustainability. *Ecol. Econ.*, **55**, 551–563.

Kruse, S., Flysjö, A., Kasperczyk, N., and Scholz, A. (2009) Socioeconomic indicators as a complement to life cycle assessment – an application to salmon production systems. *Int. J. LCA*, **14**, 8–18.

Kyläkorpi, K., Rydgren, B., Ellegård, A., Miliander, S., and Grusell, E. (2005) The Biotope Method 2005: A Method to Assess the Impact of Land Use on Biodiversity, Stockholm, Vattenfall.

Labuschagne, C. and Brent, A. (2006) Social indicators for sustainable project and technology life cycle management in the process industry. *Int. J. LCA*, **11**, 3–15.

Lapkin, A. and Constable, D. (2008) *Green Chemistry Metrics: Measuring and Monitoring Sustainable Processes*, John Wiley & Sons Ltd, Chichester.

Larsen, H. and Hauschild, M. (2007) Evaluation of ecotoxicity effect indicators for Use in LCIA. *Int. J. LCA*, **12**, 24–33.

Linkov, I., Varghese, A., Jamil, S., Seager, T., Kiker, G., and Bridges, T. (2004). In I. Linkov, & A. Ramadan, *Comparative Risk Assessment and Environmental Decision Making* (pp. 15–54). Dordrecht: Kluwer Academic Publishers.

Lippiatt, B. (1999) Selecting cost-effective green building products: BEES approach. *J. Constr. Eng. Manag.*, **125**, 448–455.

Lippiatt, B. and Boyles, A. (2001) Using BEES to select cost-effective green products. *Int. J. LCA*, **6**, 76–80.

Lou, H., Kulkrani, M., Singh, A., and Hopper, J. (2004) Sustainability assessment of industrial systems. *Ind. Eng. Chem. Res.*, **43**, 4233–4242.

Lundie, S., Ciroth, A., and Huppes, G. (2007) Retrieved from *http://lcinitiative.unep.fr/includes/file.asp?site=lcinit&file=1DBE10DB-888A-4891-9C52-102966464F8D*. (accessed 2011).

Makishi, C., Barthel, L., and Fischer, M. (2006) Retrieved from *http://www.lcacenter.org/InLCA2006/Makishi2-presentation.pdf*. (accessed 2011).

Massachusetts Institute of Technology (2006) Retrieved from *http://web.mit.edu/environment/pdf/chem_alt_wiz_faq.pdf*. (accessed 2011).

Milà i Canals, L., Bauer, C., Depestele, J., Dubreuil, A., Knuchel, R., Gaillard, G. *et al.* (2007) Key elements in a framework for land use impact assessment within LCA. *Int. J. LCA*, **12**, 5–15.

Milà i Canals, L., Chenoweth, J., Chapagain, A., Orr, S., Antón, A., and Clift, R. (2009) Assessing freshwater use impacts in LCA: part I – inventory modelling and characterisation factors for the main impact pathways. *Int. J. LCA*, **14**, 28–42.

Molander, S., Lidholm, P., Schowanek, D., Recasens, M., Fullana i Palmer, P., Christensen, F., Guinée, J.B., Hauschild, M., Jolliet, O., Carlson, R., Pennington, D.W., and Bachmann, T.M. (2004) OMNIITOX – operational life-cycle impact assessment models and information tools for practitioners. *Int. J. LCA*, **9**, 282–288.

Narodoslawsky, M. and Krotscheck, C. (1995) The sustainable process index (SPI): evaluating processes according to environmental compatibility. *J. Hazard. Mater.*, **41**, 383–397.

Narodoslawsky, M. and Krotscheck, C. (2000) Integrated ecological optimization of processes with the sustainable process index. *Waste Manage.*, **20**, 599–603.

Nazarkina, L. and Le Bocq, A. (2006) Social aspects of Sustainability assessment: Feasibility of Social Life Cycle Assessment (S-LCA), EDF, Moret-sur-Loing, France.

NEEDS (2011) Retrieved from *http://www.needs-project.org*. (accessed 2011).

Norris, G. (2006) Social impacts in product life cycles. *Int. J. LCA*, **11**, 97–104.

Odum, H. (1988) Self-organization, transformity, and information. *Science*, **263**, 1243–1260.

Odum, H. (1996) *Environmental Accounting: EMERGY*, John Wiley & Sons, Inc., New York.

van Oers, L. and Huppes, G. (2001) LCA normalisation factors for the Netherlands, western Europe and the world. *Int. J. LCA*, **6**, 256.

Öko-Institut (2007) Product Sustainability Assessment (PROSA) Guideline, Öko-Institut, Freiburg.

Pearce, D. and Turner, R. (1990) *Economics of Natural Resources and the Environment*, Johns Hopkins University Press, Baltimore.

Pennington, D. (2001) Current issues in the characterisation of toxicological impacts. *Int. J. LCA*, **6**, 89–95.

Pennington, D., Potting, J., Finnveden, G., Lindeijer, E., Jolliet, O., Rydberg, T. et al. (2004) Life cycle assessment Part 2: Current impact assessment practice. *Environ. Int.*, **30**, 721–739.

Peters, G. and Hertwich, E. (2006) A comment on "functions, commodities and environmental impacts in an ecological–economic model". *Ecol. Econ.*, **59**, 1–6.

Potting, J. and Hauschild, M. (2005) Background for Spatial Differentiation in LCA Impact Assessment – The EDIP2003 Methodology, The Danish Environmental Protection Agency, Danish Ministry of Environment, Denmark.

Ritthoff, M., Rohn, H., and Liedtke, C. (2002) Calculating MIPS; Resource Productivity of Products and Services, Wuppertal Institute for Climate, Environment & Energy, Wuppertal, Germany.

Rosenbaum, R., Bachmann, T., Gold, L.H., Jolliet, O., Juraske, R., Koehler, A. et al. (2008) USEtox – the UNEP-SETAC toxicity model: recommended characterisation factors for human toxicity and freshwater exotoxicity in life cycle impact assessment. *Int. J. LCA*, **13**, 532–546.

Saaty, T. (1990) How to make a decision: the analytic hierarchy process. *Eur. J. Oper. Res.*, **48**, 9–26.

Saling, P. (2010) Retrieved from *http://www.greenchemistrynetwork.org/pdf/PeterSalingBASF.pdf*. (accessed 2011).

Saling, P., Kicherer, A., Dittrich-Krämer, B., Wittlinger, R., Zombik, W., Schmidt, I. et al. (2002) Eco-efficiency analysis by BASF: the method. *Int. J. LCA*, **7**, 203–218.

Sandén, B. and Kalström, M. (2007) Positive and negative feedback in consequential life-cycle assessment. *J. Clean. Prod.*, **15**, 1469–1481.

Schmidt, J. (2008) Development of LCIA characterisation factors for land use impacts on biodiversity. *J. Clean. Prod.*, **16**, 1929–1942.

Schmidt, A., Poulsen, P., Andreasen, J., Fløe, T., and Poulsen, K. (2004) The Working Environment in LCA. A New Approach. Guidelines from the Danish Environmental Agency, The Danish Environmental Protection Agency, Danish Ministry of Environment, Denmark.

Sciubba, E. and Ulgiati, S. (2005) Emergy and exergy analyses: complementary methods or irreducible idealogical options? *Energy*, **30**, 1953–1988.

Seppälä, J. and Hämäläinen, R. (2001) On the meaning of the distance-to-target weighting method and normalisation in life cycle impact assessment. *Int. J. LCA*, **6**, 211–218.

Singh, A., Lou, H., Yawsa, C., Hopper, J., and Pike, R. (2007) Environmental impact assessment of different design schemes of an industrial ecosystem. *Resour. Conserv. Recycl.*, **51**, 294–313.

Singh, R., Murty, H., Gupta, S., and Dikshit, A. (2009) An overview of sustainability assessment methodologies. *Ecol. Indic.*, **9**, 189–212.

Smit, M. (2007) Framework for the environmental impact factor for drilling discharges. Proceedings of the ERMS Seminar 2007, Oslo.

Soares, S., Toffoletto, L., and Deschênes, L. (2006) Development of weighting factors in the context of LCIA. *J. Clean. Prod.*, **14**, 649–660.

Sopheon Corporation (2011) Retrieved from *www.isustain.com*. (accessed 2011).

Spillemaeckers, S., Vanhoutte, G., Taverniers, L., Lavrysen, L., and van Braeckel, D. (2004) Integrated Product Assessment – The Development of the Label "Sustainable Development" for Products Ecological, Social and Economical Aspects of Integrated Product Policy, Belgian Science Policy, Belgium.

Steen, B. (1999) A Systematic Approach to Environmental Priority Strategies in Product Development (EPS): Version 2000 – General System Characteristics. Centre for Environmental Assessment of Products and Material Systems (CPM), Chalmers University of Technology, Gotheburg.

Steen, B. (2006) Abiotic resource depletion; different perceptions of the problem with mineral deposits. *Int. J. LCA*, **11** (1) special issue, 49–54.

Stuttgart University (2011) Retrieved from http://ecosenseweb.ier.uni-stuttgart.de. (accessed 2011).

Suh, S. (2009) *Handbook of Input–Output Economics in Industrial Ecology*, Springer-Verlag, Dordrecht.

Suh, S., Lenzen, M., Treloar, G., Hondo, H., Horvath, A., Huppes, G. *et al*. (2004) System boundary selection in life-cycle inventories using hybrid approaches. *Environ. Sci. Technol.*, **38**, 657–664.

Swarr, T. (2009) Societal life cycle assessment – could you repeat the question? *Int. J. LCA*, **14**, 285–289.

The Ecoinvent Centre (2011) Retrieved from http://www.ecoinvent.ch/. (accessed 2011).

The Water Footprint Network (2011) Retrieved from http://www.waterfootprint.org/?page=files/FAQ_Technical_questions. (accessed 2011).

Thiesen, J., Christensen, T., Kristensen, T., Andersen, R., Brunoe, B., Gregersen, T. *et al*. (2008) Rebound effects of price differences. *Int. J. LCA*, **13**, 104–114.

Thomassen, M., Dalgaard, R., Heijungs, R., and de Boer, I. (2008) Attributional and consequential LCA of milk production. *Int. J. LCA*, **13**, 339–349.

Toffoletto, L., Bulle, C., Godin, J., Reid, C., and Deschênes, L. (2007) LUCAS – a New LCIA method used for a CAnadian-specific context. *Int. J. LCA*, **12**, 93–102.

Udo de Haes, H., Jolliet, O., Finnveden, G., Hauschild, M., Krewitt, W., and Mueller-Wenk, R. (1999) Best available practice regarding impact categories and category indicators in life cycle impact assessment: part 1. *Int. J. LCA*, **4**, 66–74.

Ukidwe, N. and Bakshi, B. (2007) Industrial and ecological cumulative exergy consumption of the united states via the 1997 input–output benchmark model. *Energy*, **32**, 1560–1592.

UN (2005) Human Development Report. International Coopertaion at Crossroads: Aid, Trade and Security in an Unequal World, Hoechtstetter Printing Co, New York.

UNEP SETAC Life Cycle Initiative (2009) Guidelines for Social Life Cycle Assessment of Products, Opgehaald van http://lcinitiative.unep.fr/includes/file.asp?site=lcinit&file=524CEB61-779C-4610-8D5B-8D3B6B336463. (accessed 2011).

USEPA (2011) Retrieved from http://www.pbtprofiler.net/methodology.asp. (accessed 2011).

Utne, I. (2009) Life cycle cost (LCC) as tool for improving sustainability in the Norwegian fishing fleet. *J. Clean. Prod.*, **17**, 335–344.

Van Aken, K., Strekowski, L., and Patiny, L. (2006) EcoScale, a semi-quantitative tool to select an organic preparation based on economical and ecological parameters. *Beilstein J. Org. Chem.*, **2**, 3.

Van Leeuwen, C.J. and Hermens, J.L.M. (1995) *Risk Assessment of Chemicals: an Introduction*, Kluwer Academic Publishers, Dordrecht, Netherlands.

Wackernagel, M. and Rees, W. (1996) *Our Ecological Footprint: Reducing Human Impact on the Earth*, New Society Publishers, Philadelphia, PA.

Wegmann, F., Calvin, L., MacLeod, M., Scheringer, M., and Hungerbühler, K. (2009) The OECD software tool for screening chemicals for persistence and long-range transport potential. *Environ. Model. Softw.*, **24**, 228–237.

Weidema, B. (2003) Market Information in Life Cycle Assessment, The Danish Environmental Protection Agency, Danish Ministry of Environment, Denmark.

Weidema, B. (2006) The integration of economic and social aspects in life cycle impact assessment. *Int. J. LCA*, **11**, 89–96.

WHO (2011) WHO Global Burden of Disease Metrics, Retrieved from *http://www.who.int/healthinfo/global_burden_disease/metrics_daly/en/*. (accessed 2011).

World Business Council for Sustainable Development (WBCSD) (2011) Retrieved from *http://www.wbcsd.org/web/publications/ee_module.pdf*. (accessed 2011).

World Commission on Environment and Development (WCED) (1987) *Our Common Future*, Oxford University Press, Oxford.

Wrisberg, N. and Udo de Haes, H. (2002) *Analytical Tools for Environmental Design and Management in a System Perspective*, Kluwer Academic Publishers, Dordrecht.

Wynstra, F. and Hurkens, K. (2005). In M. Eßig, *Perspektiven des Supply Management* (pp. 463–482). Berlin, Heidelberg: Springer.

Xu, F., Zhao, S., Dawson, R., Hao, J., Zhang, Y., and Tao, S. (2006) A triangle model for evaluating the sustainability status and trends of economic development. *Ecol. Model.*, **195**, 327–337.

Young, D., Scharp, R., and Cabezas, H. (2000) The waste reduction (WAR) algorithm: environmental impacts, energy consumption, and engineering economics. *Waste Manage.*, **20**, 605–615.

Zamagni, A., Buttol, P., Porta, P., Buonamici, R., and Masoni, P. (2008) Retrieved from *fr1.estis.net/includes/file.asp?site=calcas&file=E8D08559-D042-4D5B-A935-395A140E2606*. (accessed 2011).

Zhang, X., Li, C., Fu, C., and Zhang, S. (2008a) Environmental impact assessment of chemical process using the green degree method. *Ind. Eng. Chem. Res.*, **47**, 1085–1094.

Zhang, Y., Bakshi, B., Rathman, J., and Zakin, J. (2008b). Ecologically-Based LCA: an Approach for Quantifying the Role of Natural Capital in Product Life Cycles. The Ohio State University, Ohio.

7
Integrated Business and SHESE Management Systems
Kathleen Van Heuverswyn and Genserik L.L. Reniers

7.1
Introduction

The chemical sector has a long tradition of working with management systems to improve performance. Traditionally, two broad categories of management systems can be distinguished: business management systems and risk management systems. The former are concerned with developing, deploying, and executing business strategies, while the latter focus on reducing SHESE (Safety, Health, Environment, Security and Ethics) risks.

Business management systems specifically aim at improving the quality or business performance of an organization, through the optimization of stakeholder satisfaction, with a focus on clients (e.g., by using the ISO Standard 9001:2008) or extended to other stakeholders such as employees, society, and shareholders (e.g., by using the EFQM 2010 Model for Business Excellence or the ISO 9004:2009 Guidelines).

Some of the most popular generic examples of risk management systems are the international standard for environmental management, ISO 14001:2004; the European Eco-Management and Audit Scheme, EMAS2; the internationally acknowledged specification for occupational safety and health, OHSAS 18001:2007; and the international standard for integrity management, SA 8000.

The boundaries between those two categories seem to have faded in recent years: first, through the increase of alignment and compatibility between the specific management systems, as a result of successive revisions (Van Heuverswyn, 2009); secondly, by the introduction of a modular approach with a general, business-targeted framework and several sub-frameworks, dealing with specific issues such as risk management, as is shown by the development of the EFQM risk management model in 2005. Thirdly, this evolution is illustrated by the emergence of integrated risk management models, which emphasize both sides of risks: the threat of danger, loss, or failure (the typical focus of risk management) as well as the opportunity for increased business performance or success (typical focus of business management). Examples of this last category are the Canadian Integrated Risk Management Framework (2001) and the Australian/New Zealand Standard

Management Principles of Sustainable Industrial Chemistry: Theories, Concepts and Industrial Examples for Achieving Sustainable Chemical Products and Processes from a Non-Technological Viewpoint,
First Edition. Edited by Genserik L.L. Reniers, Kenneth Sörensen, and Karl Vrancken.
© 2013 Wiley-VCH Verlag GmbH & Co. KGaA. Published 2013 by Wiley-VCH Verlag GmbH & Co. KGaA.

AS/NZS 4360:2004, which served as the basis for the development of the generic ISO risk management standard 31000:2009.

The aim of developing an integrated management system within an organization is to implement a continuous improvement cycle throughout the company's activities. By synchronizing different concepts, processes, and methodologies within an organization, integration leads to a sustainable improvement in the organization's overall performance.

The device "People–Planet–Profit" (P–P–P) reflects the idea of sustainability actually being considered as one of the results of integration because it clearly refers to societal, ecological, and economic dimensions that the management seeks to reconcile by integrated management. Within one company, sustainability's P–P–P dimensions are obviously represented in different disciplines, and are often the concern of separate departments. Their alignment within an organization is exactly what an integrated management approach attempts to achieve.

In this chapter, we take a closer look at the added value of integrated management systems and the steps required to successfully implement an integrated approach.

In the following, we discuss the requirements for integrating management systems (Section 7.2), the obstacles and advantages of an integrated approach (Section 7.3), examples of integrated models (Section 7.4), and the characteristics and added value of integrated management systems (Section 7.5).

7.2
Requirements for Integrating Management Systems

Specific management systems, as opposed to integrated management systems, focus on the improvement of one specific discipline, for example, quality, environmental performance, occupational safety and health, financial results, ethics, and so on. Despite the specific focus, the main characteristics of these specific models are to a large extent similar. Some basic requirements can be drawn from their comparison, offering a common platform for their integration and for the establishment of a global, integrated management system (Van Heuverswyn, 2009).

Integrated management requires so-called "systems thinking" due to existing interactions between different organizational processes and due to the many common aspects of these processes. A systems thinking approach allows visualizing the various existing relationships between individual processes, as well as analyzing and dealing with them. Such an approach encompasses four elements (Van Heuverswyn, 2009):

1) **Policy determination**: Define the core values and the engagement of top management; define the organization's policy (objectives and strategy); implement a global management system having "risk management" as one part thereof.
2) **Resources management**: Designate responsibilities and assign resources for achieving the desired objectives regarding personnel, infrastructure, planning, materials, technology, finances, training and education, and so on.

3) **Policy implementation**: Ensure adequate organization of the different processes, supporting measures, and procedures (such as communication, document management, operational control, emergency management, etc.).
4) **Measurements to learn**: Monitor and measure by registrations, audits, and feedback mechanisms, in order to assess and evaluate and eventually make changes to the policy.

To be able to elaborate the four systems thinking elements within an organization, one has to have a notion of the meaning of "an organizational process." A *process* can actually be defined as "any activity whereby input is transformed into output, by a number of chronological and functional interdependent phases" (Van Heuverswyn, 2009). One important process within any organization is the risk management process, containing three phases:

1) **Analyze the context**: This involves the identification (and making an inventory) of weaknesses, threats, hazards, dangers, and risks as well as of strengths and opportunities. The evaluation of all the identified elements is related to the strategic goals and objectives of the organization. Risk analyses, vulnerability assessments, SWOTs,[1] environmental impact assessments, and so on can be used to this end.
2) **Risk policy and strategy**: The results of the previously conducted analyses serve as the basis for designing a risk policy and implementation strategy, with clearly described targets and desired results, as well as responsibilities, tasks, time frames, and so on.
3) **Implementing the risk policy**: In order to achieve the targets and results set forward in the policy, an appropriate process configuration is elaborated.

As already mentioned, an effective management process requires the application of the principle of continuous improvement. The Deming loop of continuous improvement can be employed to explain how this principle can be achieved. The following are the different steps of the Deming loop, also called the *PDCA cycle*:

1) **Plan**: Develop a policy and determine the goals and processes necessary for achieving certain objectives based on the risk analyses carried out and the subsequent action programs.
2) **Do**: Execute the control measures and realize the policy objectives; implement the processes (e.g., the risk management process).
3) **Check**: Monitor, measure, and analyze the realization of the aims, targets, action programs, and so on and their effects, by means of inspections, audits, and so on. The result of these measurements and analyses is the definition of new corrective and/or preventive improvement actions, aiming to improve organizational processes and products in relation to the policy and its objectives.
4) **Act**: Take measures to continuously improve process achievements. The company policy is reviewed and eventually revised.

The chronological elements of systems thinking and the Deming cycle are very similar. The main difference between them is that systems thinking represents

1) (internal) Strengths-Weaknesses – (external) Opportunities and Threats.

```
┌─────────────────────────────┐     ┌─────────────────────────────┐
│           Plan              │     │            Do               │
│                             │     │                             │
│ – Policy development        │     │ – Structures and responsi-  │
│ – Risk identification and   │ ──► │   bilities                  │
│   evaluation                │     │ – Training and awareness    │
│ – Planning of measures      │     │ – Carry out measures        │
│   and actions to be taken   │     │ – Management of incidents,  │
│                             │     │   accidents, etc.           │
└─────────────────────────────┘     └─────────────────────────────┘
              ▲                                   │
              │                                   ▼
┌─────────────────────────────┐     ┌─────────────────────────────┐
│           Act               │     │          Check              │
│                             │     │                             │
│ – Carry out improvement     │     │ – Measurements and          │
│   actions                   │     │   analyses                  │
│ – Revise and adjust         │ ◄── │ – Registrations and audits  │
│   organizational risk policy│     │ – Controls and improvement  │
│                             │     │   actions                   │
└─────────────────────────────┘     └─────────────────────────────┘
```

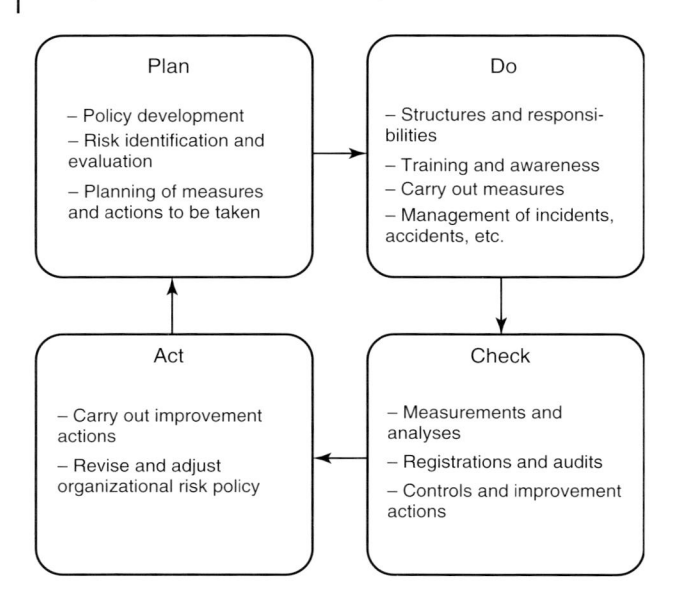

Figure 7.1 Basic philosophy of PDCA for specific risk management systems.

an approach of organizational theory, emphasizing the cyclic sequence of steps starting with the explicit definition of the company's policy and resulting in the achievement of the desired objectives. The Deming cycle describes the quality management principle of continuous improvement that needs to be applied to all aspects, processes, and activities throughout the company (Van Heuverswyn, 2011). They thus represent complementary approaches, sharing the same vision and the same objective, that is, of efficient and good businesses.

The PDCA cycle can be illustrated in different ways and it can be used for different types of activities. Figure 7.1 illustrates the basic philosophy of the PDCA cycle for SHESE risks. Figure 7.2 shows the systems thinking and PDCA approaches to be complementary, as well as the added value of both approaches for managing risks in an integrated way within an organization.

7.3
Integrating Management Systems: Obstacles and Advantages

When comparing different types of management systems, the parallelisms between them are remarkable from different points of view: philosophically, structurally, as well as with regard to the envisioned objectives. It seems therefore rather surprising that a discussion concerning the usefulness of an integrated management system has taken such a long time. It is especially difficult to find a solution for a number of structural obstacles. An overview of the pros and cons of an integrated approach are discussed in this section (Van Heuverswyn, 2009).

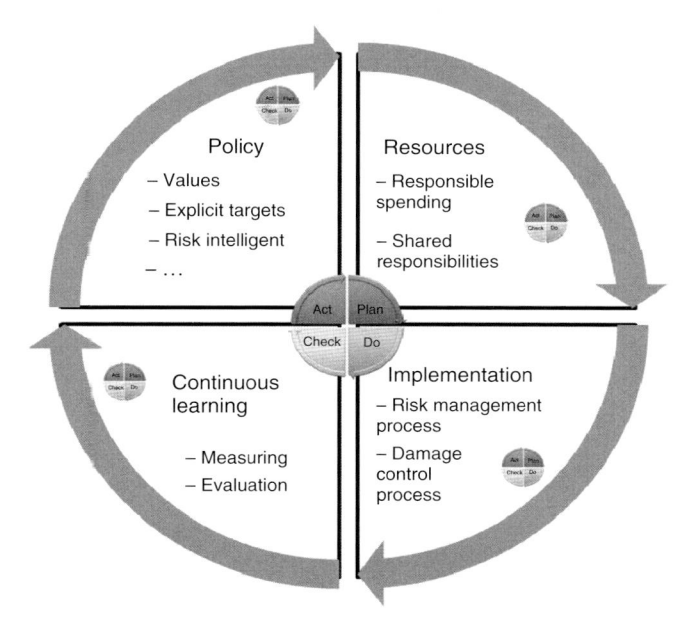

Figure 7.2 PDCA applied to integrated management. Source: Van Heuverswyn, 2009.

Every organization has a number of strategic objectives to ensure its existence and continuation. Each of the domains safety, health, environment, security, ethics, and quality form an essential contribution to this end. Balancing the importance of each of these management domains and their share in accomplishing the strategic goals can only be carried out at the top managerial level. This requires a global, holistic approach.

The most important goals are common to the different domains. The goals consist of into two parts: (i) to achieve the organizational objectives, including legal compliance, and (ii) to continuously improve overall performance. The previous section indicates that to attain both these goals, an organization has to set up systems, to designate people who are responsible, to develop procedures for monitoring regulations, and so on, and this, in a way that organizational systems become ever more effective and efficient. Elaborating such optimized organizational systems requires the introduction of analogous components concerning structures (organograms, determining tasks and responsibilities, etc.), processes (drafting procedures and instructions, planning and control of processes, etc.), personnel (training, education, competence management, communication, etc.), and documents (handbooks, working procedures, spreading of information, registrations, etc.).

By extending and integrating existing procedures, important savings can be made in time as well as in resources. Integration also requires high levels of transparency and communication, leading to better knowledge and enhanced dealing with (all types of) risks and business processes.

As a result, common values, shared targets, and joint responsibilities encourage the distinctive parts of an organization to work together to achieve the corporate goals instead of becoming each other's competitors in achieving their (department's) goals.

On the other hand, in many organizations, the specific management domains are traditionally separated owing to different points in time when they were installed and introduced in the company, and owing to different people (different functions, different profiles) being responsible for the implementation of the respective management systems. Integration of these management systems at a later stage in time is not an easy task in such case. Moreover, integrating the management systems is a complex project since all levels and all departments of the organization are affected. The domains sometimes require different academic disciplines and are subject to specific legislation. As a result, terminology and methodology can be very different, making interdisciplinary cooperation and integration difficult. Besides practical problems, integration often leads to tension due to psychological barriers. People need to think out of their reference framework, there may be a "global manager" imposed on a management domain, there is anxiety for the unknown, there is fear of losing power and influence, and so on.

An overview of advantages and disadvantages of integrating management systems is given in the following.

- Advantages of management integration:
  - Employees are confronted with one single procedure or instruction
  - Double work is avoided
  - All aspects of one management problem are considered simultaneously
  - Continuous improvement of all risk and business domains in one go
  - Structural components are converged (organograms, determination of tasks and responsibilities, audits, reviews, determination of key processes, crafting of organizational and technical procedures and instructions, planning and control of processes, monitoring, measuring, training, education, competence management, qualification, communication, documentation, handbooks, registrations, etc.)
  - All management systems have a structure based on the philosophy of continuous improvement
  - Risk control/limitation due to uniformity of procedures and documents
  - Risk control/limitation due to transparency and familiarity
  - Risk control/limitation due to job descriptions (responsibilities and liabilities) for several domains
  - Legal compliance becomes much more transparent and easier because of the existence of one integrated system for identification, monitoring, and dissemination of legislation, regulations, and guidelines for several domains (e.g., safety, security, environment)
- Disadvantages of management integration:
  - Separate procedures/instructions are preferred by some auditors for safety reasons (to avoid accidents)
  - Dependence on one coordinating person: risk of person leaving the organization and dependence on (risk of) single judgment

- In the current situation, a separate responsible person is assigned for every management domain, having his/her own policy accents
- Different management systems have been introduced at different points in time, and hence their maturity may differ a lot
- Dominance of one domain due to psychological factors or due to historical reasons (respect and value perception) within a company
- Holistic approach may lead to an unmanageable organization if not carried out well
- Different jargon of every domain
- Insufficient insight into integration objectives due to fear, resistance to changes, and so on.

Summarizing, there are arguments for as well as against integration resulting both from the complexity of such an undertaking and from the fragmentation of expertise, legislation, and mentality. The link between a company's complexity and the maturity of the risk management practices was also one of the conclusions of the fifth edition of the FERMA Risk management benchmarking survey in 2010 (FERMA, 2010).

7.4
Integrated Risk Management Models

Despite the many obstacles, and as a result of the many advantages, integrated management systems have been elaborated in industrial practice. This section presents some examples of such integrated management systems.

7.4.1
FERMA Risk Management Standard 2003

FERMA is the Federation of European Risk Management Associations, being the umbrella organization of 12 national risk management organizations. Its aim is mainly to exchange information between its members. The standard describes aspects such as terminology, process approach for risks, structural issues, and risk management goals.

The FERMA Risk Management Standard considers risk management as a crucial part of a corporate strategy, describing it as follows:

> It is the process whereby organizations methodically address the risks attaching to their activities with the goal of achieving sustained benefit within each activity and across the portfolio of all activities.

Risk management is seen as a continuous and continuously evolving process. Contrary to other standards, the RM Standard emphasizes much more the importance of a process-driven approach than the principle of continuous improvement. The Standard indeed makes no reference to the Deming's PDCA cycle, but gives an extensive description of the constituting steps of the risk management process:

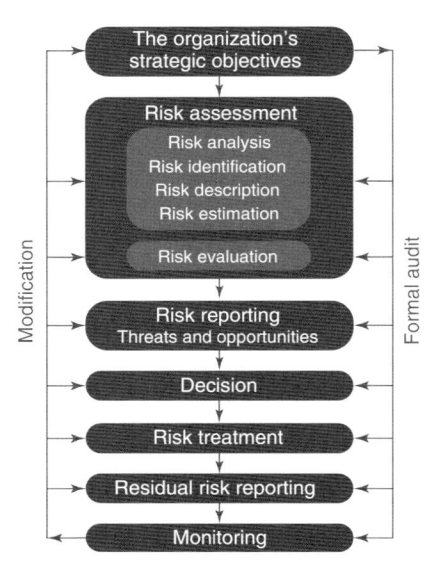

Figure 7.3 The FERMA risk management process.

from risk analysis (composed of identification, description, and estimation), to risk decisions based on informed reporting and risk treatment by implementing the chosen policy (Figure 7.3).

7.4.2
Australian/New Zealand Norm AS/NZS 4360:2004

The AS/NZS 4360:2004 norm was drafted as a general framework to establish the context, identification, analysis, evaluation, treatment, monitoring, and communication of risks. It does not strive for uniformity, but it explains the risk management process such that it may be implemented in different industrial settings.

AS/NZS 4360 aims to formulate guidance for public as well as private organizations that aim for a more reliable and rigorous basis for decision making and planning, better identification of opportunities and threats, taking advantage of uncertainty and variability, solid proactive management, more efficient use of resources, limitation of costs and losses, increase of stakeholders' trust, increase of compliance, and corporate governance.

The idea behind this norm is that risk management should be an integrated part of adequate management practices and of corporate governance. Risk management is an essential domain of an organization's culture, and is everyone's concern.

Risks are looked upon from a very wide angle, allowing the norm to be used in different organizational contexts. The process approach and the principle of continuous improvement are essential elements of – and used throughout – the norm. Figure 7.4 describes the risk management process within such a loop of continuous improvement.

7.4 Integrated Risk Management Models

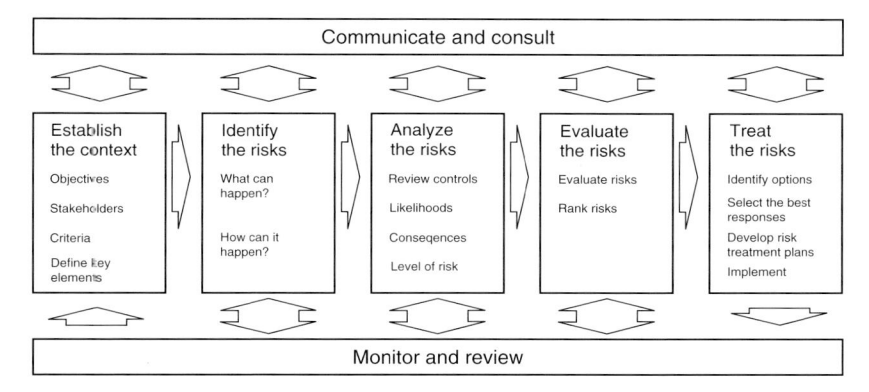

Figure 7.4 Overview of the risk management process within AS/NZS 4360.

7.4.3
ISO 31000:2009

ISO 31000:2009 was drafted for managing operational risks. The norm integrates general management principles of strategic management, offering an approach that is systems- and process oriented, and based on continuous improvement.

Figure 7.5 illustrates the integral risk approach of ISO 31000:2009.

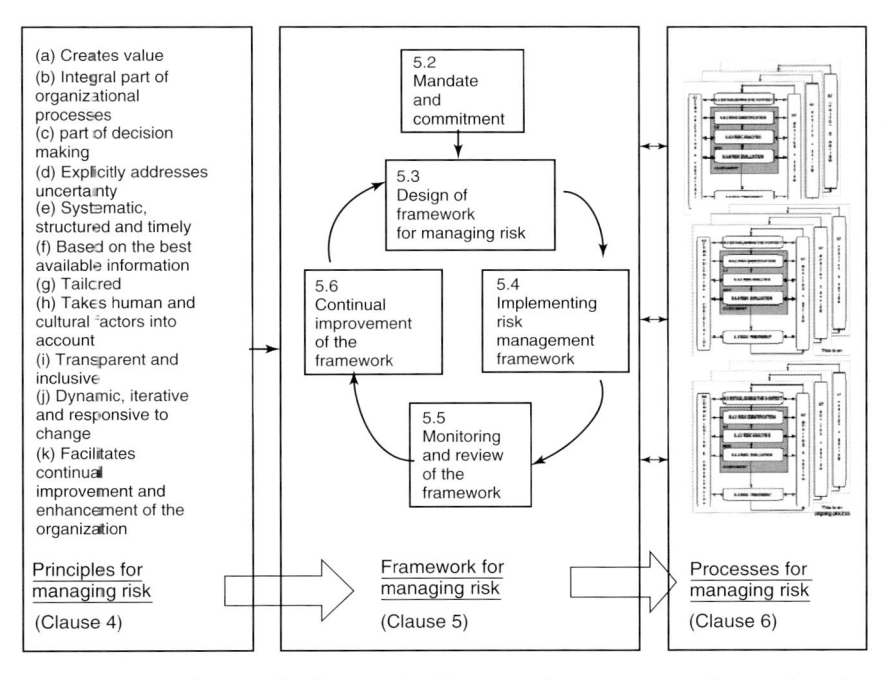

Figure 7.5 Integral approach of ISO 31000:2009: principles, management framework, and risk management process.

This ISO norm can actually be seen (and can be employed) as an "umbrella norm" to elaborate a management system for various operational risks and continuously and effectively improve an organization. As an example, ISO 31000:2009 stresses the need to apply the PDCA loop on all organizational processes. Nonetheless, although ISO 31000:2009 provides generic guidelines, it is not intended to promote uniformity of risk management across organizations. The design and implementation of risk management plans and frameworks will still need to take into consideration the specific needs of an organization, its particular objectives, context, structure, operations, processes, functions, projects, products, services, or assets, and specific practices employed.

It is intended that ISO 31000:2009 be utilized to harmonize risk management processes in existing and future standards. It provides a common approach in support of standards dealing with specific risks and/or sectors, and does not replace those standards.

7.4.4
The Canadian Integrated Risk Management Framework (IRM Framework)

Although the Canadian IRM Framework (2001) is specifically developed for implementation in the public sector, its characteristics are so generic that its possible applications are much larger and valuable for any type of organization.

The general purpose of the IRMF is to support all divisions, departments, and so on of an organization to think more strategically from an organization-wide perspective and to improve their ability in identifying common goals and priorities. At individual level, the IRMF helps employees to develop new skills and capabilities for anticipating, evaluating, and managing risks.

The main characteristics of this Standard are in line with the four management commitments it intends to support: (i) a better response to the needs of citizens/clients through sound, more informed decision making and greater emphasis on consultation and communication; (ii) respect on all levels for key values such as client-driven services, honesty, probity, and integrity; (iii) better performance, as a result of a proactive and smart risk approach; and (iv) a holistic, whole-of-government view based on rational priority setting and the principles of responsible spending.

The IRMF offers support in achieving these management commitments, through guidelines encouraging a more holistic and systematic approach for managing risks. In the IRFM's concept of risk management, it is closely linked to strategic management, because it is "about making decisions that contribute to the achievement of an organization's objectives." The IRFM wants to provide a mechanism that leads to more result-based management and improved results, through better-informed strategies and based on operational decisions that contribute to the achievement of the desired corporate goals.

Therefore, a cultural shift is needed toward a risk-smart workforce and environment, embracing both innovation and responsible risk taking, while ensuring precaution where necessary to protect the public's interests and to maintain public trust.

The IRMF proposes a set of risk management practices that the end users (agencies, departments) can adopt and adjust wherever necessary to their own requirements (mandate, circumstances). The common framework with its coordinated initiatives and shared responsibilities guarantees an organization-wide coherence in managing risks. On the other hand, the framework allows flexibility and adaptations to specific or changing circumstances.

Because of the integrated risk management approach, the Canadian Framework applies to all kinds of risks across all levels and departments of an organization. The IRMF focuses on minimizing the possible negative outcomes of risks, but also emphasizes the need to find an optimal balance between both sides of any risk, including positive opportunities. Therefore, the Framework provides guidelines to find a balance between stimulating innovation (which includes taking risks) and maintaining stability and continuity (a requirement that often leads to immobility, especially in the public sector).

Other characteristics of the Framework are a process-based approach and the principle of continuous improvement.

Figure 7.6 offers an overview of the different features of the IRM Framework.

The IRM Framework distinguishes four constituting elements, which can be summarized as follows:

- **Element 1. Developing the corporate risk profile** Integrated risk management starts with – and relies continuously on – internal and external environmental scanning of the organization, which includes (i) the identification of risks and opportunities and a description of the context and (ii) the assessment of the current status of risk management within the organization. From this knowledge,

Figure 7.6 The IRM Framework approach.

the organization's risk profile is drawn: keys risk areas, abilities and capabilities, and so on.
- **Element 2. Establishing an integrated risk management function** This refers to the whole "risk management infrastructure" enabling the risk policy to be implemented. There needs to be clear direction and demonstration of the support of senior management so that the policy is understood and applied. Then clear roles and responsibilities and performance reporting ensure the operationalization. Organizational capacity is built and learning plans and tools support the implementation throughout the whole organization.
- **Element 3. Practicing integrated risk management** The practical implementation follows the two previous steps and (i) is focused on the achievement of the desired goals, (ii) takes into account the risk profile, and (iii) relies on the appropriate infrastructure.

  Every segment of the organization consistently applies the common risk management process at all levels. The common process promotes a better understanding and facilitates risk communication. Risk management activities take place horizontally (in all policies, plans, and practices across the entire organization) and vertically (in programs and initiatives in every functional department). This approach guarantees a continuous alignment between the corporate policy and its implementation on every lower level.

  Tools and methods, such as risk maps, modeling tools, and so on are provided to support decision making and efficient practices.

  As an ongoing process, there is regular communication with – and consultation of – internal and external stakeholders.
- **Element 4. Ensuring continuous risk management learning** Continuous learning is considered fundamental. Learning from experience by sharing lessons, best practices, and so on are to be valued to continuously improve and innovate.

These four elements are to be applied throughout the organization and its activities.

7.5
Characteristics and Added Value of an Integrated Model; Integrated Management in Practice

When comparing a specific management system (designed for one domain, e.g., safety or environment or quality) with an integrated management system (designed for all domains), five common aspects can be found. First, they both provide guidelines on how to develop management systems without explicitly prescribing the how-exercise in detail. Second, they consider a risk management system to be an integral part of the overall company management system. This approach guarantees the focus and ability to realize the company's general and strategic objectives. Third, they all have two common aims: (i) to realize the organization's objectives taking compliance into full consideration and (ii) continuous improvement of an organization's achievements and performances. Fourth, the process approach is employed. Fifth, the models can be applied to every type of industrial organization.

The comparison also leads to the identification of several differences. First, integrated management systems recognize the positive as well as the negative possible outcomes of risks. Hence, both damage and loss on the one hand, and opportunities and innovation on the other, are simultaneously considered. Second, all types of risks are considered: operational, financial, strategic, juridical, and so on and hence, a balanced equilibrium is strived for. Third, the objectives of integrated management systems surpass compliance and continuous improvement.

The question that needs to be answered is how the integration of management systems can be realized in industrial practice, taking the similarities and differences of specific management systems into account. A suggested option is to integrate all management systems according to the scheme illustrated in Figure 7.7.

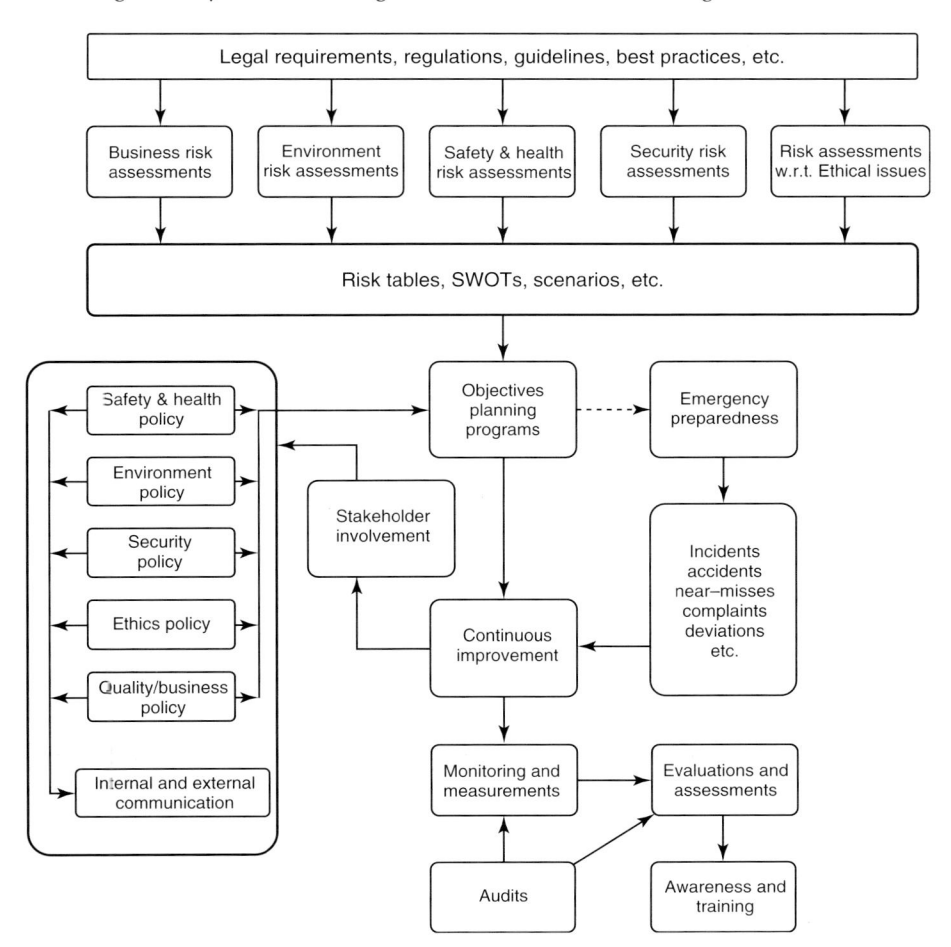

Figure 7.7 Scheme of integrated management. Source: Based on Bourmanne (2009).

7 Integrated Business and SHESE Management Systems

It is essential that an organization initially formulates a vision with respect to the system that is used to build the integrated management system (Figure 7.7 actually starts from the quality management system, but other choices are possible). Such a vision shows the intentions of the organization concerning its management system in the long term.

For a concrete elaboration of an integrated management system, the starting point can be the results of an organization, that is, its overall objectives and improvement measures. When these are known, instruments need to be thought of for carrying out daily operations and realizing the established aims (e.g., procedures, instructions, standard forms, letters). In a subsequent phase, instruments need to be figured out to assess the effectiveness and efficiency of goal realization within the company (e.g., employing risk analyses, threat assessments, SWOTs, internal and external audits, survey results, scoreboards). Finally, the analyses' results on the different domains (safety, health, environment, ethics/integrity, security, quality, etc.) as well as documents of related meetings (e.g., safety meetings, security meetings, quality meetings, management improvement meetings) need to be used for defining new objectives and improvement actions for the next time period. Figure 7.8 illustrates the entire integration process.

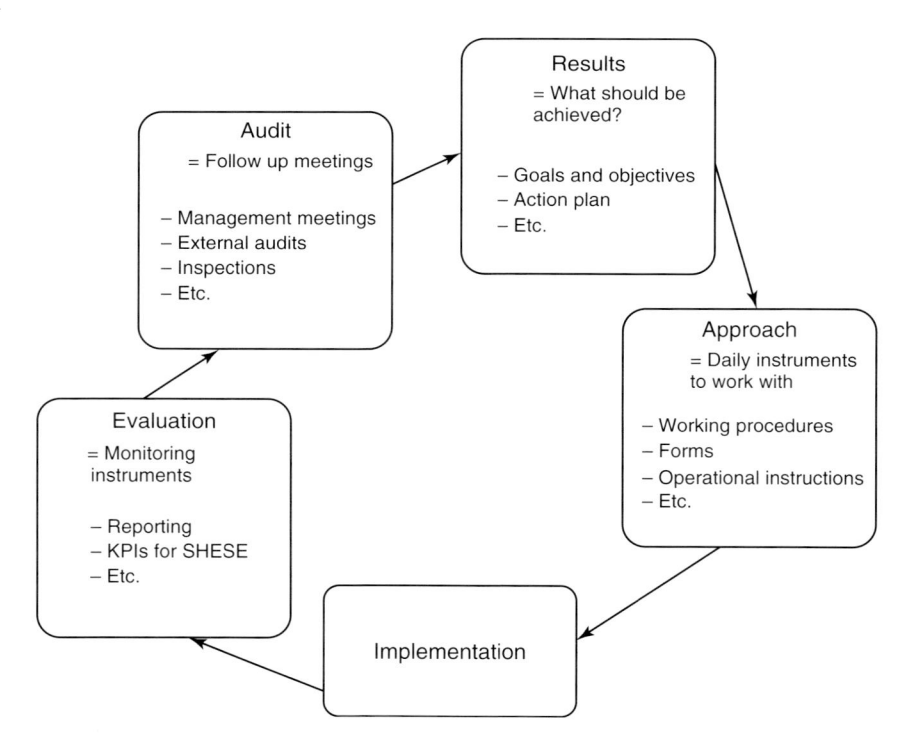

Figure 7.8 Integrated management in practice: an approach. KPIs, key process indicators. Source: Based on Bourmanne (2009).

7.6
Conclusions

Sustainability and integration both serve the same target: creating the conditions for several disciplines, divisions, and departments, each with their – possibly conflicting – objectives to work together in the interest of the long-term overall performance and success of an organization.

Both the sustainability and the integration principle are characterized by a holistic, organization-wide perspective and require a modular approach, anchored in a generic framework, creating space for specificity where needed and creating the indispensable global insight into the adequate working of an organization.

A generic integrating framework encompassing different disciplines leads to separate departments of an organization (e.g., H&S department, environment department) being aligned with each other, especially concerning an organization's policy and strategy. Common values, shared targets, and joint responsibilities can prevent the distinctive parts of an organization to strive for their own objectives or even becoming each other's competitors. Furthermore, an integrating framework allows to make information explicit, within some departments and hidden for others, even so as information is shared between, across, and beyond the distinctive departments. It is this extra information on causal insights that allows an organization to conduct a proactive policy in detecting weaknesses and strengths and creates opportunities to lift an organization to excellence. Sustainability and integration thus mutually reinforce each other in achieving overall organizational performance and success.

References

AS/NZS 4360:2004. *Australian-New Zealand Standard Risk Management*, 31 August 2004.

Bourmanne, R. (2009) *Practische Handleiding Voor Geïntegreerd Management*, Politeia, Brussels.

EFQM (2009) European Foundation for Quality management Excellence Model 2010, Excellent Organisations achieve and sustain superior levels of performance that meet or exceed the expectations of all their stakeholders, EFQM Publications, 36 p.

EMAS2 (2009) EC regulation N° 1221/2009 of the European parliament and of the council of 25 November 2009 on the voluntary participation by organisations in a community eco-management and audit scheme (EMAS), repealing regulation (EC) No 761/2001 and commission decisions 2001/681/EC and 2006/193/EC. *Off. J.*, L342.

FERMA A Risk Management Standard, AIRMIC, ALARM, IRM: 2002.

FERMA (2010) FERMA Risk Management Benchmarking Survey, 5th edn, Government of Canada Treasury Board, Integrated Risk Management Framework, Canada, 42 p., April 2010, http://www.ey.hu/Publication/vwLUAssets/FERMA_risk_management_benchmarking_survey_2010_5th_Edition/$FILE/EY_FERMA_risk_management_benchmark_2010.pdf.

ISO 9001:2008. *Quality Management Systems – Requirements*.

ISO 9004:2009. *Managing for the Sustained Success of an Organization – A Quality Management Approach*.

ISO 14001:2004. *Environmental Management – Requirements*.

ISO 31000:2009. *Risk Management Standard – Principles and Guidelines*.

OHSAS 18001:2007. *Health and Safety Standard.*

SA8000:2008. *Social Accountability Standard.*

Van Heuverswyn, K. (2011) Van prestatiemetingen naar geïntegreerde performantie, Deel I: Reeks Performance Management, Brugge, Die Keure, 52 p.

Van Heuverswyn, K. (2009) Leven in de Risicomaatschappij, Deel I: Negen Basisvereisten Voor Doeltreffend Risicomanagement, Antwerpen, Maklu/Garant, 365 p.

8
Supporting Process Design by a Sustainability KPIs Methodology

Alessandro Tugnoli, Valerio Cozzani, and Francesco Santarelli

8.1
Introduction

Achieving sustainable plants in the chemical and process industries requires the development of a balanced and responsible method to conduct production processes, so that they can be maintained indefinitely, reducing the negative impacts as far as possible. All the stages of the plant life cycle (R&D, design, construction, operation, decommissioning, etc.) play a role in the sustainability performance. However, the main determinants of the impacts during the plant life cycle are actually defined in the process and plant design stage. Therefore, design activities should take into account sustainability issues. In particular, the early phases of the process design are strategic for sustainability improvement, since a higher number of degrees of freedom is present and modifications can be implemented at relatively limited costs.

The theoretical principles of sustainability provide conceptual support and inspiration, but they alone are not sufficient to answer the problems frequently encountered in decision making during design (e.g., how much has a design modification improved the sustainability performance of the system? Is there any significant impact shift from one aspect to another?). Adequate metrics are required for the effective implementation of sustainability in process design. Sustainability performance of a system is not easy to measure, specifically in the early design stages. Applicable metrics have to take into account the specific factors of design activities, such as the availability of data, which may be limited. Moreover, suitable metrics should be able to directly quantify the impacts that can be effectively modified during a given design phase. For example, metrics requiring the evaluation of the "benefits of the employees" or of the "number of legal actions" are not applicable in the early design phase of the life cycle.

The Sustainability Key Performance Indicator (KPI) methodology (Tugnoli, Santarelli, and Cozzani, 2008, 2011) described in this chapter was specifically developed in order to meet the issues of supporting early process design. As a matter of fact, the analysis of the previous literature demonstrates several valuable efforts in the definition of metrics for evaluation of sustainability performance. However, none of these methods has a goal and a structure that makes it fully suitable for its application in early design. For example, some authors (Gonzales and Smith, 2003; Sikdar, 2003b) identified sustainability metrics in the evaluation of a set of critical parameters directly related to the process (consumption of nonrenewable raw materials, waste production, water use, CO_2 emission, yield, etc.). These parameters indeed define the sustainability profile of the process. However, the connection with the actual impact of the process is not straightforward. This hinders the evaluation of the magnitude of positive effects on sustainability of design improvements. A few proposed methodologies specifically address single contributors to the sustainability problem (e.g., material and energy flow analysis (Sikdar, 2003), exergy flows (Heui-seok et al., 2004; Jorge and Bhavik, 2004), specific potential impacts through midpoint environmental indicators (Allen and Shonnard, 2002; EC, European Commission, 2006; Bare et al., 2003), endpoint environmental indicators derived from Life Cycle Analysis (Allen and Shonnard, 2002; Chen et al., 2002; Goedkoop and Spriensma, 2001; ISO 14040–43; Pennington et al., 2000)), but do not feature a complete and comprehensive assessment of the system.

Conventional tools devoted to life cycle assessment (LCA) (ISO 14040) are not practically applicable to support process design. In fact, a large amount of data on the entire life cycle (well apart from the process section to be designed) is required and the procedure can become significantly time consuming. Moreover, in conventional LCA no site-specific considerations are usually included and the assessment is limited to environmental impacts, though some approaches for including life cycle costs and societal issues were proposed (Barthel and Albrecht, 2007; Dhillon, 1989; LCWE, 2010).

In process design activities, comprehensive indices are usually preferred to a fragmented set of metrics that cover all of the individual contributions. Numerical aggregation is usually proposed to obtain an overall quantitative indicator of process performance (Chen et al., 2002; Goedkoop and Spriensma, 2001; ISO, 14040–43; Mallick et al., 1996; Othman et al., 2010; Young, Scharp, and Cabezas, 2000; Saling et al., 2002; Shadiya and High, 2012). However, the need to compare indicators of different nature, the completeness of their set, and the subjective judgment in the aggregation process are critical and still need to be thoroughly investigated (Sikdar, 2003). Analytic hierarchy process (Satty, 1980), multiattribute utility theory (Keeney and Raiffa, 1976), and other methods were used to support aggregation activities (Chen and Shonnard, 2004). The problems relevant to the aggregation of indicators of different nature are only formally avoided in other approaches, which use a common metric for all the elements involved (e.g., economic measures, exergy, emergy, footprint, etc. (Figge and Hahn, 2004; Heui-seok et al., 2004; Krotscheck and Narodoslawsky, 1996; Narodoslawsky and Krotscheck, 2000; Jorge and Bhavik, 2004)), since a criteria must be stated to measure the different issues through the common unit of measurement applied.

Several assessment tools were promoted by USEPA for supporting the development of greener processes. Their Tool for the Reduction and Assessment of Chemical and Other Environmental Impacts (TRACI) (Bare et al., 2003) presents a wide set of environmental indicators for the analysis of industrial processes.

However the assessment is limited to environmental aspects and aggregation is avoided. Similarly, IChemE proposed a set of sustainability metrics for process plant management (IChemE, Institution of Chemical Engineers, 2002). The set encompasses all the aspects of the triple bottom line of sustainability (environment, economy, and society), but it is difficult to adapt in early design activities and still yields nonaggregated indicators.

BASF developed an eco-efficiency analysis tool that accounts for several categories of environmental and economic impacts. It is based on the aggregation of the indices in a modular scheme (Landsiedel and Saling, 2002; Saling et al., 2002; Shonnard, Kicherer, and Saling, 2012). The methodology was later extended to consider societal impacts (SEE-Balance). However, the societal upgrade considers indicators mainly linked to the facility management (e.g., working conditions, local community) and to the entire life cycle (e.g., child labor), which are poorly related to process design choices. Moreover, the use of internal normalization in the procedure may introduce biases in the assessment, especially considering highly result-sensitive applications such as design optimization.

Shonnard and coworkers, Chen et al. (2002, 2003), and Chen and Shonnard (2004) proposed and discussed the optimization of chemical processes in relation to economic and environmental performance. Their approach resorted to a multi-criteria aggregation approach similar to the KPI method, which will be discussed in the following. However no site-specific values are defined in the normalization of the indicators. The final aggregation of the economic and environmental indices is again based on internal normalization.

The Sustainability Key Performance Indicators Sustainability KPIs (Tugnoli, Santarelli, and Cozzani, 2008, 2011) is a system of indicators specifically oriented toward describing the sustainability footprint in preliminary design phases. The method provides specific approaches for the selection of relevant indicators and the introduction of site-specific factors in the normalization and aggregation of the results. This methodology constitutes an advancement of previous techniques, particularly with reference to the assessment of preliminary process flow diagrams (PFDs). It overcomes the limitations of nonspecific assessment tools and the over-simplifications of methods based, for example, on the sole reaction stoichiometry. This method will be described in the following.

8.2
Quantitative Assessment of Sustainability KPIs in Process Design Activities

Sustainability performance can be quantified during process design activities by a set of KPIs. These indicators evaluate the impact fingerprint expected from the future operation of the facility under design. The KPIs support different types of analysis in process design:

- **Analysis of impact criticality**: identifies the expected critical issues in the sustainability fingerprint of a proposed process;

- **Analysis of impact cause**: identifies the determinants of an impact in a given design option;
- **Selection of best option**: selects the best technology among a set of alternative design options;
- **Performance analysis**: monitors the performance improvement achieved by design modifications.

Besides supporting sustainability-oriented decision making, KPIs provide a simple and effective tool to communicate the results of design activities in terms of performances.

As discussed above, the limited availability of input data is one of the main characteristic features that differentiates KPI analysis of design options from sustainability assessment in other stages of the life cycle. The information typically available in the stage of preliminary design is summarized in Table 8.1. The Sustainability KPI method was specifically developed to allow its application using the input data available from a PFD (Tugnoli, Santarelli, and Cozzani, 2008).

The flow diagram reported in Figure 8.1 summarizes the assessment procedure of the Sustainability KPI method.

In the first step, the reference basis and the battery limits of the process are defined. An effective comparison of different process alternatives requires a common criterion in the definition of the assessed options. All the alternatives must share the same reference basis. This is conceptually similar to the functional unit of an LCA

Table 8.1 Summary of input data for the sustainability KPI assessment. Adapted from Tugnoli, Santarelli, and Cozzani (2008).

Type	Data
Process	Simplified process flow diagram (list of units and characterization of process streams)
	Energy input/output data
	Raw materials flow rate, composition, temperature, and pressure
	Products, coproducts and by-products flow rate, composition, temperature, and pressure
	Waste streams/effluents flow rate, composition, temperature, and pressure
	Fugitive emissions estimated rates and composition
	Hazardous proprieties of substances (flammability, instability, reactivity, toxicity)
	Operating conditions of equipment units (pressure and temperature)
	Preliminary definition of equipment types and geometry (taxonomy)
	Preliminary estimation of equipment inventories (quantity and average composition)
Economic	Preliminary estimate of capital cost
	Cost of raw materials
	Value of products
	Value or cost of disposal of coproducts and by-products
	Number of employees
	Labor cost

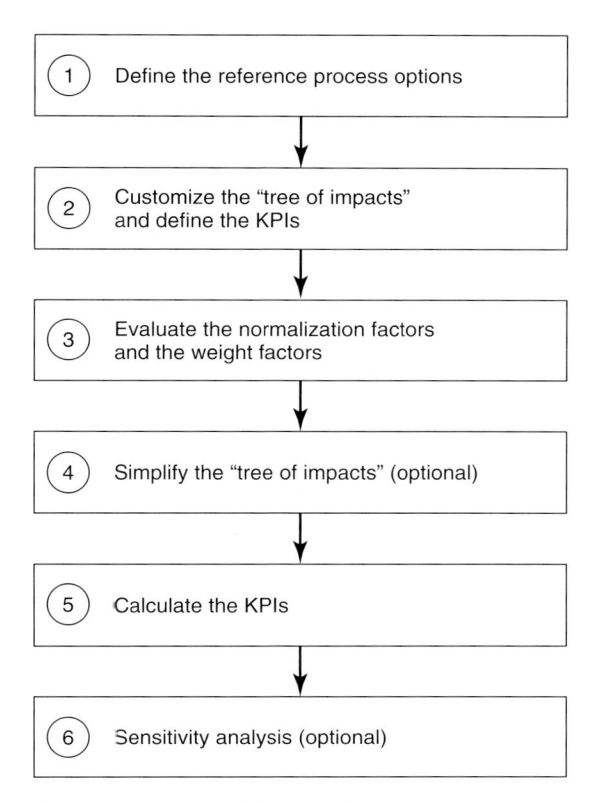

8.2 Quantitative Assessment of Sustainability KPIs in Process Design Activities | 109

1. Define the reference process options
2. Customize the "tree of impacts" and define the KPIs
3. Evaluate the normalization factors and the weight factors
4. Simplify the "tree of impacts" (optional)
5. Calculate the KPIs
6. Sensitivity analysis (optional)

Figure 8.1 Conceptual diagram of the proposed KPIs assessment method. Adapted from Tugnoli, Santarelli, and Cozzani (2011).

(International Organization for Standardization, ISO, 1997). A common reference basis may be easily introduced considering the same production potentiality for the different process alternatives. In the present framework, production potentiality is defined as (i) the production rate of the main product of the process (e.g., ammonia production, chlorine production) or (ii) the quantity of material processed per unit of time (e.g., waste treatment, LNG regasification terminal).

The process boundaries of the alternative options must be defined consistently. The goal of the assessment is to analyze different design options for a process facility at a given site. Thus the boundaries of the analysis should be limited to installations present in the site. As a consequence, they should include the main process (i.e., the core process that is the central focus of the design activity and, consequently, of the analysis) as well as the on-site utilities that are necessary for the main process to operate (e.g., boilers for steam production, nitrogen separation facilities, wastewater treatments, etc.). These must be included even if outside process design limits. However, in the analysis of the utilities, only the impacts allocated to the process should be considered (e.g., by the allocation procedures typical of LCA (Allen and Shonnard, 2002; International Organization for Standardization, ISO, 1997)).

The data required for the evaluation of the KPIs depend on the specific indicators used in the assessment. However, the typical information available in the early stages of process design should at least include the data reported in Table 8.1.

In the second step, a specific set of primary categories of impact and a correspondent set of KPIs are defined. The impact categories must cover all of the three main issues of sustainability: environmental, societal, and economical impacts. In order to grant a comprehensive and systematical analysis, a hierarchical structure, the "tree of impacts," was proposed by Tugnoli, Santarelli, and Cozzani (2011). The "tree of impacts" contains and organizes the impact categories of concern within the perspective of process sustainability assessment, reflecting both the issues identified by the sustainability policy of the site and the interests of the stakeholders. As such, the structure of the tree is generally independent from the specific process that is analyzed. Together with the tree structure, a set of suitable KPIs is identified. The construction of the tree of impacts and the definition of the KPIs is discussed in Section 8.3. A sample set of KPIs is shown in Table 8.3.

In step 3, the site-specific reference values for the normalization and aggregation of KPIs are defined. The suggested normalization approach ("external normalization") is based on the definition of a target area for each impact category (Tugnoli, Santarelli, and Cozzani, 2008). The target area is defined as the area that is significantly loaded and/or affected by the impact of concern. The external normalization factor is assessed accounting for the impacts from the industrial facilities already present in the target area. In this step, a weight factor is also defined for each branch of the tree of impacts, on the basis of the sustainability policy of the site. Further details on the definition of normalization factors and weight factors are discussed in Section 8.4.

Step 4 is an optional step, in which the event tree can be customized depending on the specific issues of the analysis. The simplification of the tree is a particular form of customization that can be applied in this step. It aims to optimize the time and resources devoted to the assessment, so it is especially valuable in early design applications where data availability is more limited. It is based on a preliminary assessment of the order of magnitude of the normalized KPIs. The definition of a cutoff criteria leads to the identification of the impact categories that are less relevant for the processes of concern. These categories can be neglected in further assessment steps, simplifying the overall assessment procedure. Further details on customization and simplification procedures of the impact tree are provided in Section 8.5.

In the fifth step of the procedure, the values of the primary KPIs defined in step 2 are calculated for each process option. These values address single key impact categories and can be directly used for the analysis of the performance of each process option. Nevertheless, non-normalized KPIs are expressed in different units and a direct comparison is possible only among the values of the same KPI. External normalization provides a reference background scale to enhance the interpretation of the primary KPIs. Normalized KPIs are calculated as

$$NI_i = \frac{I_i}{NF_i} \tag{8.1}$$

where I_i and NF_i are respectively the KPI and the normalization factor for impact category i, and NI_i is the normalized KPI.

In a site-specific external normalization approach (see Section 8.4), the normalization factors provide a quantification of the impact originated from the industrial activities present in the territory. Under these assumptions, a normalized index is a direct measure of the relative contribution of the assessed process option to the impact on the local site.

Normalized KPIs can be compared as such, also resorting to approaches as the Pareto optimization (Azapagic and Clift, 1999a,b), or can be aggregated according to the structure of the tree of impacts. A staged multi-criteria weighted summation procedure may be used to generate aggregated KPIs (Tugnoli, Santarelli, and Cozzani, 2008):

$$I_{aggr} = \sum_i W_i NI_i \tag{8.2}$$

where I_{aggr} is the aggregated index, W_i is the weight factor for the ith impact category, and NI_i the ith normalized KPI. The aggregation may produce indicators at any level of the hierarchy defined by the tree of impacts. Usually, indices for the three spheres of sustainability (second tier of the tree) and an overall impact index (first tier of the tree) are of practical interest in the comparison of alternative process options.

Clearly enough, the introduction of weight factors in step 4 is a nontechnical stage of the analysis that may introduce a bias in the results. Thus, after the calculation of the indicators, a sensitivity analysis may be carried out to understand how the above assumptions influence the final results (step 6). Details on this procedure are reported in Section 8.5.

Extensive examples of application for the Sustainability KPI method are not reported here for sake of brevity. Case studies comparing alternative options for the disposal of hazardous wastes (solvents and printed wiring boards) are presented by Tugnoli, Santarelli, and Cozzani (2008). The selection of an optimal process scheme for cyclohexanone production is discussed in detail in Tugnoli, Santarelli, and Cozzani (2011).

8.3
Identification of Relevant KPIs: the "Tree of Impacts"

The identification of a comprehensive and complete set of KPIs is of fundamental importance for a consistent assessment of the sustainability fingerprint. This requires two phases: (i) the identification of the set of impact categories to be considered in the assessment and (ii) the identification of suitable KPIs for each impact category.

A review of the literature (see Section 8.1) reveals that the selection of impact categories can be based on different practical criteria (experience, standard practice, identification of concerns, etc.). However, optimal and auditable results can be obtained only by the application of a formal procedure. In a recent study, Tugnoli, Santarelli, and Cozzani (2011) introduced a systematic procedure based on the definition of a "tree of impacts." The *tree of impacts* is defined as a hierarchical structure built according to the identification of impact categories of concern within the perspective of process sustainability assessment. Globally, the tree of impacts provides an organized representation of the concerns of the sustainability policy and of the interests of the stakeholders. The structure of the tree is generally independent from the specific process that is being analyzed, but is dependent on the macroeconomic sector (process industry). Clearly enough, the definition of the tree of impacts and of the weight factors (see step 3 of the assessment method) are intimately linked. The definition of a treelike structure was inspired by similar representations of a hierarchical structure extensively applied in safety, decision making, and computer technology (e.g., see Goodwin and Wright, 1998; Mannan, 2005; de Ville, 2006).

The conceptual flow diagram proposed for the definition of the tree of impacts in early design assessment is provided in Figure 8.2.

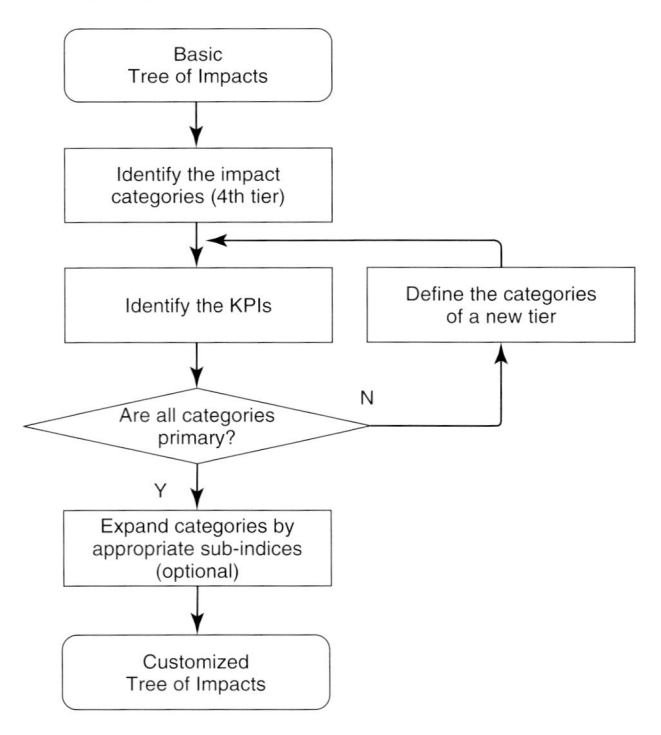

Figure 8.2 Conceptual diagram of the impact tree generation procedure. Adapted from Tugnoli, Santarelli, and Cozzani (2011).

The basic structure of the tree of impacts suitable for application in the process industry is shown in Figure 8.3. The structure was inferred combining the metrics for the classification of impacts proposed in former literature studies (Barthel and Albrecht, 2007; IChemE, Institution of Chemical Engineers, 2002; Jin and High, 2004; Saling et al., 2002; Sikdar, 2003). The root of the tree is the impact target selected for the analysis (i.e., the overall impact on sustainability). The second tier of branches reports the three domains of sustainability (IChemE, Institution of Chemical Engineers, 2002; Sikdar, 2003). At the third tier, the domains are further specified in subdomains of concern. Environmental impacts are classified according to emissions to distinct environmental media (air, water, and soil) and use of valuable resources (inclusive of materials I/O flows, energy I/O flows, land use, etc.). Impacts on the society are classified as either internal or external to the facility (i.e., the system of concern). Regarding the economic impacts, no standard subdomains are proposed, since all the relevant components for industrial processes (e.g., profit, value, investments, etc. (IChemE, Institution of Chemical Engineers, 2002)) in general can be described by a single monetary indicator (Barthel and Albrecht, 2007; Chen et al., 2002; EC, European Commission, 2006; Saling et al., 2002). The standard definition of this basic structure for the tree aims at providing a common taxonomy in the assessment and at promoting completeness in the definition of the impact categories and KPIs.

The fourth tier of the tree of impacts reports the sets of impact categories that describe the seven subdomains of the third tier. These sets are defined according to the specific sustainability concerns in the site where the system or process assessed will be located. A suitable fourth-tier impact category should reflect the potential outcomes of choices/elements implemented in the design activity of concern (i.e., preliminary PFD definition). A structured review method should be

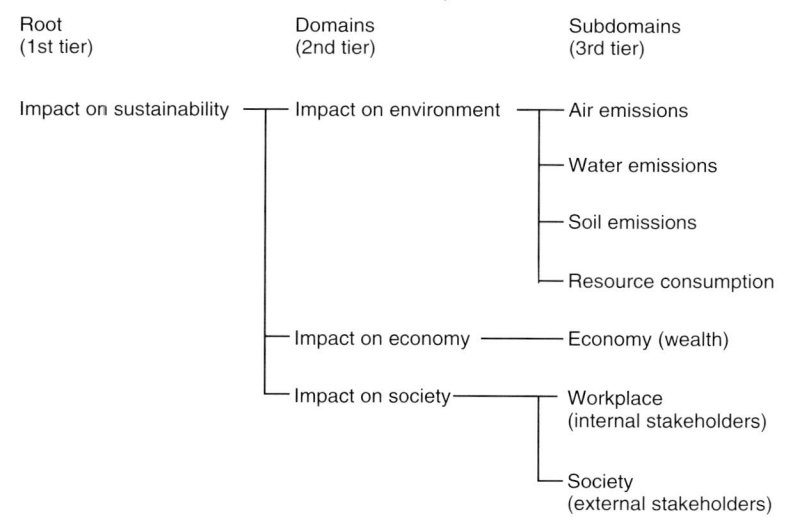

Root (1st tier)	Domains (2nd tier)	Subdomains (3rd tier)
Impact on sustainability	Impact on environment	Air emissions
		Water emissions
		Soil emissions
		Resource consumption
	Impact on economy	Economy (wealth)
	Impact on society	Workplace (internal stakeholders)
		Society (external stakeholders)

Figure 8.3 Basic structure of the "Tree of Impacts." Adapted from Tugnoli, Santarelli, and Cozzani (2011).

used for the identification of the potential impact mechanisms and, consequently, of the relevant impact categories. It should promote the flexible selection of a systematic set of impact categories representing the available knowledge and the specific features of the analyzed problem. Suitable tools include the OECD approach to safety KPI definition (Organisation for Economic Co-operation and Development, OECD, 2008), the application of system dynamics (Guest et al., 2010; Yaqin et al., 2008), and the use of guideword-based approaches adapted from process hazard identification techniques (International Organization for Standardization, ISO, 2000; Mannan, 2005). An example of application of the latter technique is provided by Tugnoli, Santarelli, and Cozzani (2011) and is reported in Table 8.2. The identification procedure consists in recognizing the potential impacts on sustainability deriving from the elements and parameters manipulated and defined in design activities. In the example, the brainstorming team considered the following components as relevant design elements/parameters: (i) materials; (ii) units/machines; (iii) utilities; and (iv) process as a network of units. The design elements were systematically screened in their impact potential considering the guidewords provided by the third-tier branches of the tree of impacts. The possible threats were identified according to the knowledge of the brainstorming team and the pertinent fourth-tier impact categories were defined accordingly. In some cases, impact mechanisms were simplified to single components (e.g., water species in lieu of the whole ecosystem) or included in other categories (e.g., soil emissions leached to water) owing to practical considerations in the definition of KPIs.

The impact categories at the fourth tier are usually suitable to be defined as "primary." A primary impact category (that may well be defined as a "leaf" of the tree of impacts) is an impact category that is not further developed, since a suitable KPI for its assessment can be defined. A suitable KPI for a primary impact category should satisfy the following attributes:

1) It introduces a quantitative metric.
2) It is directly linkable to the specific characteristics of the assessed options (e.g., midpoint indicator).
3) It has additive characteristics, thus allowing for normalization.
4) It is, as far as possible, a well-known or consolidated metric.

If no practical metric can be defined for a given impact category, a successive tier of the tree of impacts should be developed for that category, progressing recursively until a suitable KPI can be defined. Generally, the definition of these new tiers in the tree complicates the aggregation stage and, unless strictly necessary, should be avoided resorting to the more comprehensive metric available.

The definition of metrics for the primary categories may be conditioned by the availability of data for the evaluation of the normalization factors. In fact, the current methodology requires the definition of sound references for the normalization of indicators. The data sources for normalization factors (databases and statistics on industrial activities, emissions, occupation, welfare, accidents, etc.) frequently report generic and aggregated data, hindering the applicability of excessively detailed metrics.

Table 8.2 Example of a structured matrix for the identification of fourth-tier impact categories. Adapted from Tugnoli, Santarelli, and Cozzani (2011).

Element/parameter manipulated by design	Guideword	Threat/Impact mechanism	Impact category	Code
Materials	(A) Air emissions	Greenhouse gas	Global warming	A1
		Damage to stratospheric ozone layer	Ozone depletion	A2
		Acidifying agent or precursors of acidifying agent	Rain acidification	A3
		Promote/induce photochemical smog	Smog formation	A4
		Short-term damage to human beings	Toxicity in air	A5
		Long-term damage to human beings	Carcinogenicity in air	A6
		Short/long-term damage to non-human beings (consider a single critical target in the ecosystem: water species)	Account in B	See B5
	(B) Water emissions	Carbon content in water	Organic load	B1
		N/P content in water	Eutrophication	B2
		Short-term damage to human beings	Toxicity in water	B3
		Long-term damage to human beings	Carcinogenicity in water	B4
		Short/long-term damage to non-human beings (consider a single critical target in the ecosystem: water species)	Ecotoxicity	B5
	(C) Soil emissions	Carbon content in soil → leach to water	Account in B	See B1
		N/P content in soil → leach to water	Account in B	See B2
		Short-term damage to human beings → leach to water	Account in B	See B3
		Long-term damage to human beings → leach to water	Account in B	See B4

(continued overleaf)

Table 8.2 (continued)

Element/parameter manipulated by design	Guideword	Threat/Impact mechanism	Impact category	Code
		Short/long-term damage to non-human beings (consider a single critical target in the ecosystem: water species)	Account in B	See B5
		Discharge of materials (wastes) to soil	Solid waste disposal	C1
	(D) Resource consumption	Materials of nonrenewable origin (e.g., crude oil, natural gas, minerals, etc.)	Nonrenewable raw materials	D1
		Materials of renewable origin (e.g., corn, wood, etc.)	Renewable raw materials	D2
	(E) Economy (wealth)	Monetary value of the material	Economic impact	E1
		Disposal cost of materials	Economic impact	E1
	(F) Workplace	Short-term damage to human beings in the workplace	Toxicity of fugitive emissions	F1
		Long-term damage to human beings in the workplace	Carcinogenicity of fugitive emissions	F2
	(G) Society	Short-term damage to human beings by continuous emissions to air/water/soil	Account in A/B/C	See A5, B3
		Long-term damage to human beings by continuous emissions to air/water/soil	Account in A/B/C	See A6, B4
		Accidental release of hazardous/energetic (instable, reactive, etc.) material	Inherent safety	G2
Units/Machines	(omitted)	(omitted)	(omitted)	(omitted)
Utilities	(omitted)	(omitted)	(omitted)	(omitted)
Process (unit network)	(omitted)	(omitted)	(omitted)	(omitted)

Several metrics are suitable as primary KPIs. While providing an exhaustive list is beyond the scope of this section, Table 8.3 provides a few sample KPIs frequently applicable in early process design. As shown in the table, well-known and widely used "potential impact indicators" (Allen and Shonnard, 2002; Pennington et al., 2000) were proposed for the assessment of the environmental impact. The adoption of potential impact indicators is particularly useful in this context, since the indicators selected present a direct correlation among flows and impacts. Thus, their assessment is possible even in the initial stages of process or plant design, since it does not require the use of complex models of environmental fate, affected by a relevant uncertainty (Bare et al., 2000). Moreover, potential impact indicators yield conservative impact values, thus reflecting a precautionary approach. Standard procedures are available in the literature for the quantitative assessment of potential impact indicators, usually based on the reference to a benchmark substance (Allen and Shonnard, 2002; Pennington et al., 2000). However, in Table 8.3, an "ad hoc" definition of the benchmark rule has been introduced for a few indicators (toxic and carcinogenic releases, soil emissions, and resource consumption), for which poorly standardized approaches were found in the literature. The rules used to define the impact assessment criteria for these indicators are briefly described in the table. According to the benchmark rule, the indicators are expressed as a quantity of an equivalent reference material by an equivalence factor, named "potential impact factor" (PIF). The PIF represents the ability of a material to cause damage related to the reference material. The generic indicator I_i is calculated according to the general equation:

$$I_i = \sum_j \text{PIF}_{i,j} m_j \tag{8.3}$$

where $\text{PIF}_{i,j}$ is the potential impact factor for the ith impact category for jth material and m_j the mass of jth material crossing the system boundaries.

With respect to economic indicators, any suitable indicator should be able to capture at least the operative and capital costs involved in the proposed process (Chen et al., 2002; Saling et al., 2002). The index proposed in Table 8.3 is based on the net present value (NPV), which is typically adopted in the economic comparison of alternatives. The economic index used is actually NPV multiplied by -1. The change of sign was introduced since impact indices, used in the present approach, must have a higher value in the worst option.

The definition of quantitative indices for the societal impacts relevant to the process sustainability is a difficult task, as management and policy aspects are involved, that are not directly related to the process or that are not easily quantifiable. Tugnoli Santarelli, and Cozzani (2008) proposed the use of a limited number of societal indices directly related to the available choices in process design. According to the third tier of the tree of impacts, the KPIs are divided into two classes. The first concerns the potential threat to the health of stakeholders spending a significant amount of time inside the plant. The continual exposure to fugitive emissions is evaluated by a couple of indicators very similar to the ones used to assess the impact of toxic and carcinogenic air emissions. The second class concerns the

Table 8.3 Examples of primary KPIs suitable for process design applications, with definition of the process and normalization data required, and of the suitable impact metrics. Adapted from Tugnoli, Santarelli, and Cozzani (2011).

Impact category	Process data required	Reference normalization data	Suitable metrics (KPIs)
Air emissions			
Global warming	Air emissions of greenhouse gases and organic compounds, for example, from (i) boiler stacks, (ii) potential fugitive emission sources, and (iii) vents and waste incinerators	Emissions of greenhouse gases from industrial sites and power plants	Equivalents of CO_2 of the gas emissions per year (Lowe et al., 2007)
Rain acidification	Air emissions of acid compounds, for example, from (i) boiler stacks, (ii) potential fugitive emission sources, and (iii) vents and waste incinerators	Emissions of acid gases (SO_x, NO_x, etc.) from industrial sites and power plants	Equivalents of SO_2 (max H^+ release) of the gas emissions per year (Allen and Shonnard, 2002)
Smog formation	Air emissions of volatile organic compounds, for example, from (i) boiler stacks, (ii) potential fugitive emission sources, and (iii) vents and waste incinerators	Emissions of volatile organic compounds from industrial sites and power plants	Equivalents of maximum incremental reactivity of the gas emissions per year (Carter, 1994)
Air toxicity	Air emissions of toxic compounds, for example, from (i) boiler stacks, (ii) potential fugitive emission sources, and (iii) vents and waste incinerators	Air emissions of toxic compounds from industrial sites and power plants	Equivalents of toluene toxicity ($1/LC_{50'inh}$) of the gas emissions per year (Chen and Shonnard, 2004)
Air carcinogenicity	Air emissions of carcinogenic compounds, for example, from (i) boiler stacks, (ii) potential fugitive emission sources, and (iii) vents and waste incinerators	Air emissions of carcinogenic compounds from industrial sites and power plants	Equivalents of benzene cancer potency slope factor (inhalation) of the gas emissions per year (Chen and Shonnard, 2004)
Water emissions			
Organic load	Liquid emissions of organic compounds, for example, from (i) water discharges and (ii) fugitive emission sources	Water emissions of organic compounds from industrial sites and power plants	Total oxygen demand of the water emissions per year (Pennington et al., 2000)
Water toxicity	Liquid emissions of toxic compounds, for example, from (i) water discharges and (ii) fugitive emission sources	Water emissions of toxic compounds from industrial sites and power plants	Equivalents of toluene toxicity ($1/LD_{50'orl}$) of the water emissions per year (Chen et al., 2002)

Table 8.3 (continued)

Impact category	Process data required	Reference normalization data	Suitable metrics (KPIs)
Water carcinogenicity	Liquid emissions of carcinogenic compounds, for example, from (i) water discharges and (ii) fugitive emission sources	Water emissions of carcinogenic compounds from industrial sites and power plants	Equivalents of benzene cancer potency slope factor (oral) of the water emissions per year (Chen et al., 2002)
Ecotoxicity	Liquid emissions of ecotoxic compounds from (i) water discharges and (ii) fugitive emission sources	Water emissions of ecotoxic compounds from industrial sites and power plants	Equivalents of toluene aquatic ecotoxicity $(1/PNEC_{aquatic})$ of the water emissions per year (EC, European Commission, 2006)
Soil emissions			
Solid waste disposal	Production of solid and liquid waste sent to external disposal facilities	Waste production from industrial sites and power plants	Equivalents of urban solid waste (based on the disposal costs) of the waste production per year (Saling et al., 2002)
Resources consumption			
Raw materials	Raw material input of the process including, for example, (i) primary feedstock, (ii) auxiliary materials (e.g., NaOH, H_2SO_4, etc.), (iii) fuel, and (iv) valuable coproducts	Consumption of fossil fuels in industrial sites and power plants	Equivalents of crude oil (unitary costs) of the raw material resource per year
Electric power	Electrical energy consumption, for example, by main compressors, pumps, stirrers, and electro-operations (e.g., electrooxidation)	Consumption of electrical energy in industrial sites and power plants	Electrical energy use (kW h) per year
Land use	Area occupied by the plant on the long time period and not available for other activities.	Total area occupied by other similar facilities (e.g., all the industrial activities in the region/province).	Units of area permanently occupied on average year
Economic impact			
Economic impact	Annual savings (e.g., from product sale), operative costs (e.g., raw materials, fuels, auxiliaries, electrical power, employees, etc.), capital costs (e.g., plant cost)	Gross domestic product (value added) by the industrial sector	Actualized value (monetary) of costs and savings in the process life

(continued overleaf)

Table 8.3 (continued)

Impact category	Process data required	Reference normalization data	Suitable metrics (KPIs)
Workplace			
Toxicity of fugitive emissions	Fugitive emissions of toxic compounds (excluding emissions from elevated sources designed to promote safe dispersion e.g., stacks)	Air quality levels for workplace	Equivalents of toluene toxicity ($1/LC_{50,\,inh}$) of the fugitive emissions per year (Chen et al., 2002)
Carcinogenicity of fugitive emissions	Fugitive emissions of carcinogenic compounds (excluding emissions from elevated sources designed to promote safe dispersion e.g., stacks)	Air quality levels for workplace	Equivalents of benzene cancer potency slope factor (inhalation) of the fugitive emissions per year (Chen et al., 2002)
Society			
Occupation	Work hours required for process operation (e.g., control room crew) and regular maintenance operations.	Employees in the industrial sector	Equivalent number of employees (total standard work hours)
Inherent safety	Preliminary design of the units, operative conditions, inventories in the units, flows of the main lines	Average risk to human life from major industrial accident	Equivalent area potentially affected by accidents per year (Tugnoli, Cozzani, and Landucci, 2007).

$LC_{50,inh}$: lethal concentration for 50% of the sample, inhalation; $LD_{50,orl}$: lethal concentration 50% of the sample, oral; $PNEC_{aquatic}$: predicted no effect concentrations for plants and animals that live in that aquatic environment.

impact on external stakeholders. An occupational index assesses the contribution of the facility to local employment statistics. The index is based on the evaluation of the equivalent number of workers necessary for the activity, multiplied by −1. Further, in this case, the change of sign was necessary because impact indices must have a higher value in the worst option. An inherent safety index was proposed to assess the impact of process activities on the safety of the population. The inherent safety approach (see Hendershot, 1997; Koller, Fischer, and Hungerbuhler, 2001; Khan, Sadiq, and Amyotte, 2003; and references cited therein) was identified as particularly suitable to capture the process safety performances in the sustainability framework, especially in the early phases of design activities, where other safety indicators are not applicable (Tugnoli, Santarelli, and Cozzani, 2008). A specific index based on the credible accident consequences was proposed in Table 8.3 (Tugnoli, Cozzani, and Landucci, 2007; Tugnoli et al., 2012).

8.4
Criteria for Normalization and Aggregation of the KPIs

Reference values are required for the normalization and aggregation of the KPIs. Normalization is a necessary step both for the aggregation of indices and for the interpretation of primary KPIs. In the current method the internal normalization (i.e., using one of the alternatives as normalization reference) is avoided, since it is likely to introduce biases in the calculations and it requires a definition of weight factors specific for each set of the considered alternatives. Hence a site-specific external normalization is adopted. This allows for the comparison of single indicators to a reference that is independent of the process alternatives assessed, but is dependent on the specific site. This reference represents the impact burden on the site generated by the other industrial activities. Thus the normalized indicators measure the relative contribution of the alternatives to the local impact loads.

A characteristic target area is defined for each impact category. The target areas are defined in accordance to the spatial dimensions that are significantly influenced by the impacts. These are evaluated for each impact category by suitable approaches (e.g., characteristic distances calculated by models for pollutant dispersion, administrative division of the land, etc.). In some cases, practical trade-offs should be applied: for example, global warming is an impact category acting on a global scale; however, the use of a national reference may be preferred, since national emission balances and reduction plans may exist as a consequence of shared international agreements (e.g., the Kyoto Protocol). Data from national references are generally easier to collect and closely reflect the emission policy in the chosen site, supporting the aggregation step. Table 8.4 reports typical scales that can be assumed for a preliminary evaluation of the characteristic target area.

Several reliable databases may be used for gathering data on the emissions, consumptions, economics, and social issues within the pertinent target area (see examples in Table 8.3). The proper selection of data from databases is fundamental to obtain significant results. In particular the data considered must refer to homogeneous groups of activities in order to avoid biases (e.g., industrial facilities, power plants, and waste treatment plants).

All the facilities considered within the pertinent target area contribute to the normalization value for a given impact category. The normalization factors are straightforwardly calculated from the data available for the area of interest by the same rules used in the assessment of primary KPIs. Thus, the consistency of the normalization factor and of the indices for the assessed alternatives is granted. Since the impact scales are different among the impact categories, the normalization values are expressed per unit of area:

$$NF_i = \frac{\sum_{s \in At} E_{s,i}}{At_i} \qquad (8.4)$$

where $E_{s,i}$ is the total KPI index for the ith impact category from the sth source belonging to the target area (At) and At_i is the size of the target area for the ith

Table 8.4 Examples of target areas applicable to the normalization of KPIs. Adapted from Tugnoli, Santarelli, and Cozzani (2008).

Domains	Subdomains	Impact categories	Target area
Environment	Air emissions	Global warming	Nation
		Ozone depletion	Nation
		Rain acidification	1500 km radius
		Smog formation	100 km radius
		Toxicity in air	100 km radius
		Carcinogenicity in air	100 km radius
	Water emissions	Eutrophication	Hydrographical basin
		Organic load	Hydrographical basin
		Toxicity in water	Hydrographical basin
		Carcinogenicity in water	Hydrographical basin
		Ecotoxicity	Hydrographical basin
	Soil emissions	Solid waste disposal	Region
	Resource consumption	Non renewable materials	Nation
		Renewable materials	Nation
		Electrical power	Nation
		Land use	Region
Economic		NPV	Nation
Societal		Inherent safety index	Region
		Occupational index	Region

impact category (usually in km^2). The formula is valid under the assumption of background emission sources uniformly distributed in the target area.

The weight factors, later used for aggregation of the KPIs, represent the relative importance of load reduction among the impact categories. Their definition requires to reference a sustainability policy for the site of concern. The sustainability policy belongs to a macro-scale level that is different than the technical domain of process design. Weight factors must be defined within the macro-scale policy level, eventually creating a link between the general suitability policy and the process design. An important consequence of the shift to a different scale in the definition of weight factors is that the weight factors should be independent from the assessed alternatives, and should depend only on local conditions and on sustainability policy.

A possible approach for the numerical determination of weight factors is based on the distance of present impacts from future target values. Weight factor can be evaluated as:

$$W_i = \frac{r_i^\alpha}{\sum_i r_i^\alpha} \tag{8.5}$$

$$r_i = f\left(IM_i\,(t=0), IM_i\,(t=n), n\right) \tag{8.6}$$

where W_i is the weight factor of the ith impact category, r_i is the reduction rate, and α is an aversion factor (generally assumed equal to 1). The function f describes how impact loads are planned to be reduced from the present $IM_i(t = 0)$ to the future value $IM_i(t = n)$ in the span of n years. The same structure for function f must be adopted for all the impact categories, in order to yield consistent rates. A simple example of function f is the linear reduction:

$$r_i = \frac{IM_i(t=0) - IM_i(t=n)}{n \times IM_i(t=0)} \tag{8.7}$$

Reliable values for the impact targets may come from large scale (i.e., regional/national) ecosystem modeling of sustainability. Targets and trends are fixed for many categories by national policies or international agreements (e.g., Kyoto Protocol, Goteborg Protocol, etc.). An alternative approach for weight estimation may be based on the judgment by a panel of experts. Resorting to an expert panel for weighting estimation is an approach common to several assessment methods in the literature (e.g., BASF method (Saling et al., 2002), the analytic hierarchy process (AHP) procedure (Chen et al., 2002), etc.). The panel should be composed of experts with different backgrounds and, preferably, be the same that had defined the local sustainability policy. Proper consistency checks must be made in order to limit the subjectivity introduced by the decisions of the expert panel. No matter the approach chosen for the weights definition, a level of uncertainty always affects these values. As a consequence, a sensitivity analysis of the results is suggested (see Section 8.5).

8.5
Customization and Sensitivity Analysis in Early KPI Assessment

The tree of impacts defined in step 2 of the KPI assessment procedure may be customized, both to cut impacts not relevant to the specific problem that is assessed and to tune the level of detail of the KPIs. Though the holistic perspective of sustainability calls for a widely inclusive assessment, time and resources can be saved in the evaluation of KPIs if the nonrelevant impacts are removed from the assessment (i.e., simplification or "pruning" of the tree). In many practical cases, the impact categories that are negligibly affected by all the assessed options can be easily identified by a simplified pre-analysis. Tugnoli, Santarelli, and Cozzani (2011) proposed a shortcut approach for the identification of relevant impact categories, based on the estimation of the order of magnitude of the metrics. The order of magnitude of the impact KPIs is evaluated for the option or the options that are expected to yield the worst performance for each impact category. As a matter of fact, while assessing the actual value of a KPI requires an effort of data collection and estimation that may be cumbersome, the evaluation of its order of magnitude is usually swift, even in the early phases of design, since it can be based on less specific criteria (rules of thumb, similarities to other processes, expertise of the analyst, technical standards, emission limits, etc.). The difference between the order of magnitude of the KPIs and the order of magnitude of the normalization factors yields the expected order of magnitude of the normalized KPIs.

Differences up to several orders of magnitude may be present in the actual values of impact KPIs, depending on the process or system assessed. In the customization of the set of KPIs selected for the assessment, a relevance criterion can be applied to screen the categories having a higher order of magnitude of the expected impact and, therefore, having the higher relevance in the assessment. A practical cutoff criterion retains for further assessment only the categories within the upper 3 or 4 orders of magnitude. This wide range is believed to compensate the uncertainties introduced in the swift preliminary estimation of the order of magnitude in the indicators.

Table 8.5 reports a working example of the approach. The table concerns the assessment of alternative process schemes for the production of cyclohexanone (Tugnoli, Santarelli, and Cozzani, 2011). The cutoff criterion adopted was to consider in the further assessment the KPIs having normalized values falling in the upper 3 orders of magnitude. Impact categories as occupation, ecotoxicity, and so on were thus excluded in the detailed analysis of the specific case considered, reducing the effort required for the estimation of the indicators.

It is worth noticing that the weight factors defined in step 3 of the assessment procedure are not rescaled to sum up to one after the elimination of impact categories from the analysis: the fraction allocated to the impact categories, which are significant for the load from other industrial processes in the area (but are negligible for the processes analyzed), is maintained coherently with the definition of a dynamic set of impact indicators.

The set of KPIs can be further customized introducing relevant subindices. This is usually aimed at increasing the level of detail of a primary KPI, allowing for a better interpretation of the results in impact cause analysis. The subindices usually assess specific contributions to the primary KPI. In order to limit the revision of the KPI set and of the aggregation strategy, the subindices should be subparameters or addends of the primary KPI metric. For example, submetrics may be defined for selected environmental indices in order to explore the contribution of the process section (reaction, separation, water treatment, boilers, etc.) to the overall index.

Sensitivity analysis is another optional resource that can be implemented in the sustainability assessment. Although all of the parameters contributing to the evaluation of KPIs may be affected by uncertainties and approximations, the weight factors, owing to their evaluation procedure and to the direct influence on the values of aggregated KPIs, require specific attention (Goedkoop and Spriensma, 2001; Steen, 1997). A statistical approach is usually proposed to analyze the sensitivity of the aggregated KPIs to the weight factors assumed in the assessment. Owing to the mathematical structure of the procedure, large variations in the values of weight factors may significantly change the numerical value of the aggregated indices. However, in early design assessment, the relative performance (or, more simply, the rank of the compared options) is of interest, rather than the absolute values of the aggregated indices. Thus, in the framework of the methodology discussed herein, the sensitivity analysis can be specifically oriented to explore, for

Table 8.5 Customization of the set of indicators: example of diagram for the exclusion of nonrelevant impact categories. Adapted from Tugnoli, Santarelli, and Cozzani (2011).

Impact category	Unit	O indicator	O normalization factor	O normalized indicator	N/A	−5	−4	−3	−2	−1	0	1	2	3
Global warming	$kg_{eq}\,yr^{-1}$	8	6	2	—	—	—	—	—	—	—	—	•	—
Ozone depletion	$kg_{eq}\,yr^{-1}$	N/A	−1	N/A	•	—	—	—	—	—	—	—	—	—
Rain acidification	$kg_{eq}\,yr^{-1}$	5	4	1	—	—	—	—	—	—	—	•	—	—
Smog formation	$kg_{eq}\,yr^{-1}$	6	4	2	—	—	—	—	—	—	—	—	•	—
Toxicity in air	$kg_{eq}\,yr^{-1}$	8	6	2	—	—	—	—	—	—	—	—	•	—
Carcinogenicity in air	$kg_{eq}\,yr^{-1}$	4	2	2	—	—	—	—	—	—	—	—	•	—
Organic load	$kg_{eq}\,yr^{-1}$	3	2	1	—	—	—	—	—	—	—	•	—	—
Eutrophication	$kg_{eq}\,yr^{-1}$	1	1	0	—	—	—	—	—	—	•	—	—	—
Toxicity in water	$kg_{eq}\,yr^{-1}$	4	2	2	—	—	—	—	—	—	—	—	•	—
Carcinogenicity in water	$kg_{eq}\,yr^{-1}$	0	3	−3	—	—	—	•	—	—	—	—	—	—
Ecotoxicity	$kg_{eq}\,yr^{-1}$	0	5	−5	—	•	—	—	—	—	—	—	—	—
Solid waste disposal	$kg_{eq}\,yr^{-1}$	5	5	0	—	—	—	—	—	—	•	—	—	—
Nonrenewable raw materials	$kg_{eq}\,yr^{-1}$	9	6	3	—	—	—	—	—	—	—	—	—	•
Renewable raw materials	$kg_{eq}\,yr^{-1}$	N/A	0	N/A	•	—	—	—	—	—	—	—	—	—
Electrical power	$kW\,h\,yr^{-1}$	8	6	2	—	—	—	—	—	—	—	—	•	—
Land use	$m^2\,yr^{-1}$	−1	0	−1	—	—	—	—	—	•	—	—	—	—
Economy	¤	9	8	1	—	—	—	—	—	—	—	•	—	—
Population safety	$km^2\,yr^{-1}$	−3	−5	2	—	—	—	—	—	—	—	—	•	—
Occupation	People	2	2	0	—	—	—	—	—	—	•	—	—	—

O: order of magnitude; N/A: category nonapplicable.

any given aggregated index, the change in the relative ranking of the options due to uncertainties in the weight factors.

The statistical approach to sensitivity analysis requires the association of a value distribution to each of the weight factors used in the aggregation. Ideally, the shape and the range of these distributions should be proposed during the evaluation of the weight factor in step 4 of the method. If distance from a future target is used to assess a KPI, the limits of a triangular distribution can be easily defined by the identification of a "best case" and of a "worst case" scenario for the weight value. However, the uncertainty that affects target values is frequently unknown or difficult to assess in practical applications. Thus Tugnoli, Santarelli, and Cozzani (2011) propose an alternative pragmatic approach: a symmetric triangular probability distribution is assumed for all the weight factors and the maximum range is fixed for all values as an arbitrary percentage of the central value (see Figure 8.4a). Clearly, enough, this associates a higher range of uncertainty to the more relevant impact categories (i.e., the ones having higher value of the weight factor). The probability distributions for each impact category are then associated at the assumed shape in order to satisfy the consistency condition:

$$\int_{-\infty}^{+\infty} P(x)\,dx = 1 \tag{8.8}$$

Once the value distributions are defined, a Monte Carlo simulation may yield a straightforward assessment of the distribution of the aggregated indices. The consistency of the weight factor should be maintained in each run: therefore the randomly generated values of the weight factors should be rescaled before calculating the aggregated index, so that their sum is one.

The relative performance of the alternative options for an aggregated category is evaluated as a "difference of the indices" ($DI_{i,k1-k2}$) (Goedkoop and Spriensma, 2001; Steen, 1997):

$$DI_{i,k1-k2} = I_{i,k1} - I_{i,k2} \tag{8.9}$$

in which (i) refers to the aggregated index of concern (e.g., overall, environmental, etc.) and ($k1$) and ($k2$) refer to a couple of alternative options. The alternative options are labeled progressively ($k1$, $k2$, $k3$, ...) according to the decreasing value of the ith index as obtained applying the central values of the weight factors. A change in the rank of the options produces a change of sign in the index difference ($DI_{i,k1-k2}$). The application of Monte Carlo runs allows an easy assessment of the effect of the distributions assumed on the differences in the aggregated indices (Equation 8.9) and, thus, of the possibility of rank change among the options.

Moreover, the sensitivity of the index results on the uncertainty of weight factors can be explored repeating the procedure for several values of the maximum range in weight factor distribution (φ). Figure 8.4b–d illustrates an example (Tugnoli, Santarelli, and Cozzani, 2011). The figure shows that, for the system under assessment, changes in the weight factors until an order of magnitude ($\varphi = 100\%$) negligibly affect the rank of the three compared process options (A, B, and C).

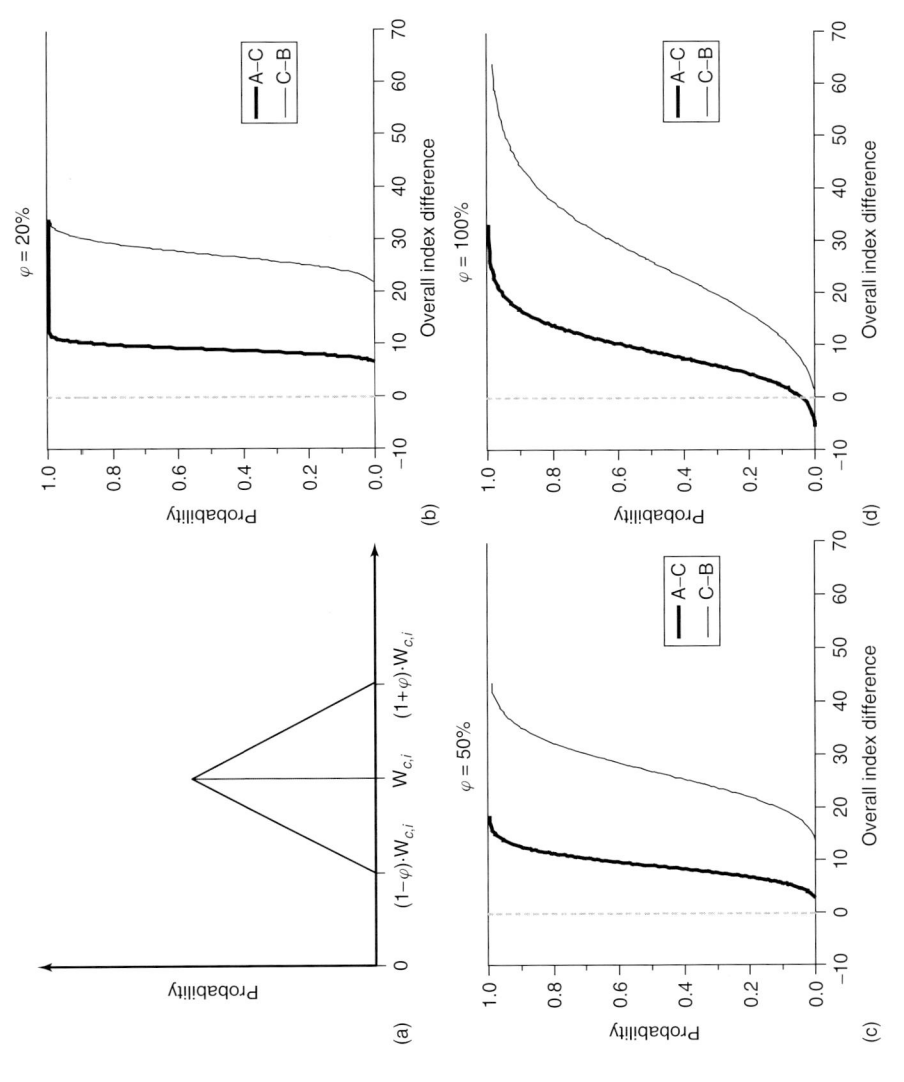

Figure 8.4 Example of result from sensitivity analysis of weight factors: (a) range and shape of the distribution assumed for weight factors and (b–d) difference of the values of the overall impact index for options A, B, and C with different maximum range of the weighting factor (b: $\varphi = 20\%$; c: $\varphi = 50\%$; d: $\varphi = 100\%$). Adapted from Tugnoli, Santarelli, and Cozzani (2011).

This proves a stability of the conclusions with respect to the uncertainty in the assumed weight factors.

8.6
Conclusions

In this chapter an approach to the assessment of the expected sustainability performance of alternative options during process design is presented. The analysis may be applied in the early stages of process development (conceptual and basic design), in order to identify, evaluate, and improve the sustainability of future industrial facilities or production processes. The method was based on the systematic development of a tree of impacts that yielded a comprehensive set of KPIs. Reference procedures were proposed for the identification of the relevant impact categories and suitable KPIs. A site-specific normalization approach was introduced to provide a background reference for the interpretation and aggregation of KPIs. The hierarchic structure of the tree of impacts allowed for the definition of weight factors and aggregated KPIs. A Monte Carlo analysis was proposed to prove the robustness of the ranking results. Altogether, the methodology can be an effective support tool for decision making in the design for sustainability of process industries.

References

Allen, D.T. and Shonnard, D.R. (2002) *Green Engineering: Environmentally Conscious Design of Chemical Processes*, Prentice Hall PTR, Upper Saddle River.

Azapagic, A. and Clift, R. (1999a) Life cycle assessment and multiobjective optimisation. *J. Cleaner Prod.*, **7**, 135–143.

Azapagic, A. and Clift, R. (1999b) The application of life cycle assessment to process optimisation. *Comput. Chem. Eng.*, **23**, 1509–1526.

Bare, J.C., Hofstetter, P., Pennington, D.W., and Udo de Haes, H.A. (2000) Midpoints versus endpoints: the sacrifices and benefits. *Int. J. Life Cycle Ass.*, **5** (6), 319–326.

Bare, J.C., Norris, G.A., Pennington, D.W., and McKone, T. (2003) TRACI: the tool for the reduction and assessment of chemical and other environmental impacts. *J. Ind. Ecol.*, **6** (3-4), 49–78.

Barthel, L. and Albrecht, S. (2007) The Sustainability of Packaging Systems for Fruit and Vegetable Transport in Europe Based on Life-Cycle-Analysis. Final Report n. 070226, Stiftung Initiative Mehrweg, Michendorf, http://www.plasticsconverters.eu/uploads/Final-Report-English-070226.pdf. (accessed November 2012).

Carter, W.P.L. (1994) Development of ozone reactivity scales for volatile organic compounds. *Air Waste*, **44**, 881–899.

Chen, H., Rogers, T.N., Barna, B.A., and Shonnard, D.R. (2003) Automating hierarchical environmentally-conscious design using integrated software: VOC recovery case study. *Environ. Prog.*, **22** (3), 147–160.

Chen, H. and Shonnard, D.R. (2004) Systematic framework for environmentally conscious chemical process design: early and detailed design stages. *Ind. Eng. Chem. Res.*, **43** (2), 535–552.

Chen, H., Wen, Y., Waters, M.D., and Shonnard, D.R. (2002) Design guidance for chemical processes using environmental and economic assessments. *Ind. Eng. Chem. Res.*, **41** (18), 4503–4513.

Dhillon, B.S. (1989) Life Cycle Costing: Techniques, Models and Applications, OPA, Amsterdam.

EC (European Commission) (2006) Integrated Pollution Prevention and Control – Reference Document on Economics and Cross-Media Effects, Sevilla, Spain, http://eippcb.jrc.es/reference/BREF/ecm_bref_0706.pdf. (accessed November 2012).

Figge, F. and Hahn, T. (2004) Sustainable value added - measuring corporate contributions to sustainability beyond Eco-efficiency. *Ecol. Econ.*, **48**, 173–187.

Goedkoop, M. and Spriensma, R. (2001) The Eco-Indicator 99 – A Damage Oriented Method for Life Cycle Impact Assessment, PRè Product Ecology Consultants, Amersfoort, www.pre.nl.

Gonzales, M.A.and Smith, R.L. (2003) A methodology to evaluate process sustainability. *Environ. Prog.* /vol.22, No.4, 269–276.

Goodwin, P. and Wright, G. (1998) *Decision Analysis for Management Judgment*, 2nd edn, John Wiley & Sons Ltd, Chichester.

Guest, J.S., Skerlos, S.J., Daigger, G.T., Corbett, J.R.E., and Love, N.G. (2010) The use of qualitative system dynamics to identify sustainability characteristics of decentralized wastewater management alternatives. *Water Sci. Technol.*, **61** (6), 1637–1644.

Hendershot, D.C. (1997) Inherently safer chemical process design. *J. Loss Prev. Proc.*, **10** (3), 151–157.

Heui-seok, Y., Jorge, L.H., Nandan, U.U., and Bhavik, R.B. (2004) Hierarchical thermodynamic metrics for evaluating the environmental sustainability of industrial processes. *Environ. Prog.*, **23** (4), 302–314.

IChemE (Institution of Chemical Engineers) (2002) The Sustainability Metrics, Rugby, http://www.icheme.org/communities/special-interest-groups/sustainability/resources/sustainability%20tools.aspx. (accessed November 2012).

International Organization for Standardization (ISO) (1997) ISO 14040 Series. *Environmental Management–Life Cycle Assessment*, International Organization for Standards (ISO), Geneva.

International Organization for Standardization (ISO) (2000) ISO 17776:2000. *Petroleum and Natural Gas Idustries–Offshore Production Installations–Guidelines on Tools and Techniques for Hzard Identification and Risk Assessment*, 1st edn, International Organization for Standards (ISO), Geneva.

Jin, X. and High, K.A. (2004) A new conceptual hierarchy for identifying environmental sustainability metrics. *Environ. Prog.*, **23** (4), 291–301.

Jorge, L.H.and Bhavik, R.B. (2004) Expanding exergy analysis to account for ecosystem products and services. *Environ. Sci. Technol.*, **38**, 3768–3777.

Keeney, R.L.and Raiffa, H. (1976) *Decisions with Multiple Objectives: Preferences and Value Trade-Offs*, John Wiley & Sons Inc., New York.

Khan, F.I., Sadiq, R., and Amyotte, P.R. (2003) Evaluation of available indices for inherently safer design options. *Process Saf. Prog.*, **22** (2), 83–97.

Koller, G., Fischer, U., and Hungerbühler, K. (2001) Comparison of methods suitable for assessing the hazard potential of chemical processes during early design phases. *Trans IChemE*, **79** Part B, 157–166.

Krotscheck, C. and Narodoslawsky, M. (1996) The sustainable process index – a new dimension in ecological evaluation. *Ecol. Eng.*, **6** (4), 241–258.

Landsiedel, R.and Saling, P. (2002) Assessment of toxicological risks for life cycle assessment and eco-efficiency analysis. *Int. J. Life Cycle Ass.*, **7** (5), 261–268.

LCWE (2010) Documentation of LCWE data in GABI 4, University of Stuttgart, http://www.gabi-software.com/fileadmin/Documents/lcwe.pdf. (accessed November 2012).

Lowe, D.C., Myhre, G., Nganga, J., Prinn, R., Raga, G., Schulz, M., and Van Dorland, R. (2007) Changes in atmospheric constituents and in radiative forcing, in *Climate Change 2007: The Physical Science Basis–Contribution of Working Group I to the Fourth Assessment Report of the Intergovernmental Panel on Climate Change* (eds S. Solomon, D. Qin, M. Manning, Z. Chen, M. Marquis, K.B. Averyt, M. Tignor, and H.L. Miller), Cambridge University Press, Cambridge and New York.

Mallick, S.K., Cabezas, H., Bare, J.C., and Sikdar, S.K. (1996) A pollution reduction methodology for chemical process simulators. *Ind. Eng. Chem. Res.*, **35** (11), 4128–4138.

Mannan, S. (2005) *Lees' Loss Prevention in the Process Industries*, 3rd edn, Elsevier, Oxford.

Narodoslawsky, M. and Krotscheck, C. (2000) Integrated ecological optimization of processes with the sustainable process index. *Waste Manag.*, **20**, 599–603.

Organisation for Economic Co-operation and Development (OECD) (2008) Guidance on Developing Safety Performance Indicators, Series on Chemical Accidents, Vol. 19, 2nd edn, OECD, Paris.

Othman, M.R., Repke, J., Wozny, G., and Huang, Y. (2010) A Modular Approach to Sustainability Assessment and Decision Support in Chemical Process. *Design. Ind. Eng. Chem. Res.*, **49** (17), 7870–7881.

Pennington, D.W., Norris, G., Hoagland, T., and Bare, J.C. (2000) Environmental comparison metrics for life cycle impact assessment and process design. *Environ. Prog.*, **19** (2), 83–91.

Saling, P., Kicherer, A., Dittrich-Kramer, B., Wittlinger, R., Zombik, W., Schmidt, I., Schrott, W., and Schmidt, S. (2002) Eco-efficiency analysis by BASF: the method. *Int. J. Life Cycle Ass.*, **7** (4), 203–218.

Satty, T. (1980) *The Analytic Hierarchy Process*, McGraw-Hill, Book Company, New York.

Shadiya, O.O. and High, K.A. (2012) Sustainability Evaluator: Tool for evaluating process sustainability. *Environ. Prog. Sustainable Energy*, doi: 10.1002/ep.11667.

Shonnard, D.R., Kicherer, A., and Saling, P. (2003) Industrial applications using BASF eco-efficiency analysis: perspectives on green engineering principles. *Environ. Sci. Technol.*, **37** (23), 5340–5348.

Sikdar, S.K. (2003a) Sustainable development and sustainability metrics. *AICHE J.*, **49** (8), 1928–1932.

Sikdar, S.K. (2003b) Journey toward sustainable development: a role for chemical engineers. *Environ. Prog.*, **22** (4), 227–232.

Steen, B. (1997) On uncertainty and sensitivity of LCA-based priority setting. *J. Cleaner Prod.*, **5** (4), 255–262.

Tugnoli, A., Cozzani, V., and Landucci, G. (2007) A consequence based approach to the quantitative assessment of inherent safety. *AICHE J.*, **53** (12), 3171–3182.

Tugnoli, A., Santarelli, F., and Cozzani, V. (2008) An approach to quantitative sustainability assessment in early process design. *Environ. Sci. Technol.*, **42** (12), 4555–4562.

Tugnoli, A., Santarelli, F., and Cozzani, V. (2011) Implementation of sustainability drivers in the design of industrial chemical processes. *AICHE J.*, **57** (11), 3063–3084.

Tugnoli, A., Landucci, G., Salzano, E., and Cozzani, V. (2012) Supporting the selection of process and plant design options by Inherent Safety KPIs. *Journal of Loss Prevention in the Process Industries*, **25** (5), 830–842.

de Ville, B. (2006) *Decision Trees for Business Intelligence and Data Mining*, SAS Institute Inc., Cary, NC.

Yaqin, F., Qiuwen, Z., and Cheng, W. (2008) Utilizing UML and system dynamics to optimize the Eco-environmental impacts evaluation and prediction models. Proceedings of the 2008 IEEE International Conference on Information and Automation, ICIA 2008, pp. 238–243.

Young, D., Scharp, R., and Cabezas, H. (2000) The waste reduction (WAR) algorithm: environmental impacts, energy consumption, and engineering economics. *Waste Manag.*, **20** (8), 605–615.

Part III
Managing Horizontal Interorganizational Sustainability

9
Industrial Symbiosis and the Chemical Industry: between Exploration and Exploitation

Frank Boons

9.1
Introduction

Since the early 1990s, the field of industrial ecology has evolved as a systemic perspective on industry and ecological sustainability (Erkman, 1997; Ehrenfeld, 2002; Boons and Howard-Grenville, 2009). Industrial ecology is "the study of the flows of materials and energy in industrial and consumer activities, of the effects of these flows on the environment, and of the influences of economic, political regulatory, and social factors on the flow, use, and transformation of resources"[1] (White, 1994). Within the field, researchers use various system boundaries to delineate industrial and consumptive activities, of which the most prominent are product chains and clusters of colocated firms (Korhonen, 2002; Boons and Howard-Grenville, 2009). Regional clusters are also referred to as examples of industrial symbiosis, which according to Chertow (2007, p. 12) consists of "engaging traditionally separate industries in a collective approach to competitive advantage involving physical exchange of materials, energy, water, and by-products."

Over time, a number of examples of industrial symbiosis have been documented, and researchers have started to conceptualize the conditions and processes that lead geographically proximate firms to engage in symbiotic exchanges such as by-product exchange and utility sharing. While many conceptualizations are based on one, or only a few case studies (see Boons, Spekkink, and Mouzakitis, 2011 for an overview), research is now beginning to test hypotheses on larger sets of cases (Boons and Spekkink, 2012). The aim of this chapter is to build on current knowledge on industrial symbiosis to provide insight into the possibilities for chemical industry to develop sustainable regional clusters of activity through industrial symbiosis.

While building on recent empirical analyses, a conceptual addition is made to the symbiosis literature by introducing the notions of exploitation and exploration.[2]

The following section provides a critical assessment of three common ideas about industrial symbiosis. This serves as an introduction to novices, while giving readers

---

1) For a business-centered overview of the field of industrial ecology, see Lifset and Boons, (2012).
2) This distinction is derived from the work of James March (1991) on organizational learning.

who are familiar with the concept an insight into recent developments. Section 9.3 discusses two crucial dimensions of systems that need to be balanced in some way: exploration and exploitation. Section 9.4 contains illustrations of how these dimensions occur in three chemical clusters in the Netherlands: the industrial port of Terneuzen, the petrochemical complex in the harbor of Rotterdam, and the chemically oriented industrial park Moerdijk. The chapter concludes with a generic discussion of exploitative and explorative forms of industrial symbiosis.

9.2
Understanding Industrial Symbiosis

Industrial symbiosis has been one of the icons of the industrial ecology field, especially as realized in the network of by-product exchanges among firms located in the vicinity of the village of Kalundborg, Denmark (Lifset and Boons, 2012; Boons and Janssen, 2004) (see Figure 9.1).

This empirical case, which was introduced into the field after a field visit by researchers in the early 1990s, was important first of all because it showed that developing close linkages among the production processes of a diversity of firms

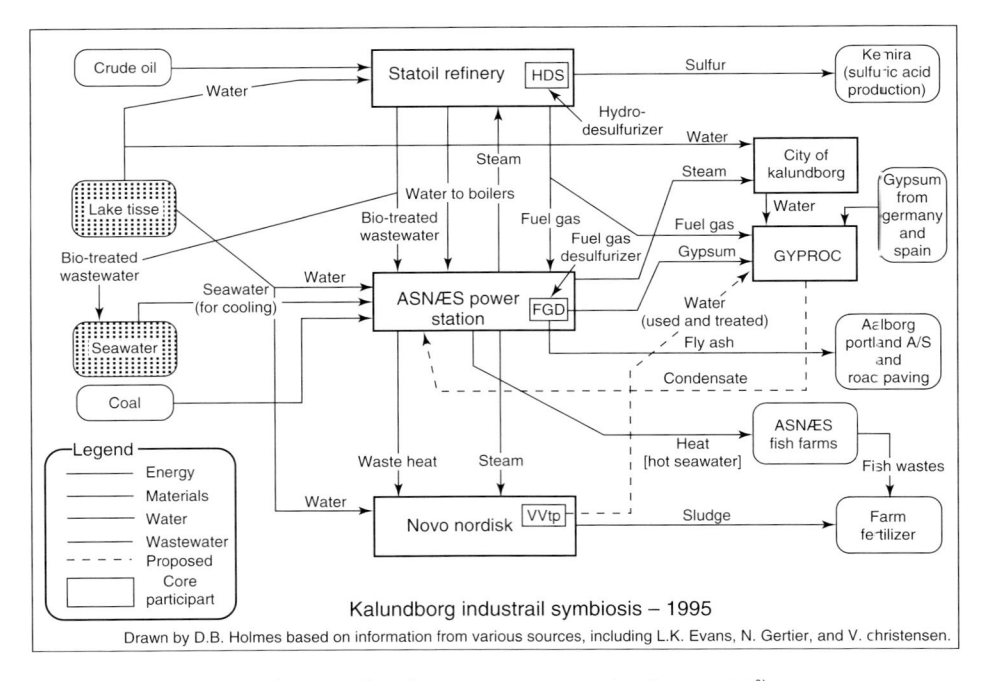

Figure 9.1 Visualization of symbiotic exchanges at Kalundborg, 1995.[3]

---

3) Source: http://www.indigodev.com/Kal.html (accessed 30 March 2012).

was practically possible. Secondly, it made clear that such linkages, at least as they appeared there, required the support of a social infrastructure that took years to evolve (Ehrenfeld and Gertler, 1997).

Subsequently, more cases of industrial symbiosis have been either uncovered (Chertow, 2007), or developed, inspired by the Kalundborg example (Baas and Boons, 2007). As the number of cases increased, it also became clear that not all of these went beyond envisioning symbiotic exchanges (Deutz and Gibbs, 2008). The successful examples have spawned a number of conceptual contributions that still require rigid empirical testing to assess their external validity. Empirical tests that recently have become available provide the basis for a critical discussion of three central ideas about industrial symbiosis. This sets the stage for assessing the way in which industrial symbiosis relates to a sustainable chemical industry. The chapter concludes by distinguishing between explorative and exploitative types of industrial symbiosis.

9.2.1
Industrial Symbiosis Leads to Decreased Ecological Impact

We begin with what for many is the central purpose of establishing symbiotic linkages: one implicit idea related to industrial symbiosis is that such linkages reduce the ecological impact of industrial activity. Taking a systemic perspective, a first question here is what system boundary is involved here. The rationale of industrial ecology implies that the system should be drawn not around the individual firm, but rather for the regional system as a whole. However, there are few studies available that actually assess the consequence of establishing symbiotic linkages in terms of ecological impact at the cluster level. One of the reasons is that this is difficult to establish. The impact fluctuates with production levels of firms involved, with entries and exits of firms, and the impact of activities within the cluster is often difficult to disentangle from that of other geographically proximate activities. This means that the major rationale for seeking to develop industrial symbiosis is based on an assumption rather than on an empirically established fact.

Symbiotic linkages may fail to lead to reduced ecological impact at the cluster level for several reasons. One possibility is that exchanges of by-products lead to an increased number of process stops. As processes become more inefficient, they are likely to produce more waste. Another reason that has been voiced is that by-product exchanges reduce the incentive for the firm that produces the by-product to seek ways to prevent it from occurring at all. In this sense, industrial symbiosis and pollution prevention may be at odds with each other (Oldenburg and Geiser, 1997).

Again, these are conceptual speculations that require systematic empirical tests to assess their validity. But until such tests are done, they are equally valid as the claims that industrial symbiosis does lead to reduced ecological impact at the cluster level.

9.2.2
Industrial Symbiosis Requires a Highly Developed Social Network

On the basis of the analysis of a number of successful examples, most conceptual frameworks build on the assumption that establishing symbiotic linkages among firms requires a well-developed social infrastructure at the level of the cluster as a whole. This structure takes the form of a dense network of relations among (environmental) managers of firms, where information is exchanged and a common vision is developed on how to develop symbiotic projects. Thus, the bilateral exchanges among firms, which are governed by contracts or other organizational arrangements, are complemented by a clusterwide structure that enables and facilitates the bilateral relationships. This structure may be decentralized, but it has also been argued that a centralized network around an anchor company or even a governmental agency can perform this function.

Recent evidence challenges this assumption. In a set of 233 symbiotic projects in the Netherlands, we found that high levels of institutional capacity (the theoretically based conception of social structure) are not associated with an increase in perceived opportunities to engage in symbiosis projects (Boons and Spekkink, 2012). Also, in a comparison of symbiotic networks in the United Kingdom and in the Netherlands, it becomes clear that symbiotic exchanges occur both in networks with high and low institutional capacity (Stift and Boons, unpublished manuscript).

These results make clear that it is important to distinguish among types of symbiotic exchanges. The comparison between the United Kingdom and the Netherlands shows that several bilateral exchanges that fall under the header of by-product exchange are similar to "normal" economic market transactions. While new for the firms involved, and thus a case of increased coupling, they do not require any additional institutional support; the activities of a facilitating organization mainly serve to increase market transparency. Symbiosis projects that move beyond such exchanges occur only in the context of more developed institutional capacity.

This finding is consistent with a conceptualization on how symbiotic clusters evolve over time (Baas and Boons, 2004). Initially, firms may establish symbiotic linkages that are based on direct economic returns for both parties, similar to a market exchange. Such interactions may facilitate the development of trust among firms, so that in later phases more advanced symbiotic linkages occur, over longer time frames and with less direct and equal pay-off structures.

9.2.3
The Regional Cluster Is the Preferred Boundary for Optimizing Ecological Impact

An implicit assumption for advocates and practitioners of industrial symbiosis is that the regional cluster is a useful, or even the preferred boundary for optimizing industrial systems in terms of ecological impact. In fact, other system boundaries are available (Boons and Baas, 1997; Boons and Wagner, 2009). Without seeking to resolve this issue here (it may only be resolvable in specific contexts), it is important to be sensitive to the fact that individual firms are not only part of regional clusters but also are simultaneously connected to (often global) chains of production and consumption. In addition, local production plants are often part of larger (multinational) firms. Each of these systems provides a possible context for optimizing activities in terms of minimized ecological impact. Until now, there has been no research that considers these as alternatives in a comparative analysis. At the same time, managers in production facilities within regional clusters are required to make decisions that concern these alternatives, and their decisions to engage in a symbiotic exchange are often constrained by pressures from a parent company, or from suppliers or buyers in their product chain. One important point here is that such boundaries have consequences not by being objectively given, but instead because they are perceived by managers and policy makers to be important (Ulrich, 2003).

9.3
Resourcefulness

The issues discussed above can be summarized in two main questions:
1) To what extent does system integration enable or preclude actions to make systems more sustainable?
2) How can we incorporate different system boundaries (as adopted by both researchers and practitioners) into industrial symbiosis?

In order to answer these questions, it is useful to look at two dimensions that systems in general need in order to be adaptive. According to March (1991), systems need to balance exploitation and exploration. The former stands for qualities such as efficiency, routinization, and goal directedness, while the latter refers to the creation of variation, diversity, and problem solving. As becomes clear in the industry life cycle, a system may move from an explorative phase (in developing new products, conquering new markets) to a phase of exploitation (gaining economies of scale through standardization, profits based on efficiency improvements of given designs) (Nooteboom, 1999). The previously mentioned hypothesis that industrial symbiosis evolves through different phases is another example of this sequential display of both qualities, where, in this case, exploitation precedes exploration.

In contrast with such sequential balancing, many systems display both qualities simultaneously. Consider the growing number of product chains where Western firms are responsible for the design of new products that are then produced by suppliers based in China. This has become a dominant strategy for western firms in the product chain of electronic products.

The conceptual pair of exploration and exploitation provides a way to conceptualize the question about the extent to which industrial symbiosis prevents systems from becoming more sustainable. If the regional cluster, together with its symbiotic exchanges, is characterized by exploitation, then it may become locked into a path where there is no incentive or infrastructure to look for more fundamental improvements (Green and Randles, 2006).

According to March (1991), adaptive systems over time have a tendency to evolve toward exploitation because the feedback loops from the effect of actions back to managerial decisions are usually much shorter for exploitative actions than for explorative actions. In that case, effects take longer to materialize and the risks of failure are higher.

The pair is also helpful in addressing the second question. As system boundaries are a construct of the observer, balancing exploration and exploitation becomes an issue in variously delineated systems. Even if in a specific case, industrial symbiosis consists of efficiency-increasing linkages, it requires exploration (for instance, because firms need to learn about the ways in which to organize a by-product exchange with colocated firms). The question is thus not about the preferred level of optimization, but instead at balancing exploration and exploitation at each system level. This could be termed *resourcefulness*.

9.4
Putting Resourcefulness to the Test

These are all abstract observations that require an empirical context in order to prove their usefulness. In this chapter, the empirical context is provided by the chemical industry, and this section provides case vignettes that illustrate the role that industrial symbiosis could play in moving this industry toward sustainability.

The chemical industry has been well studied with regard to its ecological impact, as well as the ways in which firms have sought to deal with that impact (see, for instance, Hoffman, 1999). It is generally considered to have a high ecological impact, which results from substantial negative contributions to three categories of ecological effects (Commoner, 1971; 1997): the introduction of new substances into natural ecosystems, the interference with ecological cycles (such as the water cycle and atmospheric cycles), and the intensive exploitation of natural resources. Given the fact that many of its processes are based on the use of either oil or chlorine, it is seen by many analysts as inherently unsustainable. As a result, the sector has since long been subject to regulatory efforts by governments.

At the same time, the production of chemical substances takes place to a considerable degree in industrial complexes where processes are highly integrated, using production technology with long life cycles. These complexes could be seen as prototypical examples of industrial symbiosis.

A final relevant characteristic of the industry is that at least parts of it are highly innovative, giving rise to new production processes and products as well as new risks, and the consequences of introducing these are difficult to assess.

9.4.1
Petrochemical Cluster in the Rotterdam Harbor Area

The harbor industrial complex in Rotterdam includes a large number of petrochemical activities. Traditionally, in this sector of industry different production processes that are colocated are linked together in order to increase production efficiency. When production facilities are owned by the same firm, such linkages occur within the firm boundary and thus do not fall under the industrial symbiosis umbrella. Since the early 1990s, firms in the Rotterdam cluster have started to work together to develop linkages among different firms. After a period where they developed a successful network in which they exchanged experiences on implementing environmental management systems within their facilities, firms looked for new inspiration, which they found in the Kalundborg example. On the basis of a visit from Kalundborg representatives, they started to define possible projects, and develop a social infrastructure for exploring the feasibility of linkages such as heat exchanges. The industry association Deltalinqs has been a constant factor in this process that has evolved in the last 20 years in facilitating knowledge development and acting as an agent that made sure that sensitive information was not diffused among competitors (which is always a potential impediment to collaboration among firms with similar processes). While firms have been innovative in the sense that they have acquired and applied new locally relevant knowledge on linking their production processes, the main innovation in this cluster seems to reside in learning how to actually implement available technological possibilities. This is then a case where firms collaborate and learn how to exploit existing resources more efficiently (Table 9.1 provides an overview; a more detailed description can be found in Baas and Boons, 2007).

9.4.2
Terneuzen

The industrial development of the port of Terneuzen is closely intertwined with that of the canal that links it to the port of Ghent in Belgium. Terneuzen is located in the southwest of the Netherlands, a region that is characterized by agricultural activities and tourism. Over the years, regional and local authorities have developed several initiatives to support industrial development and increase demand for local labor. The harbor authority of Terneuzen has been quite successful in doing so; a major chemical complex of Dow has been the core of the harbor since 1962. During the 1990s, the first attempts were made to infuse industrial development in the port of Terneuzen and adjacent areas with ideas of industrial symbiosis (Van de Laak and Goedman, 1999). To some extent, such interfirm exchanges are the outcome of a process in which Dow Chemical decided to outsource noncore activities. This forced plant managers to develop their skills in managing such relationships with proximate firms, skills that could then be used to manage new relationships (De Valk, 2011).

A larger initial initiative was that firms wanted to develop a market-based waste-exchange network, but eventually other initiatives took priority. Several plans of governmental authorities, developed in collaboration with firms, resulted in increased symbiotic exchanges among existing firms, as well as the exchange of CO_2 between incumbent firms and newly developed greenhouse activities. In 2005, a producer of bioethanol announced the construction of a large production plant,

Table 9.1 Characteristics of the development of industrial symbiosis in the Rotterdam harbor complex.

Issue\period	1990–1993	1994–1997	1999–2002	2003 to ongoing
Theme	Development of company's environmental management systems	INES-project: exploration of industrial ecology	INES mainport project: implementation of industrial ecology projects	R3: sustainability in the Rotterdam harbor and industry complex; exploration of transitions
Actors	Deltalinqs, environmental coordinators, and consulting firm (project facilitator)	Deltalinqs, environmental coordinators, consulting firm (project facilitator), and academic researchers	Deltalinqs (project manager), representatives of industry, government (national, regional EPA, provincial, and port management), academia, and environmental advocacy organization	ROM-Rijnmond program (including project management), representatives of industry, government (national, regional EPA, and provincial), port management, academia, and environmental advocacy organization
Relationship	Industry partners	Industry plus academic researchers	Industry lead plus all stakeholders	Government lead plus all stakeholders
Process	Knowledge dissemination and quarterly information exchange	Academic research and decision making in INES platform led by staff member of Deltalinqs	Feasibility research plus implementation of INES projects render an account to a stakeholders platform	Energy transition research and reflection of a strategy platform of stakeholders
Financing	Subsidy of government	Subsidy of government	50% subsidy of government + 50% time input of industry	Subsidy of government plus project input industry
Type of learning	Learning by doing and information exchange by partners	Traditional research projects plus feasibility study	Learning of variety in stakeholders' perspectives and information in strategic platform	Reflexive learning by evaluation of projects in strategy platform of high-level stakeholders
Institutionalization	Development of network of environmental coordinators for EMS development	Representatives of industry and academics in a garbage can model for industrial ecology development	Strategic decision-making stakeholder approach for industrial ecology development	Strategy platform of stakeholders for sustainability transition policy development

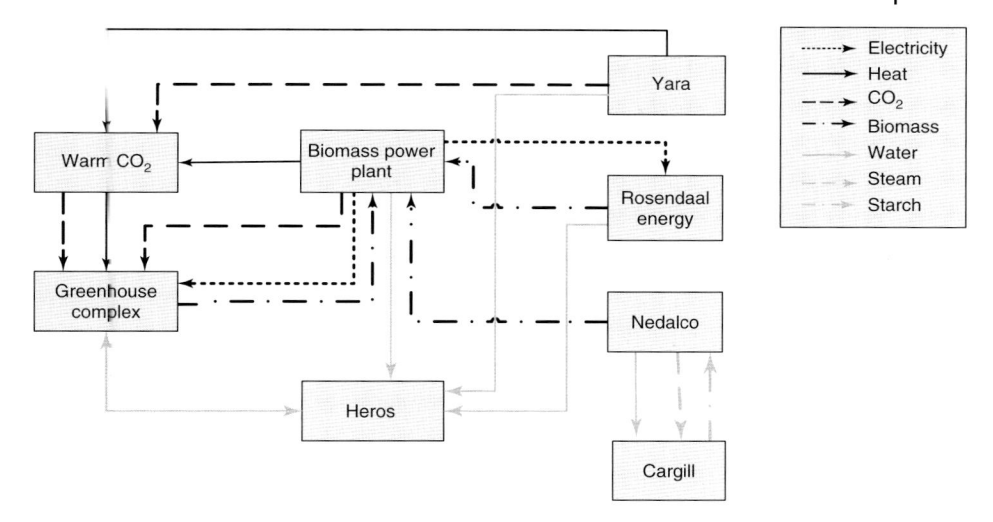

Figure 9.2 Pattern of exchanges in BioparkTerneuzen, 2009.[4]

initiating the development of a bioenergy cluster. In 2007, the BioparkTerneuzen was officially inaugurated, a collaboration among firms, regional authorities, and (local and national) knowledge institutes. The further development of this cluster has gained momentum through funding obtained from the European Union in collaboration with partners in the Ghent harbor area, another region where bio-based energy activities are developed (Figure 9.2 presents an overview of exchanges).

9.4.3
Moerdijk

The Moerdijk industrial park initially was created to give place to firms for which there was no room in the Rotterdam harbor. Over time, it has developed into a mixed industrial park with harbor access, where a large number of chemical activities take place. There are also pipelines that connect some of the activities in Moerdijk with those in Rotterdam. Also, Moerdijk has a history in the public debate of being seen as a park with substantial ecological impact. The provincial authority of Brabant, in which Moerdijk is situated, initiated a process in which consultants were asked to evaluate the situation and propose ways of improving the sustainability of the park. This resulted in the development of a vision by firms in collaboration with governmental authorities (including the municipality and the Water Board), and over 20 initiatives from firms who developed symbiotic exchanges, including multilateral exchanges among a petrochemical facility, a waste incinerator, and an electricity producer; CO_2 exchange by four firms with diverse production processes; the use of fly ash by a phosphate processor; and the

---

4) Source: Van Waes and Huurdeman (2009).

Table 9.2 Comparative overview of industrial symbiosis at three industrial parks in The Netherlands.

	Rotterdam	Moerdijk	Terneuzen
Type of cluster	Industrial complex dominated by petrochemical activities	Industrial park with diversity of chemical production processes	Industrial park focusing on bio-based chemistry
Type of symbiotic exchanges	—	Steam exchange, CO_2 exchange, use of fly ash, waste gasses, and heat exchange	CO_2 exchange and utility sharing
Nature of social networks	Emergent network of firms facilitated by industry association; linked to other initiatives over time	Bilateral exchanges, mostly initiated by governmental action to increase sustainability, resulting in a shared vision on symbiosis by business and governmental stakeholders	Networks based on vision development by regional authorities and involvement of large corporate players
Exploitation and exploration	Exploitation dominant	Exploitation dominant	Exploitation and exploration in parallel

production of construction materials (Stift, N. 2011; see Figure 9.3 for an overview of exchanges).

Table 9.2 provides a comparative overview of the three cases.

9.5
Conclusions

To the extent that the chemical industrial activities that take place in regional clusters are decentralized production facilities that operate scaled up processes of mature products, we may expect exploitation to be a dominant preoccupation of the local managers, as is indicated by the findings in Table 9.2. By themselves, and pressured by headquarters, managers at production plants will aim to maximize the efficient utilization of their production facilities. The current focus on short-term shareholder value reinforces this efficiency drive. Also, downtime will be minimized, thus further decreasing the possibility for experimentation.

To the extent that local (environmental) managers are interested in developing symbiotic linkages, they will experience limited freedom because they are part of social systems that also prioritize exploitation. This has been found for production plants in regional clusters of various industrial sectors. If symbiotic linkages occur here, they will tend to be exploitative in nature, seeking to reap benefits of increased system integration. Within the chemical industry, there is a tradition of seeking improvements through establishing linkages, which provides fertile ground for this type of symbiosis. It will be difficult to counteract this strong bias

9.5 Conclusions | 143

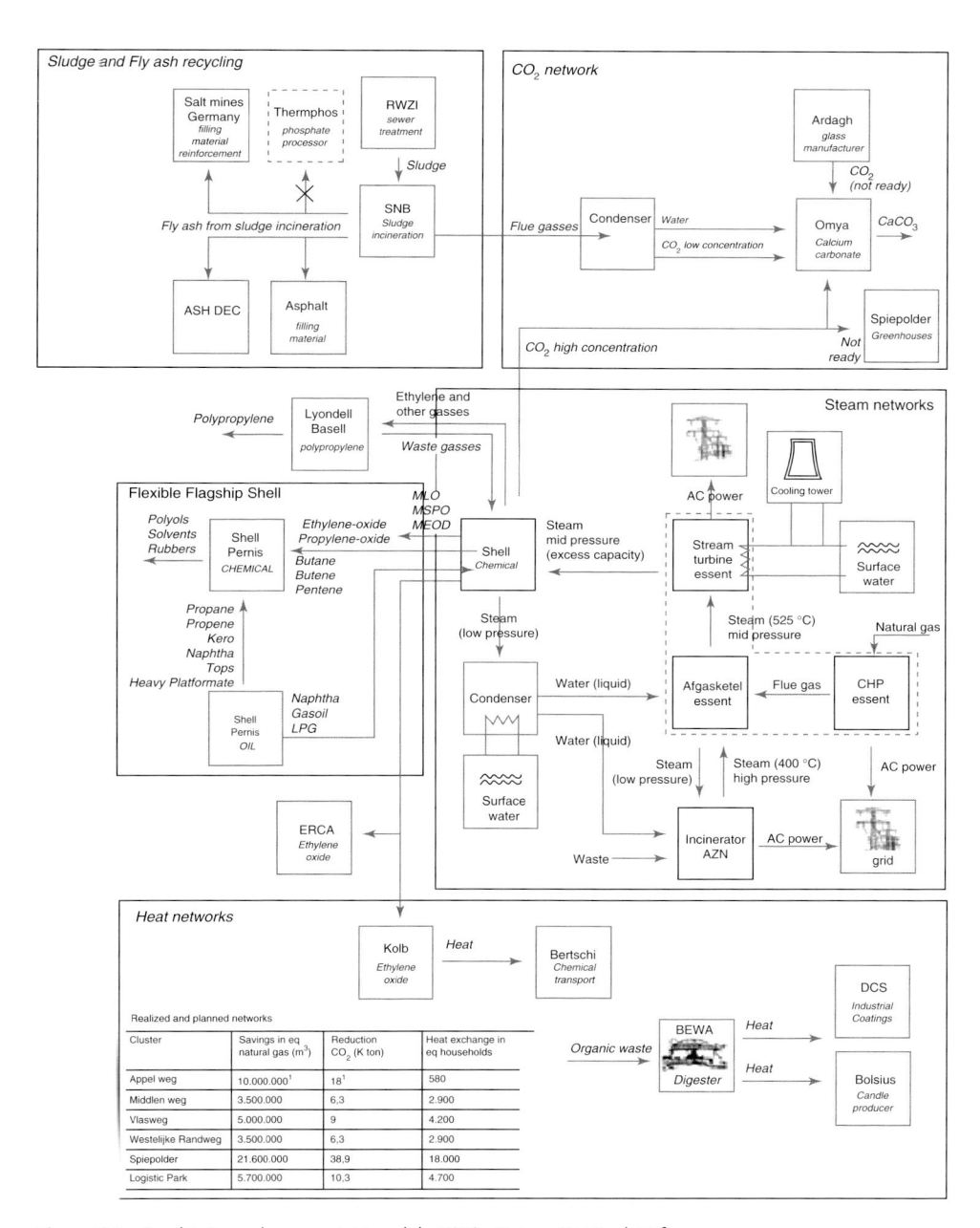

Figure 9.3 Symbiotic exchanges at Moerdijk, 2010. Picture © Nigel Stift.

toward exploitation. The author considers the Rotterdam petrochemical cluster and Moerdijk to be mainly of this nature.

If the activities of chemical firms in a regional cluster include research and development of new products, there is potential for more explorative types of symbiosis.

Here the main condition is the extent to which exploration takes place within the firm boundary, or whether instead a network of firms and knowledge institutes engage in this exploration. In the latter case, the cluster acts as a regional innovation system (Cooke, 2001). The Terneuzen case is an example of such a system (see Table 9.2). There, the continuous efforts of regional authorities (in close collaboration with anchor firms in the industrial park) have shaped the evolution of a set of integrated activities with bio-based solutions to sustainability problems as a focus.

References

Baas, L.W. and Boons, F.A.A. (2004) An industrial ecology project in practice: exploring the boundaries of decision-making levels in regional industrial systems. *J. Clean. Prod.*, **12** (8), 1073–1085.

Baas, L.W. and Boons, F.A.A. (2007) The introduction and dissemination of the industrial symbiosis projects in the Rotterdam Harbour and industry complex. *Int. J. Environ. Technol. Manag.*, **7** (5/6), 551–577.

Boons, F.A. and Baas, L.W. (1997) Types of industrial ecology: the problem of coordination. *J. Clean. Prod.*, **5** (1–2), 79–86.

Boons, F.A. and Howard-Grenville, J. (2009) *The Social Embeddedness of Industrial Ecology*, Cheltenham, Edward Elgar.

Boons, F.A.A. & Janssen, M. (2004). In J. van den Bergh & M. Janssen (Eds.), *Economics of Industrial Ecology* (pp. 337–355). Cambridge: MIT Press.

Boons, F.A.A. and Spekkink, W.A.H. (2012) Levels of institutional capacity and actor expectations about industrial symbiosis: evidence from the Dutch stimulation program 1999–2004. *J. Ind. Ecol.*, **16** (1), 61–69.

Boons, F.A.A., Spekkink, W.A.H., and Mouzakitis, Y. (2011) The dynamics of industrial symbiosis: a proposal for a conceptual framework based upon a comprehensive literature review. *J. Clean. Prod.*, **19**, 905–911.

Boons, F.A.A. and Wagner, M.A. (2009) Assessing the relationship between economic and ecological performance: distinguishing system levels and the role of innovation. *Ecol. Econ.*, **68**, 1908–1914.

Chertow, M.R. (2007) 'Uncovering' Industrial symbiosis. *J. Ind. Ecol.*, **11** (1), 11–30.

Commoner, B. (1971) *The Closing Circle: Nature, Man, and Technology*, Knopf, New York.

Commoner, B. (1997) The relation between industrial and ecological systems. *J. Clean. Prod.*, **5** (1–2), 125–129.

Cooke, P. (2001) Regional innovation systems, clusters, and the knowledge economy. *Ind. Corp. Chang.*, **10**, 945–974.

Deutz, P. and Gibbs, D. (2008) Industrial ecology and regional development: eco-industrial development as cluster policy. *Reg. Stud.*, **42** (10), 1313–1328.

De Valk, E. (2011) Evolution dynamics of Eco-industrial parks. Master thesis. TU Delft.

Ehrenfeld, J. (2002) Industrial ecology—becoming a new field? Paper prepared for presentation at AIChE 2002 Annual Meeting, Sustainable Engineering, November 3–8, 2002.

Ehrenfeld, J.R. and Gertler, N. (1997) Industrial ecology in practice: the evolution of interdependence at kalundborg. *J. Ind. Ecol.*, **1** (1), 67–79.

Erkman, S. (1997) Industrial ecology: an historical view. *J. Clean. Prod.*, **5** (1–2), 1–10.

Green, K. and Randles, S. (2006) *Industrial Ecology and Spaces of Innovation*, Edward Elgar, Cheltenham.

Hoffman, A. (1999) Institutional evolution and change: environmentalism and the

U.S. Chemical industry. *Acad. Manag. J.*, **42** (4), 351–371.

Korhonen, J. (2002) Two paths to industrial ecology: applying the product-based and geographical approaches. *J. Environ. Plan. Manag.*, **45** (1), 39–57.

Lifset, R. and Boons, F. (2012) in *Handbook of Business and the Natural Environment* (eds P. Bansal and A.J. Hoffman), Oxford University Press, Oxford, pp. 310–326.

March, J. (1991) Exploration and exploitation in organizational learning. *Organ. Sci.*, **2** (1), 71–87.

Nootebcom, B. (1999) Innovation, learning and industrial organization. *Camb. J. Econ.*, **23** (2), 127–150.

Oldenburg, K. U. & Geiser, K. (1997) Pollution prevention and… Or industrial ecology? *J. Clean. Prod.*, **5**/1–2, 103–108.

Stift, N. (2011) Industrial symbiosis and transaction costs. Master thesis, Leiden, Delft University.

Stift, N. and Boons, F. Transaction cost economics and industrial symbiosis. *J. Ind. Ecol.*, unpublished manuscript.

Ulrich, W. (2003) Beyond methodology choice: critical systems thinking as critically systemic discourse. *J. Oper. Res. Soc.*, **54**, 325–342.

Van de Laak, P. and Goedman, J. (1999). Industrieel ecosysteem kanaalzone, in: Van de Laak, P., and Boons, F. (eds.). *Industriële Ecosystemen*, Hague, SMO-99-10.

Van Waes, M. and Huurdeman, M. (2009) Biopark Terneuzen. Building Smart Links into the Sustainability Chain. Public Report.

White, R. (1994) Preface, in *The Greening of Industrial Ecosystems* (eds B.R. Allenby and D. Richards), National Academy Press, Washington, DC.

10
Cluster Management for Improving Safety and Security in Chemical Industrial Areas

Genserik L.L. Reniers

10.1
Introduction

Collaborative relationships between organizations – from mergers and acquisitions, to joint ventures, partnerships, and alliances, to myriad forms of organizational networks and consortia – have gained increasing importance in and between organizations. Interfirm collaborative partnerships are being formed at an unprecedented pace. This is no different in the chemical industry, where different types of collaboration in chemical clusters are becoming ever more critical competitive weapons. Reniers, Dullaert, and Visser (2010) indicate that by augmenting collaborative agreements and relationships and by linking up with other firms on the same level of the market, a chemical plant may enjoy options otherwise unavailable to it, such as better access to markets, pooling or swapping of technologies and production volumes, access to specialized competencies, lower risk of R&D, enjoying larger economies of scale, and benefiting from economies of scope. This observation also explains why chemical plants form chemical clusters. However, in the case of chemical organizations, clustering not only implies profit opportunities and economic benefits of scale. A chemical cluster has a very high responsibility toward maintaining safety and security standards in the urban surroundings as well. Each additional chemical plant entering a chemical cluster might decrease the average safety and security standing of the area. Companies in chemical clusters are thus not merely linked via technological spillovers, logistics advantages, and so on. They are related through the responsibility of gaining and sustaining safety and security standards in the entire cluster as well. Hence, in chemical industrial areas, cooperating on the topics of safety and security is highly relevant. Reniers, Dullaert, and Visser (2010) further argue that collaborative arrangements lead to more sustainable situations, providing several examples (considering e.g., the supply chain, the environment and energy streams).

In this chapter, we therefore discuss the application of *cluster management* on one specific subject, namely *safety and security*.

The success of collaborative arrangements is largely determined by the cooperative efforts and coordinated activities across organizational interfaces. An approach to enhance collaborative effectiveness, creating and sustaining complex safety- and security-enhancing capabilities and collaborative tendencies in a multi-organizational environment, is elaborated in this chapter. The following are some of the questions that are addressed in this chapter:

- What is meant by *cluster management*?
- How can *cluster management be used to improve safety and security* in chemical corporations?
- Under what *conditions and circumstances is cluster management effective* for safety and security enhancement?
- What are the *implications of cluster management for people, procedures, processes, and technology* within a chemical organization?
- What are the *benefits of cluster management* compared with single-plant management?

Although answers to these questions will be formulated in this chapter, it should be clear to the reader that these answers, as well as a lot of other questions and answers, will only be revealed after putting theory to practice. This chapter is thus intended to get acquainted with cluster management for improving safety and security within chemical industrial areas from a theoretical viewpoint.

The next section describes cluster management and what it entails. Section 10.3 discusses the requirements for efficient and effective cross-organizational learning, and drafts a cluster management model for safety and security. The section then discusses the suggested model and identifies the strengths, weaknesses, opportunities, and threats (SWOT) of cluster management. Section 10.5 concludes this chapter.

10.2
Cluster Management

Cluster management can be described by a multi-organizational form in which the individual and joint approaches, methods, best practices, results, and so on of a cluster of organizations (e.g., with respect to safety and security) are discussed and debated among – and reported to – its individual members, and where for every member (thus for every individual company), a manager applies individual organizational approaches, methods, and best practices based on cross-plant information, and is responsible for individual organizational results. This way, the joint efforts by the different plant managers should optimize the individual organizational results through smart collaboration.

Successful collaboration depends on several requirements. It is evident that the different organizations should have the willingness to collaborate. Reniers, Dullaert, and Visser (2010) mention a number of hard factors, soft factors, and independent factors which provide an indication of how well companies fit with each other, or what has to be changed to make companies fit with each other. Factors include, for example, company openness, market position, knowledge spillover opportunities, financial position of companies, flexibility of companies, and innovation potential.

To maximize the gross benefits of clustering, chemical organizations have a long tradition of collaborating on many different fronts. Their cooperative strategies offer significant advantages for plants that are lacking in particular competencies or resources to secure these through links with other firms possessing complementary skills or assets. They also offer opportunities for mutual synergy and learning.

However, as Child, Faulkner, and Tallman (2005) explain, the organizational cultures can have a significant impact upon the implementation of cooperative strategies. Cultures can create serious barriers to collaboration between organizations, and yet, at the same time, the knowledge embodied in cultures can provide a valuable resource for cooperation strategies. To better understand and explain the latter reflection, Figure 10.1 conceptualizes organizational culture and puts it in contrast with two other important influences that shape the ways in which people think and behave, namely, personality and human nature.

It is widely accepted that culture is something that is learned. It is not inherited like human nature or, at least partially, personality. This has a very important practical implication for cooperation between the members of organizations coming from different organizational cultural environments, for it means that despite the way a certain company culture is absorbed by a plant's employees, it may always be possible to learn further cultural attributes through experience or training. In terms of safety and security topics in particular, the latter observation is interesting. Safety and security matters are often subject to a high degree of confidentiality. Therefore, safety and security cooperation strategies are difficult to establish. However, since clustered chemical corporations are bonded by the responsibility to keep the industrial area as a whole as safe and as secure as possible, individual plants situated next to one another should develop a safety and security cooperation strategy, bringing diverse safety and security cultures together. Shaping such a

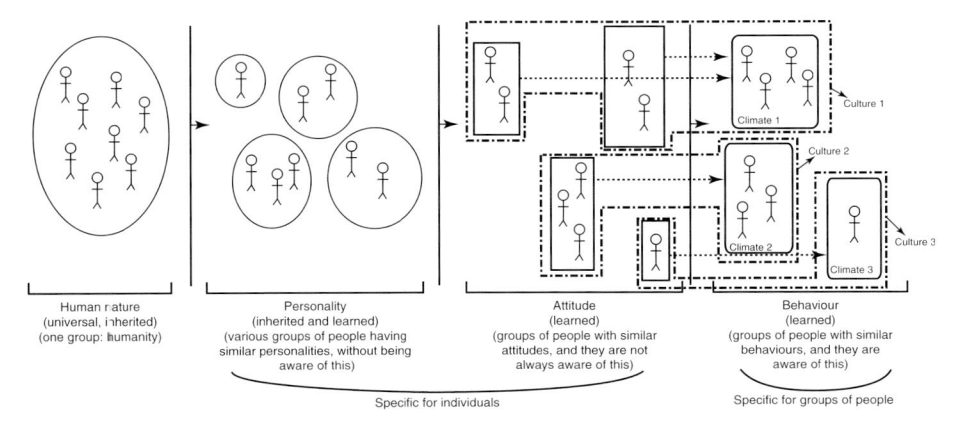

Figure 10.1 Organizational culture contrasted with other sources of human thinking and behavior.

cross-company safety and security culture can lead to significant advantages for the chemical cluster as a direct result of cross-organizational learning. Avoiding incidents and accidents and their associated direct and indirect costs can lead to substantial benefits, especially in case of averting major accidents. Although the potential benefits are obvious, cluster management for enhancing safety and security, and thereby dealing with a variety of organizational cultures, could not be described as an easy task.

As shown in Figure 10.1, an organizational culture can be roughly seen as the sum of attitude and behavior within organizations, both attributes which can be learned. Whereas attitude is not observable but rather a *set of mind* and a *way of thinking* of people, the behavior of people (/employees) can be directly observed, and can be considered as the organizational climate (see also Reniers, Cremer, and Buytaert, 2011).

Hence, it is certainly possible for different cultures to learn from each other, without one culture necessarily being assimilated by the other culture. Visions and approaches with regard to behaviors and attitudes should be shared between companies to advance individual cultures by learning. To develop an efficient way for cultures to cross-learn from one another, different aspects such as confidentiality issues and continuous improvement thinking should be taken into account. Organizing and managing these knowledge transfers between different plants belonging to the same chemical cluster, can be regarded as *cluster management*.

10.3
Cross-Organizational Learning on Safety and Security

10.3.1
Knowledge Transfer

The principles of cluster management and its accompanying knowledge transfer between companies to increase safety and security within industrial areas should be applicable for any random chemical cluster. Hence, the multi-organizational theory and the model of knowledge transfer elaborated in this chapter should be generic. If, for example, no external domino effects would be possible between two plants (thus one plant cannot suffer an accident due to the other plant), it is still possible for these plants to learn from each other. Therefore, the cluster management model proposed in this chapter can be employed for safety and security improvement within a chemical industrial area in any cluster situation.

It should be understood that knowledge is fundamentally dynamic and created through social interactions among individuals and organizations (Nonaka, Toyama, and Konno, 2000). The context-appropriate application of knowledge focuses on the transferability or application of knowledge, in terms of space, time, and mechanisms. While explicit knowledge (documents, best practices, procedures, databases, etc.) is more obviously transferable, tacit knowledge is better transmitted through practices and social experiences. Obviously, for cross-organizational learning to be effective, both types of knowledge are important and should be guaranteed in the model.

Tsai (2001) indicates that the transferability of knowledge within organizations is contingent on the network position of the transferor and the absorptive capacity of both the transferor and the recipient. Networks of business unit links (within an organization) favor access to and the exchange of knowledge between different units in a company. Absorptive capacity depends on prior related knowledge. This theory can easily be used and applied for setting up a model for effective knowledge transfer across organizations, as described in Section 10.3.3.

10.3.2
Overcoming Confidentiality Hurdles: the Multi-Plant Council (MPC)

To overcome confidentiality issues, we suggest to use an independent supra-plant body, called the *Cluster Council* or the Multiplant Council (MPC) as suggested by Reniers *et al.* (2009) and Reniers (2010), to implement the model in a chemical cluster. The MPC is divided into two parts. The first part consists of plant representatives who mainly have a counseling function, formulating recommendations as a result of brainstorming sessions. The second part consists of independent and external consultants who are responsible for gathering, assessing, and analyzing all relevant and confidential financial risk information from the chemical plants in the cluster. By dividing the MPC into two parts, a balance between confidentiality and data information is targeted. Figure 10.2 illustrates the different parts of the MPC.

The first main part consists of two sub-parts and is composed of plant safety representatives (sub-part 1) and plant security representatives (sub-part 2). It has a typical counseling function, formulating safety and security recommendations as a result of joint think tank brainstorming and communication sessions. The other part, the MPC Data Administration, consists of two sub-parts as well: one for safety

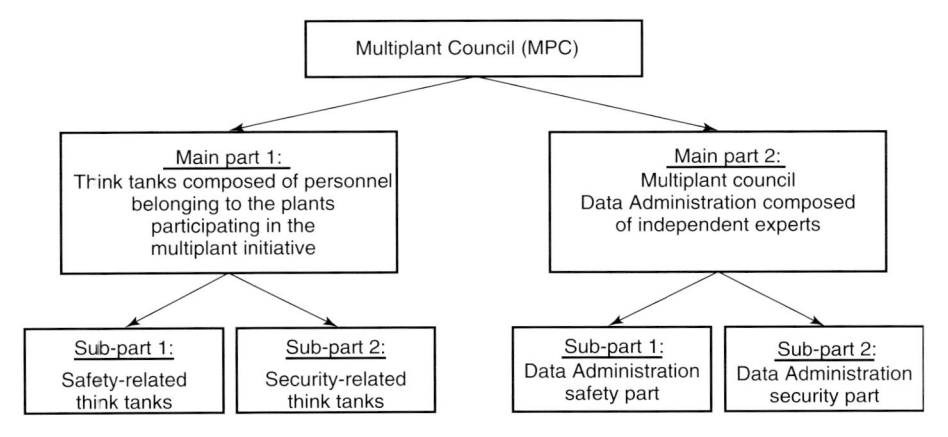

Figure 10.2 Constitution of the Multi-Plant Council. Source: Reniers (2010).

and one for security. The MPC Data Administration safety part is composed of independent consultants (i.e., impartial knowledgeable personnel) responsible for administering all necessary (confidential) safety-related information gathered from the different plants of the cluster. Although the MPC Data Administration security part has an equivalent structure as the safety sub-part, both are separated and work independently from each other owing to the extremely high confidentiality of plant security data.

10.3.3
A Cluster Management Model for Safety and Security

As mentioned in the previous section, by splitting the MPC into a part composed of plant personnel, that is, the think tanks, and a part composed of independent experts, that is, the Data Administration, balance between confidentiality and information sharing is targeted. The think tanks ensure the continuous improvement in taking safety and security measures as regards multiplant topics and, to a lesser extent, single-plant safety and security topics. The Data Administration collects the necessary (confidential) technical installations (and processes) data and information, incident and near-incident data, and other safety and security information and uses this information, for example, as input for computer-automated software, audits, and inspections. On the basis of the output, the MPC Safety Data Administration gives guidance and recommendations to the MPC safety management think tank, as does the MPC Security Data Administration to the MPC security management think tank. If calamities should occur having a possible impact beyond the originating company, the necessary data of all the plants are centralized in an MPC Data Administration data bank and may be used without any delay. A practical model for such a suggested optimized multiplant organization is given in Figure 10.3.

The procedure can be described as follows. An MPC think tank organizes brainstorming sessions for the plant departments of Production, Maintenance, Human Resources, Logistics, Environment, Management, Safety Management, and Security Management. The number of participants of such a think tank should be deliberately limited with a view to maximizing output efficiency. If the number of participating plants in the multiplant initiative is too large, a method of systematic alternation of representatives can be used. As a guideline, a maximum of eight representatives per think tank is proposed: seven department representatives (after a fixed period of time one representative after another alternating with department representatives from the other companies) and one independent consultant with expertise in the department field. The MPC safety management think tank and the MPC security management think tank aim at achieving integrated preventive multiplant safety and security respectively, by drawing and proposing standardized procedures based on plant department recommendations and added multiplant safety issues and multiplant security topics. The crucial role of the safety and security management think tanks is reflected in their composition, that is, permanent multiplant safety and security specialists are added to the small

10.3 Cross-Organizational Learning on Safety and Security | 153

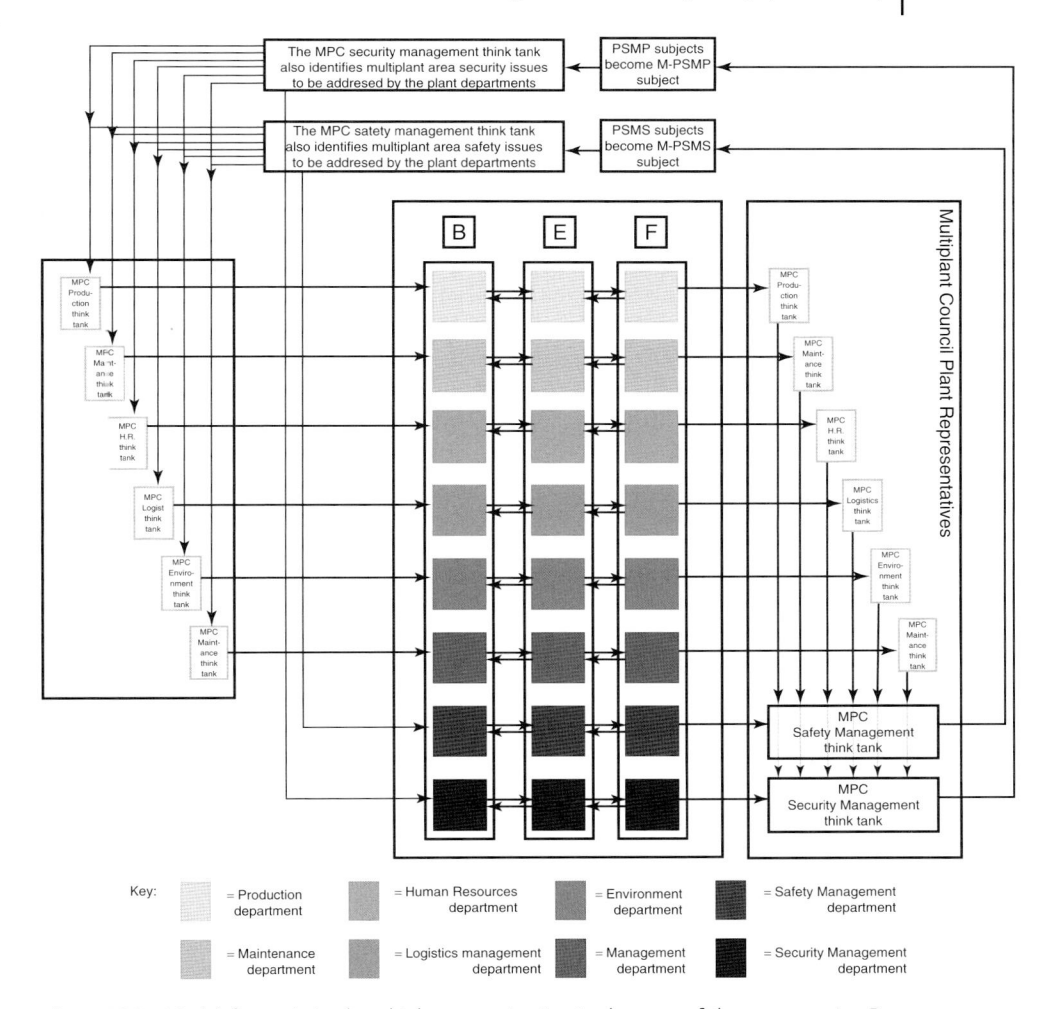

Figure 10.3 Model for optimized multiplant organization in the case of three companies B, E, and F. Source: Reniers (2010).

group of eight. The safety and security management think tanks maximum consist of 12 group members, the numbers again being limited for the same reason as given previously. The multiplant procedures are translated and implemented at individual plant level by the departments. Hence, a continuous improvement of drafting the different parts of the multiplant procedures is achieved by optimized communication and cooperation of every department of the plants participating to the MPC.

In Figure 10.3, each of the three companies (B, E, and F) is composed of eight plant departments. Plant departments, for example, communicate through e-mail and a multiplant intranet. Think tank sessions are periodically organized.

154 *10 Cluster Management for Improving Safety and Security in Chemical Industrial Areas*

Although the scheme of Figure 10.3 provides an extensive illustration of organizing the MPC plant representatives in an optimizing way, it fails to include the MPC Data Administration parts. To this end, Figure 10.4 is given.

Figure 10.4, combined with Figure 10.5a,b, illustrates the safety and security communication, and cooperation procedures, offering a loop for continuous improvement of multiplant safety and multiplant security.

Figure 10.5a,b suggest a nonexhaustive division of responsibilities of the different parts constituting the MPC.

The multiplant model and the MPC as a part thereof guarantee the performance of processes entailing planning, establishment of goals and objectives, monitoring of progress and performance, analysis of trends and development, and implementation of corrective actions to be continuously improved, both for safety and security, and both at plant and at multiplant levels.

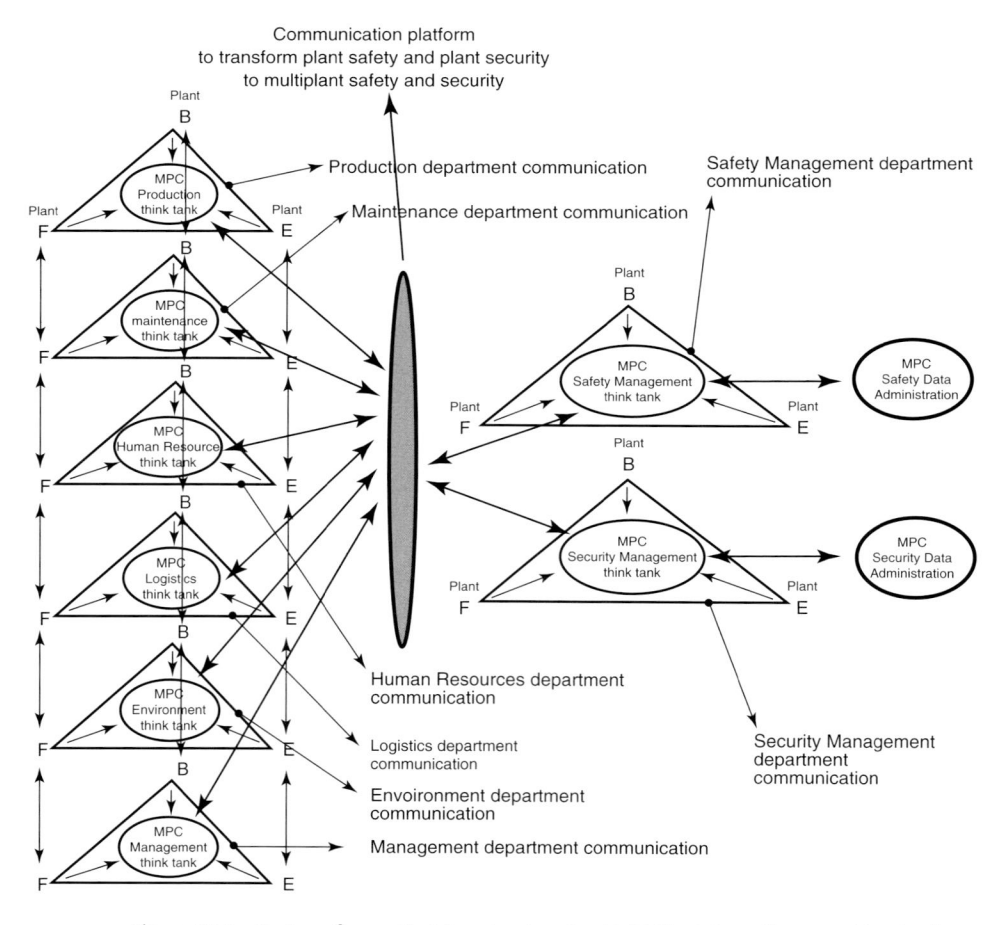

Figure 10.4 Design of an optimizing structure for Multi-Plant Council composition in the case of three plants B, E, F. Source: Reniers (2010).

10.3 Cross-Organizational Learning on Safety and Security

Plant departments

DO
Information exchange on
- Safe work practices
- Management of change
- Plant hazards and risks analyses
- Human factors
- Training and performance
- Processes
- Equipment integrity

MPC Safety Data Administration

CHECK
- Collecting confidential data
- Nonconfidential central database
- Transports of hazardous chemicals database
- Land-use planning
- Multiplant hazards and risks analyses
- Organizing multiplant safety training
- Organizing multiplant group meetings
- Incident investigations
- Audits and correctives actions
- Inspections

MPC departments think tanks

ACT
- Proposing in-house safety rules and regulation
- Recommendations for promoting safety
- Maintenance programs follow-up
- Enhancement of plant and multiplant safety knowledge
- Rigorously follow up on all deficiencies in the multiplant area identified as a result of daily observation and subsequently formulating recommendations

MPC Safety Management think tank

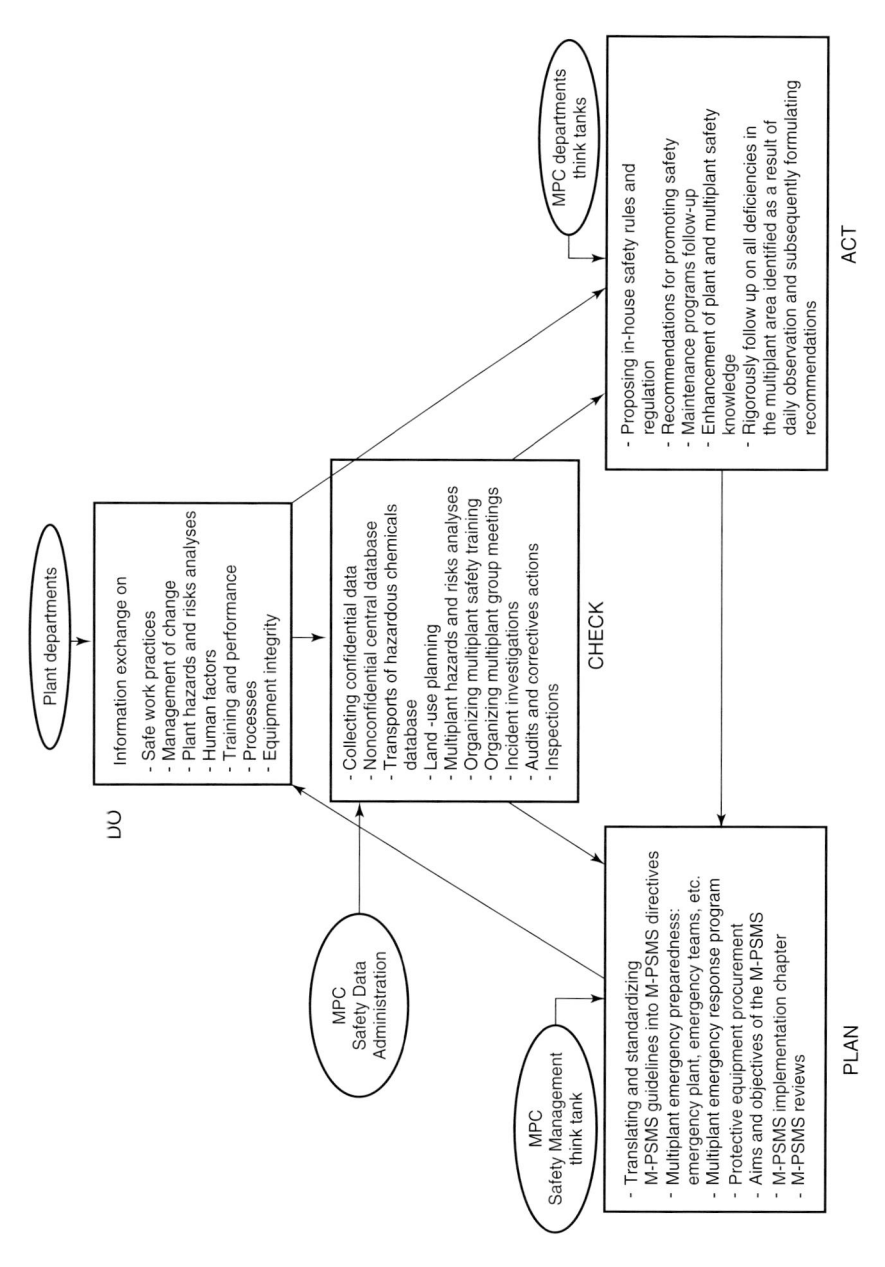

PLAN
- Translating and standardizing M-PSMS guidelines into M-PSMS directives
- Multiplant emergency preparedness: emergency plant, emergency teams, etc.
- Multiplant emergency response program
- Protective equipment procurement
- Aims and objectives of the M-PSMS
- M-PSMS implementation chapter
- M-PSMS reviews

Figure 10.5 (a) Circle of continuous improvement of multiplant safety. (b) Circle of continuous improvement of multiplant security. Source: Reniers (2010).

156 | *10 Cluster Management for Improving Safety and Security in Chemical Industrial Areas*

Plant departments

DO
Information exchange on
- Security work practices
- Points of surveillance
- Plants security vulnerability analyses
- Security training and performance
- Security promotion

Law enforcement agencies

Information exchange on
- Surveillance schemes
- Possibility to intervene
- Expectations of both stakeholders
- Organisation schedules

MPC Security Data Administration

CHECK
- Collecting confidential data
- Nonconfidential central security database
- Multiplant SVA's
- Organizing multiplant security training
- Organizing group meetings
- Security incident investigation
- Audits and Inspections

MPC departments think tanks

ACT
- Proposing in-house security rules and regulations
- Recommendations for promoting security
- Surveillance programs follow-up
- Enhancement of Plant and multiplant security knowledge
- Rigrously follow up on all security incidents in the multiplant area and formulating recommendations

MPC Security Management think tank

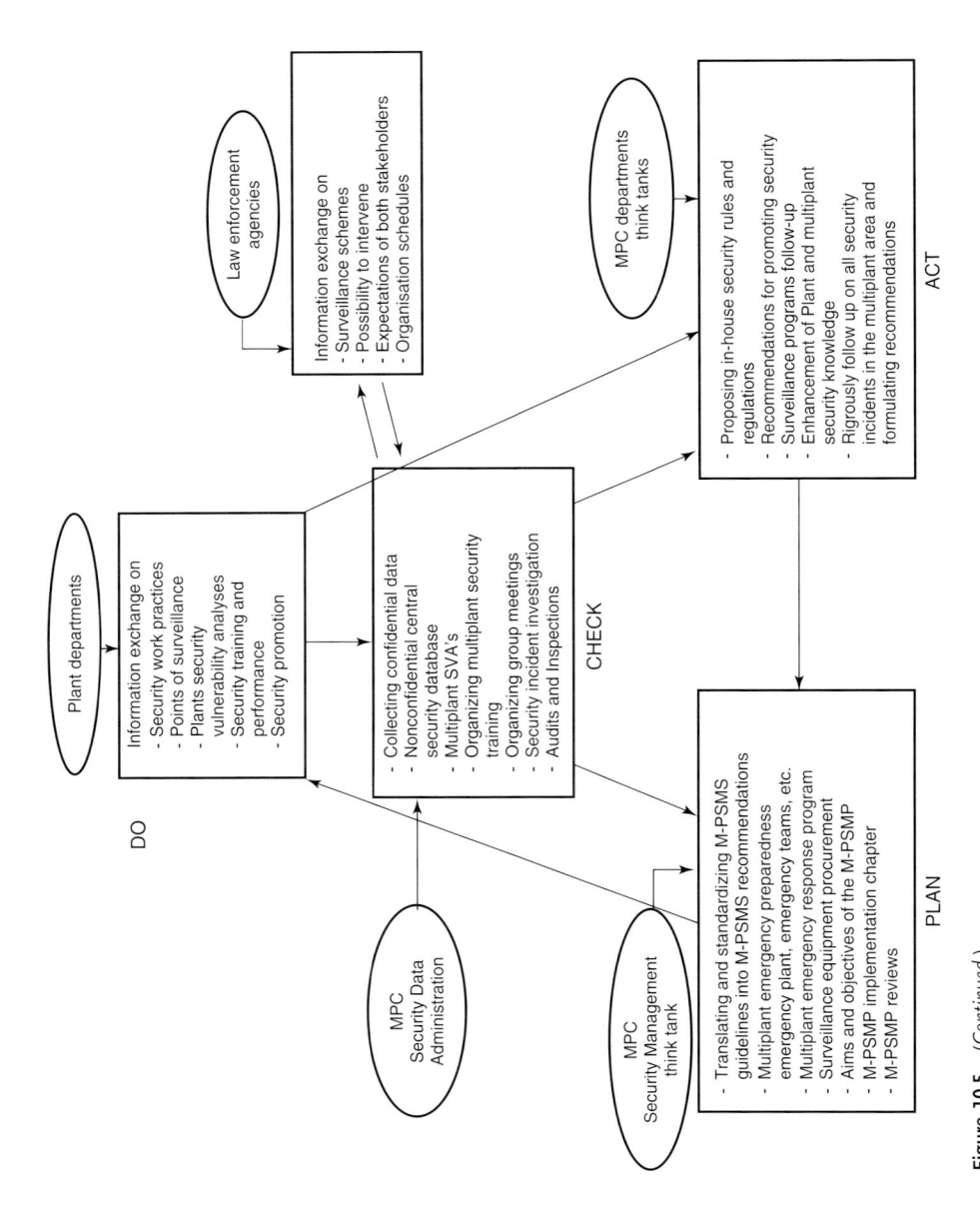

PLAN
- Translating and standardizing M-PSMS guidelines into M-PSMS recommendations
- Multiplant emergency preparedness emergency plant, emergency teams, etc.
- Multiplant emergency response program
- Surveillance equipment procurement
- Aims and objectives of the M-PSMP
- M-PSMP implementation chapter
- M-PSMP reviews

Figure 10.5 (*Continued.*)

10.4 Discussion

Instability is endemic to collaborative projects that aim to create the future. It is often more natural for collaborative ventures to come apart than to stay together. Whether a multiplant arrangement, for example, such as suggested in the previous Section 10.3, will stand the test of time hinges on its ability to learn and be flexible in the face of change. The proposed model captures this concern by having unambiguous and transparent communication protocols between the companies on one very specific topic, that is safety and security, and by using very strict procedures. Hence, the capacity to learn and the need to be flexible are built into the model.

The model is set up in a manner that documents, procedures, best practices, and so on can be exchanged in the MPC between company experts in an optimal way through structured and well-considered think tanks, and thereby guaranteeing efficient knowledge and know-how transfers on behaviors (explicit knowledge) as well as attitudes (tacit knowledge).

To have a notion of the implications of introducing the multiplant model in a chemical cluster, a so-called SWOT analysis was carried out. A SWOT analysis is a strategic planning method involving specifying the envisioned objectives of, for example, a business venture or a project and identifying the internal and external factors that are favorable and unfavorable to achieve those objectives. The SWOT analysis carried out for cluster management shows clearly the major opportunities of the collaboration model and the numerous strengths that can be built on to elaborate truly safer and more secure chemical industrial parks. The identified SWOT of cluster management for enhancing safety and security in chemical clusters can be found in Table 10.1.

The strengths of cluster management for improving safety and security within a chemical cluster can be summarized as *integration, capabilities, safety and security effectiveness, and support*. Building on these strengths, cluster management offers great opportunities such as *extra resources, safety and security knowledge and know-how spillovers*, and *Image building*. The most important weaknesses are *long-term commitment failure* and *chemical plant alignment failure*. The essential threats are *violation of mutual trust, free-riding behavior*, and *too big to manage*.

In game theory, the so-called *Tragedy of the Commons* is a well-known phenomenon. The basic idea is the following: whenever some players cooperate for mutual benefit, but others see that they could do better for themselves by breaking the cooperation, they can indeed defect (that is, cheat) until everyone else starts thinking in the same way (and starts to defect), when ultimately the cooperation collapses and everyone ends up worse off. Cluster management aims at the exact opposite and leads to the *Victory of the Few*: whenever nobody is cooperating, but some players see that they could do better for themselves by starting to collaborate, they can indeed cooperate until everybody else starts thinking identically, when ultimately everybody ends up collaborating, thereby being better off.

Table 10.1 SWOT analysis for cluster management for improving safety and security.

Strengths	Opportunities
Available safety and security knowledge from different plants	Long-term safety and security know-how increase
Available safety and security competencies from different plants	Improvement of the public image of the chemical-using industry
Unique integration of safety and security know-how and capabilities within a chemical industrial park	(Proactive) impact on safety and security legislation
Strong support from authorities and public	Attraction of "best-of-class visionary companies" (not necessarily the largest organizations)
Safety increase	Prevent major accidents
Security increase	Decrease in long-term safety and security costs due to collaboration benefits
	Spillovers on safety and security training and education
	Spillovers on safety and security research and development
Weaknesses	**Threats**
Structural financing of the Multiplant Council	Loss of "unique company hard-won knowledge"
The setup requires pioneering work where alignment of the participating plants is needed	Trust between plants gets violated
Cluster management for improving safety and security requires a long-term vision from all participating companies	Too many chemical plants in the park to manage the MPC properly
	Maintaining safety and security independence of individual plants
	Company management no longer interested in participating in the MPC
	Very large and very small companies not really interested and showing free-riding behavior

10.5
Conclusions

As could be observed in the 1970s, when Western companies struggled to close the quality gap with their Japanese competitors in global competition, the post-2010s will be characterized by chemical industrial areas trying to close the safety and security – and sustainability – gap with their chemical cluster competitors. Transferring knowledge and know-how related to safety and security between companies that form industrial areas brings them competitive advantages. More importantly, cluster management for improving safety and security leads to truly more sustainable chemical industrial clusters that are more accepted by society as a whole, thus gaining respect and a long-term license to operate from all stakeholders, including shareholders.

References

Child, J., Faulkner, D., and Tallman, S. (2005) *Cooperative Strategy. Managing Alliances, Networks, and Joint Ventures*, Oxford University Press, Oxford.

Nonaka, I., Toyama, R., and Konno, N. (2000) SECI, ba and leadership: a unified model of dynamic knowledge creation. *Long Range Plann.*, **33** (1), 5–34.

Reniers, G.L.L. (2010) *Multi-Plant Safety and Security Management in the Chemical and Process Industries*, Wiley-VCH Verlag GmbH, Weinheim.

Reniers, G.L.L., Ale, B.J.M., Dullaert, W., and Soudan, K. (2009) Designing continuous safety improvement within chemical industrial areas. *Saf. Sci.*, **47**, 578–590.

Reniers, G.L.L., Cremer, K., and Buytaert, J. (2011) Continuously and simultaneously optimizing an organization's safety and security culture and climate: the improvement diamond for excellence achievement and leadership in safety and security (IDEAL S&S) model. *J. Clean. Prod.*, **19**, 1239–1249.

Reniers, G.L.L., Dullaert, W., and Visser, L. (2010) Empirically based development of a framework for advancing and stimulating collaboration in the chemical industry (ASC): creating sustainable chemical industrial parks. *J. Clean. Prod.*, **18**, 1587–1597.

Tsai, W. (2001) Knowledge transfer in intraorganizational networks. *Acad. Manag. J.*, **44** (5), 996–1005.

Part IV
Managing Vertical Inter-Organizational Sustainability

11
Sustainable Chemical Logistics

Kenneth Sörensen and Christine Vanovermeire

11.1
Introduction

Chemicals often come in large volumes, are heavy, difficult to store and dangerous. As a result, many chemical companies spend large portions of their budget on logistics. Moreover, logistics is an increasingly complicated matter, with global sourcing and distribution being the norm rather than the exception. Most chemical companies nowadays operate in a global network of interlinked operations, as well as in a complex of legal frameworks that complicate everyday logistical operations. Coordinating the entire supply chain both in a vertical way (between companies on different levels of the supply chain, e.g., suppliers and customers) and a horizontal way (between companies on the same level of the supply chain) has become a truly daunting task.

Chemistry, and the chemical industry in general, are perceived by many people as harmful to the environment. The same is true for transportation, which is often quoted in popular news stories as responsible for producing a considerable portion of the greenhouse gases (GHGs) and other noxious substances in the atmosphere. These two perceptions combined, provide an increasingly strong incentive for chemical logistics to set the example in terms of sustainability. Broadly speaking, the chemical industry has not succeeded in convincing the public that it is part of the solution rather than the problem.

The impact of logistics on sustainability is not to be underestimated. It is trivial to see that any efforts to increase sustainability by, for example, improving chemical processes, can easily be offset by changes in transportation and logistics. A "cradle-to-grave" life cycle assessment (LCA) often reveals that a large part of the environmental impact of a (chemical) product is due to transportation performed during its lifetime.

A final reason to increase the sustainability of chemical logistics is a purely economic one. Fuel prices and energy prices in general have reached levels that make their efficient use a necessity rather than an option. Moreover, there is very little reason to assume that fuel cost will decrease anywhere in the near future. On the contrary, the increasing scarcity of fossil fuels and the increase in costs

Management Principles of Sustainable Industrial Chemistry: Theories, Concepts and Industrial Examples for Achieving Sustainable Chemical Products and Processes from a Non-Technological Viewpoint, First Edition. Edited by Genserik L.L. Reniers, Kenneth Sörensen, and Karl Vrancken.
© 2013 Wiley-VCH Verlag GmbH & Co. KGaA. Published 2013 by Wiley-VCH Verlag GmbH & Co. KGaA.

to extract crude oil and gas from difficult-to-reach oil fields will most likely result in further increases in oil and gas prices. Transportation technology that relies on renewable energy sources is still very far from reaching maturity (and just as far from being cost-competitive with transportation technology that gets its energy from fossil fuels).

Notwithstanding the fact that considerable efforts are devoted to increasing the sustainability of chemical industries worldwide, logistics is an often overlooked factor in this process. SusChem, the European Technology Platform for Sustainable Chemistry, lists "Plant Control and Supply Chain Management" as one of the topics on its strategic research agenda (SusChem, 2005), but saying that sustainable logistics is a top priority for the chemical industry is stretching the truth. Nonetheless, the strategic research agenda of SusChem states that "the field of supply chain management in chemical industry calls for significant research work for the development of new methodologies to fully exploit the potential for improvements in efficiency and service quality" and quotes the following research topics:

- Collaborative planning and control of transport and stock keeping
- Revenue management
- Inventories planning under uncertainties
- Advanced network design for production and distribution systems
- Reverse logistics strategies and operational control for the logistics of swaps.

The aim of this chapter is to present some of the challenges that the chemical industry faces when trying to improve the sustainability of its logistics operations. Some of these challenges are, in some form, indeed listed by the SusChem research agenda. The challenges that we discuss in this chapter are

1) Supply chain *optimization*, including multiobjective optimization, as a tool for sustainable decision making in the chemical supply chain
2) *Coordinated* supply chain management: managing the supply chain as a whole rather than as a set of different unrelated parts
3) *Horizontal collaboration* among companies, even competitors, at the same level of the supply chain
4) *Intermodality*: using each transport mode to its full potential.

Note that due to space restrictions several other aspects of the chemical supply chain necessarily have to be left untreated in this chapter. One aspect definitely worth mentioning is that of *reverse logistics*, in which flows in the opposite direction (from the customer back to its supplier) of used goods are explicitly taken into account in the supply chain planning. Related is the novel concept of *chemical leasing*, an approach in which the supplier sells the *function* of its product, rather than the product itself (and often remains the owner of the physical product), and which may similarly increase the sustainability of the chemical supply chain. Finally, inventory management is a factor that cannot be overlooked. A black dot between sustainable and chemical warehousing is treated in Section 4.2 of this book and chemical leasing in Section 4.3.

The next section discusses the current sustainability of transportation, mainly in the European Union. The figures given are for freight transportation in general (not only chemical products), but the conclusions drawn are equally valid for the chemical sector.

11.2
Sustainability of Logistics and Transportation

There is a broad consensus that the current evolution in both the demand and supply of transportation is inherently unsustainable. This holds true for chemical products, but also for freight in general, and even for passenger transport. Demand for freight transport in the EU-27 countries has grown at an annual rate of 2.3% in the period 1995–2008 (European Commission – Directorate General for Energy and Transport, 2008). Driving forces for this increase are general economic trends such as globalization, the speed with which modern customers expect their products to be delivered, agile manufacturing and just-in-time business practices, and integrated supply chain management (Rondinelli and Berry, 2000). Notwithstanding the increasing political pressure to move toward multimodal and intermodal transportation, this steady increase in demand for transportation has been absorbed almost exclusively by the road mode. As a consequence, the share of road freight transport in the modal split has increased from 42.1% in 1995 to 45.9% in 2008 (Figure 11.1) and does not show any sign of significant decline in the near future (Goel et al., 2009).

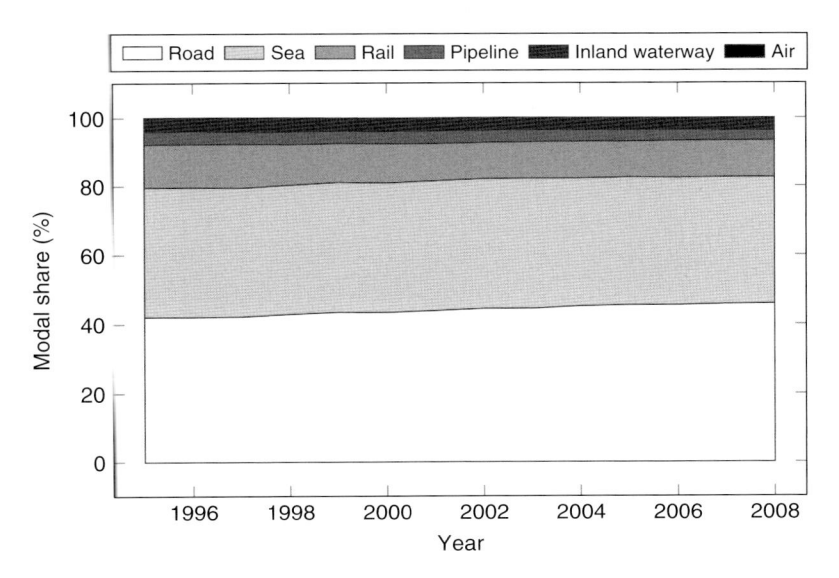

Figure 11.1 EU-27 modal split 1995–2008. (Source: European Commission – Directorate General for Energy and Transport (2008).

The increase, both absolute and relative, in the amount of tonne-kilometers that are transported on the European roads poses a threat to sustainability. The environment especially suffers, as road transport is by far the most polluting of all transport modes (EEA, 2011). While the transportation sector in general accounts for approximately 20% of all GHG emissions in the EU-27, 70.9% of this can be attributed solely to road transportation. In addition, the society and economy – the two other main pillars of sustainability – are negatively impacted by the dominant share of road transportation in the modal split as it increases the number of accidents and the congestion on the European road network. In 2005, 10% of the trans-European road network was affected by daily congestion. It is a certainty that this figure has only increased, and will keep on increasing in the coming years (Pedersen, 2005). As a result, transportation continues to negatively impact the current environmental problems in an important way (Doherty and Hoyle, 2009).

As mentioned, road is the most polluting freight transport mode for all pollutants (CO_2, NO_x, with the exception of particle matter (PM) pollutants) (EEA, 2011), whereas rail is generally the cleanest mode of transport for most pollutants. Maritime shipping is also one of the cleanest modes of freight transport, except for PM and specific sulfur oxide (SO_x) emissions, which are the highest for shipping (and aviation), mainly due to the high sulfur content of bunker fuels. The emissions generated by the maritime fleet are largely dependent on the quantity of fuel consumed; however, there are several factors besides fuel consumption that influence the emissions generated. These include the fuel quality and the engine type.

As a result of improved fuel quality and engine technology, emissions are expected to decrease, especially of sea transportation. In addition, legislation that puts stricter limits on, for example, sulfur content in fuel oils will greatly contribute to the expected emission reductions. Importantly, the emissions of all pollutants from road transport have been greatly reduced, mainly due to technological improvements and legislative measures. This is very different from other transport modes that are not subject to any emission control regulations. According to EEA 2011, similar emission control legislation could lead to equivalent reductions in the emissions of other transport modes.

11.3
Improving Sustainability of Logistics in the Chemical Sector

In this section, some ways are discussed in which chemical logistical operations can be rendered more sustainable. It should be mentioned that technological advances in transportation technology have often helped reduce the amount of pollutants exhausted by the transportation sector. Examples of important innovations include better fuels with less polluting additives, and the increasing efficiency of engines. Without playing down the importance of technological innovations, it should be equally clear that they are not a panacea for every problem in the chemical supply chain. Many supply chains are run with the latest technological advances, but are still far from being as sustainable as they could be due to inefficiencies in the way they are organized.

This chapter therefore focuses on the other types of changes to improve sustainability: *organizational* changes that do not require technological advances but stem from a better planning and coordination of the chemical supply chain. Four issues are discussed. *Optimization* is a field of study that disposes of adequate methods and tools to increase the efficiency of virtually any operation in the chemical supply chain, without requiring qualitative restructuring of this operation. In this way, the potential of any technology can be fully exploited. Coordinated supply chain management, or vertical integration of the supply chain, allows all processes in the supply chains to be managed from raw material to finished product. Horizontal collaboration is a third strategy that has recently proved to yield gains in efficiency and sustainability that no single company can achieve on its own. It does have a rather long list of factors that influence its viability, and is therefore still very far from reaching its full potential. Finally, intermodal transportation is already an important factor in the chemical industry, but to move to a more sustainable chemical logistics will certainly be accompanied by a better use of different transport modes.

11.3.1
Optimization

Even without any technological advances in transportation technology, current operations can most often be improved through the effective use of optimization technology. Optimization does not change the current manner of working in any other way than by using the current resources to their full potential. A company that employs a fleet of trucks to perform customers' deliveries, for example, can use adequate optimization techniques to minimize the number of kilometers driven by its trucks. Similarly, optimization can be used to determine the optimal schedule for ocean liners to visit their destination ports, or the optimal location of distribution facilities.

The field of optimization is part of the broader domain of Operations Research, a field that originated around WWII as a set of methods to bring the extensive military operations under tighter control (hence the name "Operations" Research). This field can be broadly defined as the use of quantitative methods for decision support. After WWII, it was realized that the methods successfully used to plan and execute the enormous logistical operations of a large-scale war could be usefully employed in a business context. Logistics was one of the first areas in which this happened, and is still one of the most active research domains in the application of civil Operations Research, especially optimization.

Optimization works in three steps, schematically represented in Figure 11.2. First, a *model* is built of the optimization problem in question. This model is essentially a simplification of reality, in which only the bare essence is left. Models can be made in varying degrees of detail, but care should be taken that all aspects of the optimization problem that might have a significant impact on the value of the optimal solution are taken into account. For example, a chemical company using

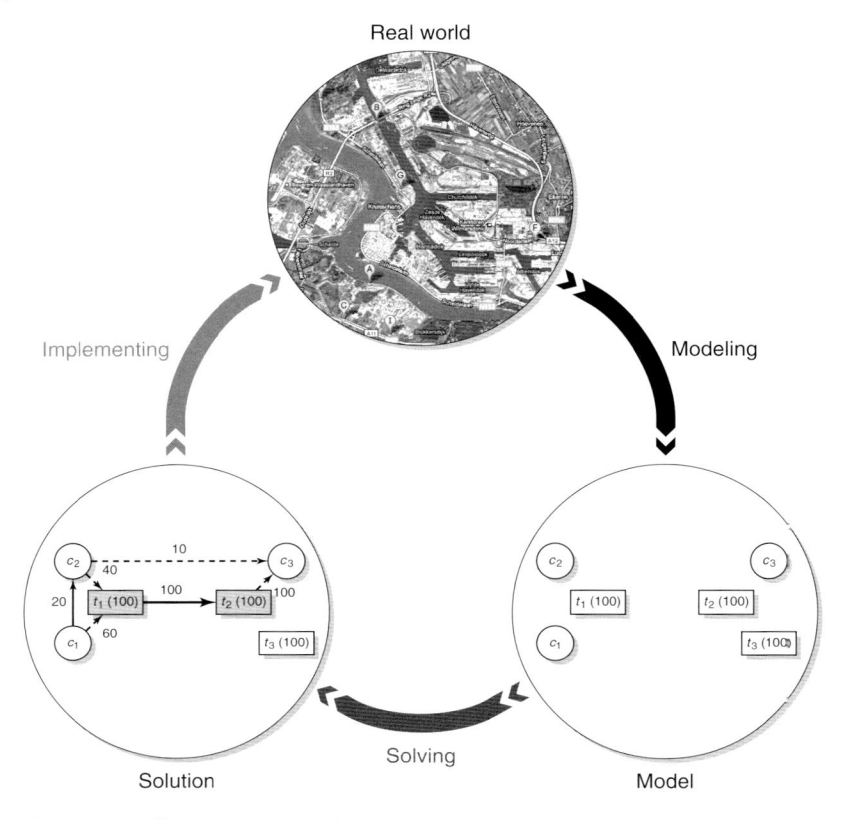

Figure 11.2 The optimization cycle.

optimization to determine the optimal location of its refineries (e.g., by minimizing the expected cost of transportation between these facilities) should take into account the fact that there may be significant differences between the transportation costs between different pairs of facilities. On the other hand, negligible aspects such as the difference in wear on the truck's tires due to the road quality in different countries, can probably be safely ignored because it will have no effect on the optimal solution.

The second step in the optimization cycle is the development and application of an optimization *method* to the model. Optimization methods come in two flavors. While *exact* methods come with a guarantee of finding the optimal (i.e., the best possible) solution, their counterparts, called *heuristics*, do not. Although exact methods may seem more appropriate because of this guarantee to find the best possible solutions out of the set of all possible solutions, there is one significant drawback to this. Most optimization problems that are relevant in practice are what in the field of complexity theory is called *NP* hard. The technical details of this term are beyond the scope of this paper, but in short, this means that the computing time of any exact method for such a problem increases exponentially with the size of the problem. This, in turn, usually means that solving any but the smallest of problems can take an extraordinarily large amount of time: days, weeks, years, and even centuries and more. It is clear that this renders most exact methods useless to solve realistic optimization problems.

Heuristics are able to overcome the *NP* hardness of optimization problems by discarding the need to find the optimal solution. Rather, such methods attempt to find a "good" solution in a "short" amount of computing time. The meaning of the terms "good" and "short" can, of course, vary enormously across optimization problems. For long-term, strategic problems such as the location of distribution or production facilities, hours or days of computing time are generally not a problem. For short-term, operational decisions such as those found in, for example, production or distribution scheduling, shorter computing times are essential.

Modern heuristics, often called *metaheuristics*, are generally able to find solutions that are close to the ones found by exact methods in a time that is much shorter. For this reason, they are the preferred way of solving most practically relevant optimization problems. Most optimization methods built into commercially available software for production planning, vehicle routing optimization, strategic decision making, and so on are therefore of a heuristic nature.

Finally, the third step in the optimization cycle is the translation of the solution found using the optimization method back into the real world. As often considerable simplifications were made in step 1, this step can be more difficult than it seems. Often, several iterations are necessary to reach the desired outcome. This is schematically represented in Figure 11.2.

Optimization has been used to successfully tackle several challenging problems in the chemical industry. A few examples of optimization problems and solution techniques found in the literature are mentioned below. Sahinidis *et al.* 1989 developed a model for the optimal selection and expansion of processes using forecasts for the demands and prices of chemicals over a long-range horizon. Lasschuit and Thijssen 2004 developed a support system for planning and scheduling production and distribution in the downstream oil and chemical industry. They argued that decision support tools must allow for consistent operational guiding of the process, while at the same time taking available (real-time) information on actual operations and market economics into account. The authors have found that the use of adequate models yields substantial benefits not only in economic terms but also in an improved understanding of the interactions between the various components of the business. Berning *et al.* 2004 considered a complex scheduling problem in the chemical process industry involving batch production. In their application, a network of production plants with interdependent production schedules, multistage production at multipurpose facilities, and chain production was studied. A production scheduling solution was obtained by a genetic algorithm, but the authors additionally described a mechanism for collaborative planning between the plants, which allowed them to work together in a transparent and flexible way.

Traditionally, most optimization algorithms have been aimed at solving so-called *uni-objective* optimization problems, in which a single objective is either maximized or minimized. Typically, the goal of such algorithms has been to minimize the cost or maximize the profit. More recently, algorithms, which specifically tackle the so-called *multiobjective* optimization problems, in which several objectives need to be simultaneously optimized, have been created. Multiobjective optimization (Gandibleux et al., 2004) is particularly useful if sustainability is an issue. Suppose a transportation company wishes to develop a multimodal transportation schedule for its products (i.e., use which mode to ship each product). The company might want to minimize the total cost of this schedule, while simultaneously minimizing the burden on the environment and the associated risk. Because different transport modes have different, and often conflicting characteristics (e.g., road transportation is often the cheapest alternative, but also the most polluting and the most risky), difficult trade-offs have to be made. Multiobjective optimization can help these companies find a schedule that simultaneously minimizes all three objectives.

For companies, it is often impossible to develop their own optimization algorithms and implement them in a decision support system. Many companies have therefore built dedicated systems for production planning, distribution planning, and other planning problems. These systems are generally called *advanced Planning and scheduling* (APS) systems and are usually tightly integrated with a company's enterprise resource planning (ERP) system.

11.3.2
Coordinated Supply Chain Management

Although it is impossible to depict the complexity of current-day supply chains, most of them look somewhat similar to the generic supply chain shown in Figure 11.3. Even this simple figure shows that most supply chains are not chains at all, but complex interconnected networks of manufacturers, suppliers of raw materials, warehouses, and customers (Thomas and Griffin, 1996). Traditionally, three stages in the supply chain are distinguished: procurement, production, and distribution, each of which may consist of different facilities around the world.

In the past, most efforts have been put on optimization and coordination *within* a facility. Typically, different facilities have been shielded from each other by

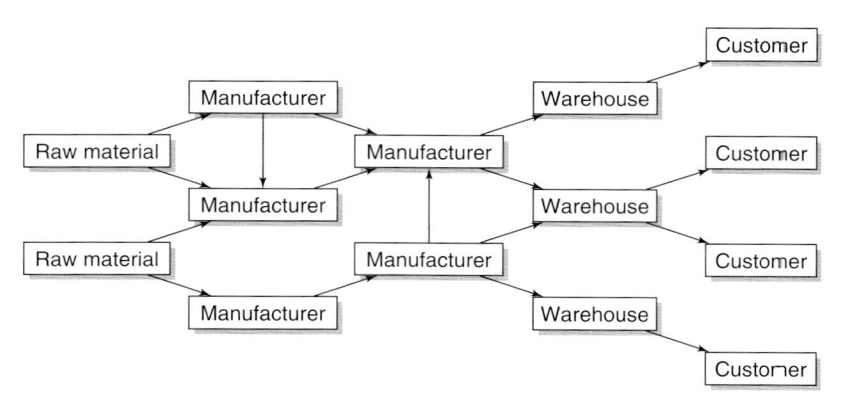

Figure 11.3 A generic supply chain (adapted from Thomas and Griffin, (1996).

large buffers of inventory. The undesirable consequences of this are legion: high inventory costs, loss of control, shrinkage, and increased susceptibility to the bullwhip effect. In recent years, an increasing focus has been placed on the so-called *coordinated supply chain management*. In this paradigm, the different stages in the supply chain network are treated as one system and optimized as such.

Thomas and Griffin 1996 distinguish three forms of supply chain coordination: buyer–vendor coordination, production–distribution coordination, and inventory–distribution coordination. In each of these forms of coordination, the aim is to reduce operational costs, generally by reducing inventories and adapting supply chain practices across the different stages of the supply chain.

Think tank sessions of EPCA-Cefic have revealed that the chemical industry indeed can find solutions to their sustainability issues in vertical supply chain management. A shortened list of opportunities is presented here (McKinnon, 2004):

- *Postponing*: Most of the packaging and customization in the chemical sector is now done at the time of production. However, opportunities exist to delay to a later point or even avoid it and thus enable more transportation in bulk as well as reduce inventory.
- *Differentiating customer service levels*: By providing the best service levels to the main customers and allowing lower service levels to the customers who do not require this, the operational planning can be further optimized.
- *Vendor-managed inventory*: VMI is an established practice in many other sectors nowadays. It consists of giving the vendor insight into and responsibility for the inventory of its customers. The vendor can use this knowledge to better plan its production and transportation in advance. This is enabled with ERP systems that are generally already in place in chemical companies.

A more recent form of coordination, however, is horizontal collaboration, between companies that operate on the same level in the supply chain. This is discussed in the following section.

11.3.3
Horizontal Collaboration

In order to improve the supply chain processes, companies first optimized internally (Section 11.3.1). The next step consisted of looking beyond the borders of their own organization and looking at the partners in the same supply chain. By vertically coordinating or even integrating the entire (or part of) the supply chain (Section 11.3.2), significant savings can be made. Nowadays, as the sustainability of supply chains still remains an issue, companies even go beyond the boundaries of their supply chains (Mason, Lalwani, and Boughton, 2007). Opportunities exist to improve the sustainability, cost, and service level of supply chains by collaborating with companies that act on the same level of the supply chain, a trend that has been coined *horizontal collaboration* (European Commission, 2011) (Figure 11.4). In practice, assets (such as warehousing, a fleet, etc.) can be shared, and orders

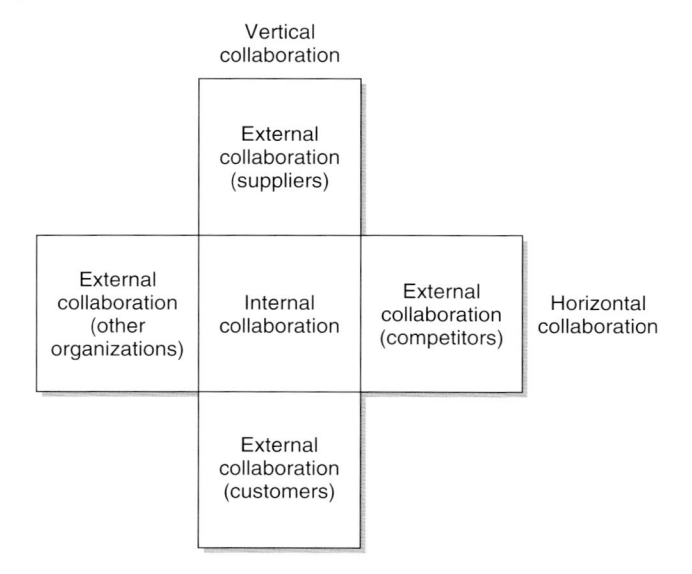

Figure 11.4 The scope of collaboration (Barratt, 2004).

of separate companies can be bundled and shipped in a common distribution effort. This results in a more efficient use of assets and a more efficient operational planning (as (half-)empty trucks and superfluous miles can be avoided).

The advantages of horizontal collaboration that are listed in the literature review of Cruijssen, Dullaert, and Fleuren 2007 are the following:

- A decrease in cost, as well as a better productivity owing to more available knowledge and better use of the available means.
- Increased service levels because of complimentary or specialization of goods. The service level can also be improved as companies can comply with more ease to the strict customer requirements. The number of drops or the frequency of delivery can be increased.
- Market position can be fortified (a faster speed to market, penetrating new markets, ...).
- Other opportunities such as developing technical standards, overcoming legal and/or regulatory barriers, accessing superior technology, and enhancing public image.

Nevertheless, there also exists a long list of hurdles to overcome. A survey among supply chain professionals by EyeforTransport 2010–2011 reveals at least 19 important factors. The results of this survey are given in Figure 11.5.

Horizontal collaboration has broken through on a more or less important scale in two sectors: air and maritime transport. As transportation in these sectors is more capital-intensive, distances are generally longer, and the number of players is smaller than in the land transportation sector, the players in these sectors have experienced bigger incentives to overcome these hurdles.

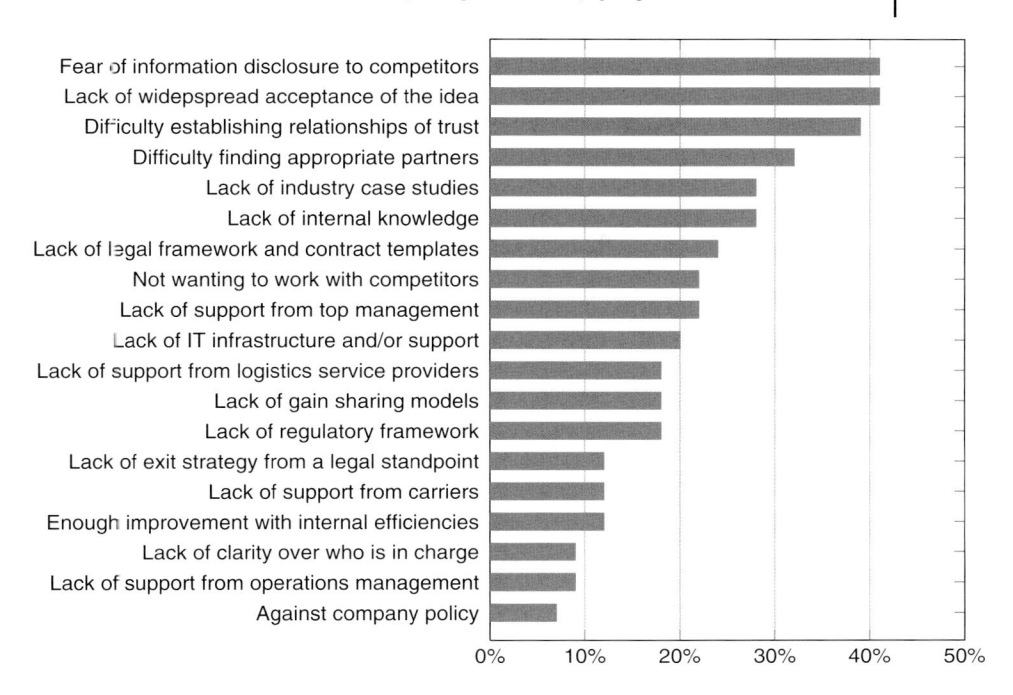

Figure 11.5 Barriers stopping further investment to horizontal collaboration from shippers (EyeforTransport, 2010–2011.)

However, as the pressure to operate in a more sustainable way grows (something that is especially true of the chemical industry), horizontal collaboration also becomes a viable option to operating in a stand-alone way. Horizontal collaboration in land transportation has indeed proved itself in several pilot projects. In all these projects, horizontal collaboration was able to reduce costs (between 10 and 35%), as well as improve sustainability (e.g., CO_2 output was decreased between 5 and 50%) (Vanovermeire and Sörensen, 2011).

The chemical industry, in particular, can certainly benefit from bundling transportation, for the following reasons.

- *Cost*: As managing the risk of transporting hazardous materials implies investing in safe transportation means, applying safety measures, choosing safer instead of cheaper solutions, and so on, transportation in the chemical industry is far more capital-intensive than in other industries. By finding partners that have the same strict requirements for the transportation of their products, these costs can be shared and an individual company can see its supply chain cost reduced. Such partners can generally be found in the same sector.
- *Cost*: Chemical companies are often located in clusters such as those in Denver or Antwerp. This can contribute to the gains that can be achieved by bundling, because less additional kilometers will have to be travelled to connect two collaborating companies and will in turn lead to more positive effects the

collaboration has on cost as well as CO_2 emissions. The same reasoning evidently applies if partners have a lot of mutual (or clustered) clients.
- *Service level*: By increasing the frequency of delivery (smaller delivery quantities, more deliveries), the amount of stock that is kept on both the shipper's as well as the client's side can be reduced. As chemical products are generally not only expensive but also hazardous, a lower stock level is preferable.
- *Other*: Because of the increased volume that is shipped in a collaborative supply chain, investments in superior technology and transportation means are easier to justify. Intermodal transportation, for example, which is nowadays highly promoted as the solution to more sustainable supply chains, is only sustainable when large volumes are transported (e.g., a train is not sustainable if less than half of its capacity is used). In the case of the chemical industry, there exists the possibility of shifting to pipelines. As pipelines do not emit any GHGs, the sustainability of using this transport mode increases significantly.
- *Other*: The green image of the chemical industry, notoriously known as a polluting industry, can be improved. For example, a pilot project of UCB and Baxter (Menedeme, 2011) has achieved 50% reduction in emission of GHGs and has therefore won the supply chain innovation award of 2011.[1]

11.3.4
Multimodal, Intermodal and Co-Modal Transportation

Although the terms "intermodal" and "multimodal" are both used to refer to the transport of goods using more than one transport mode and are often used interchangeably, most authors make a distinction between them. The Illustrated Glossary for Transport Statistics (UNECE, 2009) defines *multimodal* freight transportation as the transport of goods by at least two different modes of transport. *Intermodal* freight transportation is defined as a particular type of multimodal transportation in which the goods are transported in one and the same intermodal transport unit (e.g., a standard ISO container) by successive modes of transport without handling of the goods themselves when changing modes. Hence, while the more general term "multimodal transport" does not require any interoperability between modes, the term "intermodal transportation" puts the focus on the interaction between the modes. Given the fact that transfers play a crucial role in intermodal transportation – since they account for a sizeable portion of the total transportation journey time (Pedersen, 2005) – this distinction is important.

A relatively new term is *co-modality*, introduced by the European Commission (European Commission, 2006) to define an approach to transportation in which transport modes are not seen as competing, but rather as complementary alternatives. According to this philosophy, a more sustainable transportation situation can be reached by designing, managing, and optimizing supply chains in such

---

[1] The supply chain distinction awards website, October 2011.

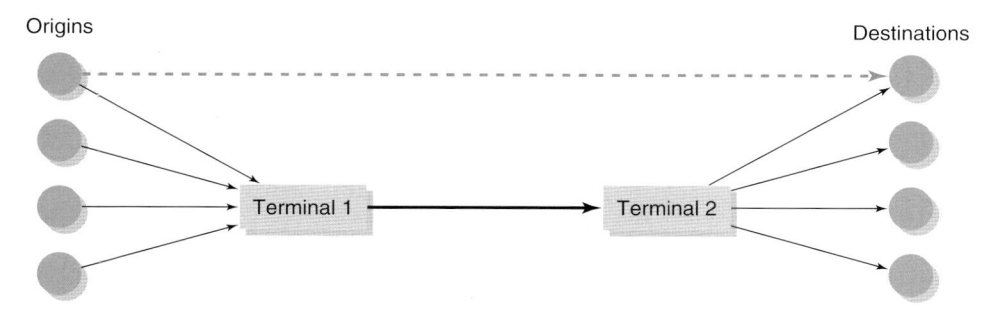

Figure 11.6 An intermodal supply chain.

a way that each transport mode is used to its full potential and synergies are exploited.

Figure 11.6 visualizes a very simple intermodal supply chain, considering only two transshipment terminals and two modes of transport, for example, rail and road. In this case, each shipper can either choose to ship his goods from origin to destination directly by truck or by using a truck–train combination. If the last option is chosen, that is, the intermodal trajectory, the goods are first transported by truck to a terminal where they are transshipped from truck to train (*prehaulage*). The train subsequently takes care of the interterminal part of the transportation journey (*long haul*). After arrival at the second terminal, the goods are once more transshipped onto a truck that delivers the goods at the receiver location (*end-haulage*). Of course, the interterminal transport can also be performed by barge, ocean liner, airplane, pipeline, or any other transport mode. One of the greatest revolutions in transport is without a doubt the introduction of the ISO-standardized containers. Goods stored in such containers can be transported by container ship, barge, trailer trucks, and cargo train without ever requiring repackaging. They can be handled by any of a number of cranes and other container-handling equipment in an efficient way.

Managing and designing an intermodal supply chain is, however, considerably more complex than a traditional unimodal one and puts heavier requirements on the models and methods used to support different decisions on different levels (strategical, tactical, operational) (Macharis and Bontekoning, 2004; Winebrake et al., 2008). Several factors contribute to this increased complexity. First, the number and diversity of stakeholders (e.g., shippers, receivers, policy makers, drayage operators, and network operators) is much larger than in a simple unimodal network. Moreover, these stakeholders have different interests (the shipper, e.g., wants his goods to arrive as quickly as possible, whereas the network operator wishes to fully utilize his capacity; this may mean that there is some resistance in the latter to send half-full freight trains, which might mean delays for the former), but they need to work in close coordination for the system to run smoothly. Another factor contributing to the complexity of intermodal supply chains is the geographical configuration of the intermodal terminal systems. Ports, rail yards, and other areas where containers are unloaded from one transport mode and loaded onto another one are all liable to failures, delays, and other adverse effects. Finally, multimodality explicitly assumes the integration of at least two different transport modes, which of course have their own requirement with respect to infrastructure and transport units. Winebrake *et al.* 2008 even concluded that *understanding intermodal freight transport is understanding a complex web of people, places, and technologies that are intertwined in a network constructed to deliver goods throughout the country in the most efficient, timely, and environmentally sustainable way possible.* As a result, intermodal transportation gives rise to new optimization problems for which the field of optimization has scrambled to develop adequate optimization methods.

In recent years, intermodal transportation has therefore evolved into one of the most rapidly emerging research fields in the transportation literature. Caris *et al.* 2008 propose a classification scheme of research on intermodal transportation on three decision-making levels: strategical, tactical, and operational. An additional division is made on the basis of the stakeholder under consideration. The different stakeholders in an intermodal supply chain, and some of the planning problems they face, are as follows:

- *Drayage operators* schedule and operate (truck) transport between shippers and terminals, and between terminals and receivers. On a strategic level, research has almost exclusively focused on the potential benefits of centrally planning the pickup and delivery trips of several drayage operators in one terminal service area. On the tactical level, the drayage operator is concerned with the optimal assignment of shippers and receivers to terminals. The most important operational problem these stakeholders need to deal with is to ensure sufficient capacity availability at the terminals to meet demand.
- *Terminal operators* manage the transshipment activities at the terminals. Strategically, terminal operators need to determine the optimal design and type of new terminals, as well as the optimal location of the terminal. Given the huge investments necessary to build, for example, a port container terminal, it is clear that such decisions may require considerable analysis. Tactical problems faced by these stakeholders are the analysis and subsequent optimization of the operating procedures at existing terminals and the terminal layout. Again, these procedures may determine, to a large extent, the efficiency and therefore the viability of the intermodal terminal and software is available for the allocation of incoming ships to berths, the layout of the terrain on which the containers are stacked, the operating procedures of the gantry cranes, and so on. On an operational level, the planning of the individual operations of the container-handling equipment needs to be planned. In a port container terminal, these decisions may include the order in which straddle carriers pickup containers from the ship and where they should put them in the yard. Again, taking such decisions optimally requires considerable analysis and is of the largest importance. Containers are usually stacked onto each other and picking up the bottom container from a stack may require considerable extra work, in moving the other containers in the stack to a different location (where they may again block the next container

that is going to be picked up). It is clear that computer algorithms for such operational planning problems may considerably enhance the working of the system.
- *Network operators* take care of the infrastructure planning and the organization (scheduling/pricing) of intermodal services. The planning problems this stakeholder faces have been studied extensively by intermodal transportation researchers. The two most prominent strategic problems are the location of new intermodal terminals and the creation of intermodal chains by ensuring the interconnectivity of transport modes. On a tactical level, the network operator needs to decide which consolidation network (point-to-point, line, hub-and-spoke, collection–distribution) and production model to use. The operational decisions relate to, for example, the load order of trains and barges and the redistribution of railcars or push barges.
- *Intermodal operators* are the actual end users of the intermodal infrastructure and services. They buy the services offered by the aforementioned stakeholders. Their main task is to determine the optimal route for their shipments through the intermodal network, which is by nature an operational decision. Hence, strategical and tactical planning problems are of no relevance for this stakeholder.

Given the nature of these planning problems, there is no doubt that Operations Research models and methods are best suited to support the stakeholders in making these decisions. Both exact as well as approximative methods have been applied for this purpose. For a review, we refer among others to Crainic and Kim 2007; Jarzemskiene 2007, and Caris *et al.* 2008.

Shifting the modal split in favor of alternative, more environmentally friendly, and safe modes of transportation would already mean a big step forward in the movement toward more sustainable transportation. It is well known that there is a large discrepancy between transport modes in terms of ecological footprint. As described in Section 11.3.4, this is inherently the objective of multimodal/intermodal transportation. Currently, however, intermodal transportation is quite far from being a viable alternative to road transportation and therefore rather exceptional in a European context. In its white paper, the European Union (European Commission, 2001) uncovers some of the main issues that explain this low level of maturity. A first element is the short shipping distances. In comparison with, for example, the United States, the sender–receiver distances in the European Union are a lot shorter, which does not only reduce the benefits of consolidation but also complicates the transportation planning significantly. Another inconvenience is the existence of different railways systems in Europe that renders it impossible to move freight from the north to south without transferring loads. Also, the intermodal transshipment infrastructure leaves much to be desired. Both the limited number and insufficient capacity of intermodal terminals and the labour-intensiveness of terminal operations significantly increase the idle time at the transshipment points.

Despite these planning problems, it is impossible to ignore the potential of intermodality in achieving more sustainable transportation systems. Therefore, it seems obvious to incorporate sustainability criteria into intermodal transportation planning. However, papers on the integration of sustainability criteria into intermodal transport planning models and methods are few and far between. Examples include the method to determine network investment priorities of Loureiro and Ralston 1996; the network analysis model to trade-off energy, environmental, cost, and time aspects in intermodal freight transport of Winebrake *et al.* 2008; and the routing and scheduling model of Bauer *et al.* 2010.

Despite the diversity of these planning problems, they do allow drawing some general conclusions. Finally, a true integration of sustainability and intermodal transportation planning gives rise to some methodological issues. In fact, it assumes that besides the traditional economic criteria, sustainability criteria are also taken into account into the decision-making process. However, the majority of the tools and techniques that have been proposed in the literature for solving transportation planning problems are not capable of handling more than one (conflicting) objective at the same time. The same observation holds for most commercial software that are designed to provide feasible solutions that will minimize economic costs (Sbihi and Eglese, 2007). This clearly provides a large incentive for the development of (multiobjective) optimization methods.

11.4
Conclusions

Both transportation and chemistry probably occupy top positions in the general public's list of factors that negatively contribute to the environment. There is some truth in this perception: notwithstanding large-scale efforts of governing organizations such as the European Union to promote a modal shift to less polluting alternatives, the share of road transportation (by far the most polluting transport mode) in the modal split is still increasing. For both the transportation and chemical sectors, sustainability therefore is and should be an important issue and improving the score on the different sustainability measures is high on the agenda.

Technological advances, for example, better engines and better fuels, can help remedy the situation, but an optimized *organization* of the chemical supply chain is equally important. This chapter has discussed four ways in which this can be achieved. First, the field of Operations Research provides *optimization* methods and software that can help increase the efficiency of virtually every step of the supply chain. Vertical integration (usually called *coordinated supply chain management*) as well as *horizontal collaboration* are also required to move beyond internal optimization possibilities. Finally, a move to more intermodal transport (co-modality, as the European Union prefers to call it) promises improvements both in terms of efficiency and sustainability.

In all these areas, a lot of work still remains to be done, both in academic research as in practical implementations. Only through an integration of new technology with efficient methods for planning and designing logistics operations can a truly sustainable chemical supply chain be achieved.

References

Barratt, M. (2004) Unveiling enablers and inhibitors of collaborative planning. *Int. J. Logist. Manage.*, **15** (1), 73–90.

Bauer, J., Bektas, T., and Crainic, T.G. (2010) Minimizing greenhouse gas emissions in intermodal freight transport: an application to rail service design. *J. Oper. Res. Soc.*, **51**, 530–542.

Berning, G., Brandenburg, M., Gürsoy, K., Kussi, J.S., Mehta, V., and Tölle, F.J. (2004) Integrating collaborative planning and supply chain optimization for the chemical process industry (i)–methodology. *Comput. Chem. Eng.*, **28** (6-7), 913–927.

Caris, A., Macharis, C., and Janssens, G.K. (2008) Planning problems in intermodal freight transport: accomplishments and prospects. *Transport. Plann. Technol.*, **31** (3), 277–302.

Crainic, T.G. and Kim, K.H. (2007) Intermodal transportation, *Handbook in Operations Researsh and Management Science*, Vol. 14, Elsevier B.V.

Cruijssen, F., Dullaert, W., and Fleuren, H. (2007) Horizontal cooperation in transport and logistics: a literature review. *Transport. J.*, **43** (2), 129–142.

Doherty, S. and Hoyle, S. (2009) Supply Chain Decarbonization: The Role of Logistics and Transport in Reducing Supply Chain Carbon Emissions. Technical report, World Economic Forum and Accenture.

EEA (2011) *Specific Air Pollutant Emissions*. European Environmental Agency TERM 028.

European Commission (2001) European transport policy for 2010: time to decide. White paper.

European Commission (2006) Keep europe moving — Sustainable mobility for our continent — Mid-term review of the European Commission's 2001 transport white paper, Communication from the Commission to the Council and the European Parliament COM/2006/0314 final.

European Commission (2011) *Guidelines on the Applicability of Article 101 of the Treaty on the Functioning of the European Union to Horizontal Co-operation Agreements*, Official Journal of the European Communities.

European Commission — Directorate General for Energy and Transport (2008) EU energy and transport in figures.

EyeforTransport (2010–2011) European Horizontal Collaboration in the Supply Chain: A Brief Analysis of Eyefortransport's Recent Survey. Technical report, EyeforTransport.

Gandibleux, X., Sevaux, M., Tkindt, V., and Sorensen K. (eds) (2004) *Metaheuristics for Multiobjective Optimisation*, Lecture Notes in Economics and Mathematical Systems, Springer, Berlin.

Goel, A. (2009) A roadmap for sustainable freight transport, in *Methods of Multicriteria Decision Theory and Applications* (eds F. Heyde, A. Lohne, and C. Tammer), Shaker, pp. 47–56.

Jarzemskiene, I. (2007) The evolution of intermodal transport research and its development issues. *Transport*, **22**, 296–306.

Lasschuit, W. and Thijssen, N. (2004) Supporting supply chain planning and scheduling decisions in the oil and chemical industry. *Comput. Chem. Eng.*, **28** (6-7), 863–870.

Loureiro, C.F.G. and Ralston, B. (1996) An investment selection model for multicommodity multimodal transportation networks. *Transport. Res. Rec.*, **1522**, 38–46.

Macharis, C. and Bontekoning, Y.M. (2004) Opportunities for or in intermodal freight transport research: A review. *Eur. J. Oper. Res.*, **153**, 400–416.

Mason, R., Lalwani, C., and Boughton, R. (2007) Combining vertical and horizontal collaboration for transport optimisation. *Supply Chain Manage. Int. J.*, **12** (3), 187–199.

Menedeme, L. (2011) Prepare your supply chain for horizontal collaboration. Presentation at the 2nd Horizontal Collaboration Summit, May 2011.

Pedersen, M.D. (2005) Optimization models and solution methods for intermodal transportation. PhD thesis. Technical University of Denmark.

Rondinelli, D. and Berry, M. (2000) Multimodal transportation, logistics, and the environment: managing interactions in a global environment. *Eur. Manage. J.*, **18** (4), 398–410.

Sahinidis, N.V., Grossmann, I.E., Fornari, R.E., and Chathrathi, M. (1989) Optimization model for long range planning in the chemical industry. *Comput. Chem. Eng.*, **13** (9), 1049–1063.

Sbihi, A. and Eglese, R.W. (2007) Combinatorial optimization and green logistics. *4OR: A Q. J. Oper. Res.*, **5** (2), 99–116.

Suschem (2005) *Innovating for a Better Future: Sustainable Chemistry Strategic Research Agenda*, European Technology Platform for Sustainable Chemistry.

Thomas, D.J. and Griffin, P.M. (1996) Coordinated supply chain management. *Eur. J. Oper. Res.*, **94** (1), 1–15.

UNECE (2009) *Illustrated Glossary for Transport Statistics*, United Nations Economic Commission for Europe.

Winebrake, J.J., Corbett, J.J., Falzarano, A., Hawker, J.S., Korfmacher, K., Ketha, S., and Zilora, S. (2008) Assessing energy, environmental, and economic tradeoffs in intermodal freight transportation. *J. Air Waste Manage.*, **58**, 1004–1013.

12
Implementing Service-Based Chemical Supply Relationship – Chemical Leasing® – Potential in EU

Bart P.A. Van der Velpen and Marianne J.J. Hoppenbrouwers

12.1
Introduction

Within the Framework becomes policy intention of the European Union (EU) to support a new industrial policy achieving a competitive and sustainable economy, three major paths on the sustainable production of goods and services are distinguished: the minimization of the use of natural resources and energy, the goal for waste-free production, and the change of production and consumption patterns (European Commission, 2009). Efforts in research and development strive to optimize life cycle aspects, especially through recovery, treatment, and the safe reuse of products and industrial waste. More efficient production systems, machines, industrial processes, and eco-friendly products have been and continue to be developed in order to minimize natural resources and energy consumption during production or/and during use. Cleaner production technologies aim at the reduction of energy consumption, gas emissions, water effluent, solid residues, and so on. These technologies contribute to climate and environmental protection.

Beyond such technical solutions, new organizational approaches are currently elaborated, such as knowledge-based supply chain and production networks (Comethe, 2008), industrial symbiosis concepts, production and procurement networks, and so on, have gained importance in industry. Accordingly, suitable business models have been developed. The chemical industry is customarily more affected by strict legal environmental specifications and occupational health and safety regulations than the manufacturing industry. Owing to the increasing direct responsibility for product safety (REACH, EC/1907/2006), pushed by EU legislation, a closer collaboration between users and producers of chemicals is stimulated. Chemical Leasing® (ChL) is a new service-oriented business model with a particular focus on the chemical industry within a business-to-business environment. The concept of ChL was first described in 2002 during the World Summit on Sustainable Development in Johannesburg. The need for sustainability of production processes, as well as the consideration of the whole life cycle of the chemical substances, was recognized as important. Resource-efficient manufacturing that takes into account the protection of human health and environment should be

*Management Principles of Sustainable Industrial Chemistry: Theories, Concepts and Industrial Examples
for Achieving Sustainable Chemical Products and Processes from a Non-Technological Viewpoint,*
First Edition. Edited by Genserik L.L. Reniers, Kenneth Sörensen, and Karl Vrancken.
© 2013 Wiley-VCH Verlag GmbH & Co. KGaA. Published 2013 by Wiley-VCH Verlag GmbH & Co. KGaA.

encouraged; especially in the chemical industry with its typically complex and long supply chains.

From an environmental perspective and as a basic principle, a supplier–customer model that aligns economic incentives toward reduced (and thus more sustainable) chemical consumption is clearly preferable. ChL is one such specific model. In this model, the chemical supplier would principally be compensated on the basis of chemical services delivered and not on the basis of the chemical volume sold. This basis for compensation is possible because customers who purchase chemicals – particularly for production-supportive use – generally see little intrinsic value in the chemicals per se. Furthermore, the customers are frequently not motivated to invest in the use or knowledge of the purchased chemicals. Linked hereto is a poor understanding of the chemical costs. These costs are perceived as relatively small (Stoughton and Votta, 2003). After all, the chemical costs are only fractions of the total operating costs. Instead, the real value of a chemical resides in the function it performs, for example, cleaning, cooling, lubricating, coloring, coating, gluing, and so on. If the supplier of chemicals can assure that these functions are delivered by managing certain aspects of the chemical use and handling, then a move to service-based compensation is possible.

For most products, chemicals are used somewhere in the manufacturing process. Chemical costs are incurred at each stage of the life cycle of the chemical in the company (procurement, delivery, inventory, internal distribution, use, waste collection, and treatment or disposal). The chemical substances are also a major source of environmental impacts and threats. Chemicals are therefore increasingly subject to national and international legal requirements. The legal advice needed for full legal compliance, including liability issues, is nearly unaffordable for big enterprises, let alone for small- and medium-sized companies. However ChL is a business model that provides solutions to several issues, in addition to those mentioned above. This article will first explain the basic principles of ChL as a business model. Then the model is compared to other service-based chemical (procurement/management) systems pinpointing the differences. Practical suggestions and economic, technical, and juridical advices related to the implementation of ChL are described in the two following paragraphs. The article ends with lessons learned and a first sight on further developments.

12.2
Basic Principles of Chemical Leasing (ChL)

Product service systems (PSSs) are a promising answer to the sustainability challenges for industrial activities. In PSS, the focus shifts from delivering products toward delivering functionality (Cowi, 2008). Life cycle thinking has to be the basis of such systems to ensure overall improvement. ChL is one such PSS business model supporting sustainable material use and considering the whole life cycle. It is a business model that converts the activity of a supplier of chemical substances from selling volume to delivering the function of the chemical, that is, a service.

The supplier guarantees an end result in line with the specifications agreed to with the customer. The objectives of ChL are to come to an integrated approach and to transfer different responsibilities from the user to the supplier (Geldermann et al., 2009). Payments are calculated on the delivered function and translated into measurable units (e.g., surface), and are no longer calculated on the volume of chemicals used (Vander Velpen et al., 2010). Since the supplier of chemicals will be paid for the functionality of the chemical products, the supplier's economic profit will depend on the effectiveness and efficiency of the usage of the chemicals. Knowing that the model strives for a transfer of liability for the chemical handling and usage from the user to the supplier of chemicals, the volume used in the process also becomes a cost for the supplier. The traditional principle "the more you sell, the more you earn" is no longer valid. The result is that the interests of the suppliers and their customers (the user of the chemicals) are similar: both aim at using fewer chemicals and reducing waste. This is a major change compared to the traditional economic relationship between chemical suppliers and their customers/users. While current business models create supply side incentives for increased chemical sales and use, this is no longer the case in the service-oriented business model ChL (Figure 12.1).

At least two actors are needed in a business-to-business environment for a ChL project the supplier and the customer (user) of the chemicals. The supplier does not need to be the manufacturer of the chemicals, but can also be a distributor or a blending company. In any case, the supplier should have sufficient knowledge of the properties and functions of the chemical substances, their use, and handling requirements. On the other hand, the customer should know the production processes, his intermediates, and end products in depth. The customer should be able to describe in detail the quality and technical specifications of the intermediates and end products.

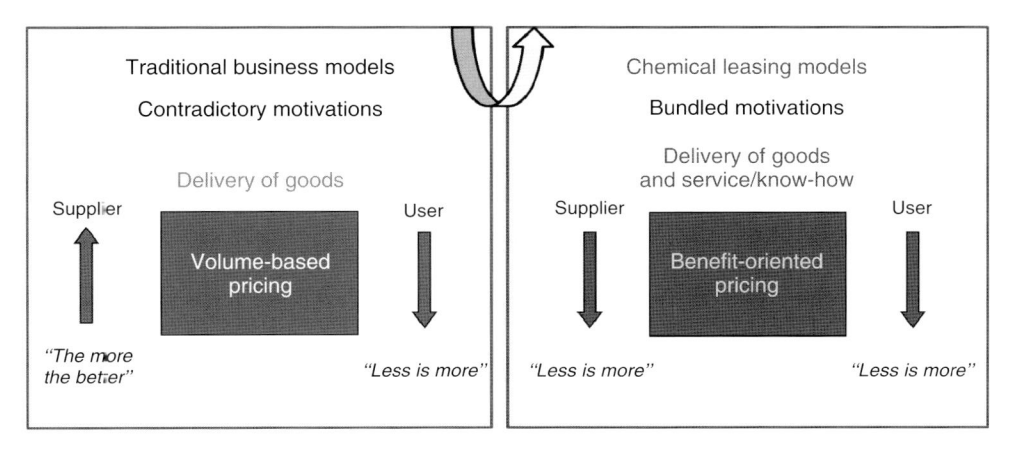

Figure 12.1 Traditional business models versus Chemical Leasing: another philosophy. Source: Jakl and Schwager (2008).

Consider the cleaning of metal parts in a company active in the automotive industry. The solvent used for cleaning purpose is a mixture of toluene, ethanol, and butane and contains traces of water and heavy residues. These substances are explosive, flammable, and harmful to health. Under a service-based compensation model, ChL, the supplier is responsible for delivering cleaned metal parts that meet the automotive industry's specifications. Since the automotive industry derives revenue from each car that leaves the facility, it is sensible to compensate the supplier on that same basis. Instead of profiting more by increased solvent use, the supplier stands to gain more by decreased solvent use. Under this service-based business model, the supplier has an incentive to work with the automotive industry to apply the most effective solvents to clean the metal parts and to be sure that as much of the solvent purchased cleans as much as possible of the metal's parts instead of disappearing in the waste drum. If, by making improvements in the chemical use and management processes, the supplier can lower solvent usage, all parties benefit: the supplier needs less raw materials and makes more money and the automotive industry has to manage less (noncore) processes and less materials (including waste). Hence system costs – including both direct procurement and indirect management costs – are reduced. Furthermore, under a gain-sharing arrangement, savings can be shared to further incentivize the supplier and the user. It is consequently desirable for the supplier to manage more of the process, in effect, to become a provider of services. The chemical service provider has thus a direct financial incentive to ensure that chemical use is minimized through both material management and process efficiency improvement. Adaptation of machinery could be necessary for efficiency improvement of the process. Thereby, it is important that sharing of knowledge between the supplier, the automotive industry, and the producer of machines is prerequisite. It is clear that with such a close interaction between the parties in a ChL project, confidential business information should also to be shared. Mutual trust between all involved is an essential condition for results.

Furthermore, the responsibilities of the supplier and/or of the customer (user) changes compared to traditional business models. For example, the supplier retaining the property of the chemical substances will have to comply with some legal requirements that were earlier the responsibility of the user of the chemicals. Depending on the arrangements made between the different parties, the management and the responsibility of the supplier could even cover the whole life cycle from production of the chemical over use up to waste. Solid agreements to support the cooperation are necessary.

Depending on the production process and the needs of the user, but also on the advice of the supplier, other stakeholders can be involved in the project. These other parties could be manufacturers or distributors of machinery and equipment, waste-recycling companies, financial institutions, engineering firms, and legal advisors. The model is essentially based on multi-stakeholder engagement with a multidisciplinary approach. The specific knowledge of every party involved determines the success of the new business model. (Figure 12.2).

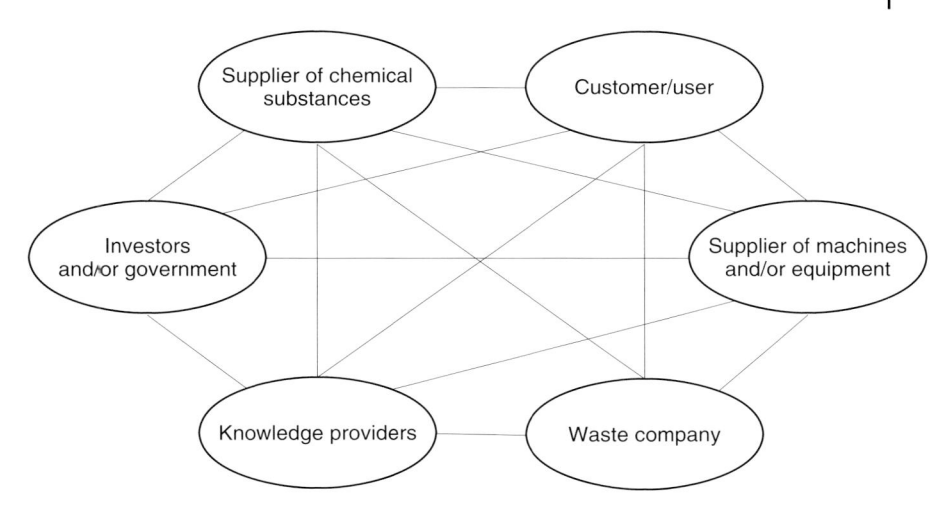

Figure 12.2 Overview of potential stakeholders involved within the business model ChL. Source: Vander Velpen et al. (2010).

Payments in ChL also differ. The quality and quantity of the end product determine the revenue for the supplier and the cost for the consumer (user). After all, the customer (user) purchases the functionality of the chemical. Pricing is typically calculated following a formula including quality measures, number of results (surface, piece, etc.), and functionality of the end product. Since this new business model (ChL) thrives on cost savings for the coynsumer (user) and greater margins for the supplier, proper profit sharing is the gist of the matter. Parties should agree upon a fair distribution of the costs and benefits of the implementation of ChL. Then right incentives are in place.

The incentives for implementing ChL can best be illustrated with the following example relating to the textile industry. Decoration and technical textiles are products in which many chemical products are used, such as fire retardant, antibacterial, and water-resistant products. Within this sector, many new innovative products are being developed. In the textile sector, ChL is mainly recognized as a means to resolve new challenges within their current processes, including

- reduction of process water during the pre- and posttreatment of textiles;
- the reuse of process water;
- avoiding the release of specific micropollutants (e.g., alkylphenol ethoxylates (APEO)) and nitrogen compounds.

There are opportunities to improve the cooperation between the textile industry and suppliers of chemical products by using ChL. By applying this business model, the innovation pressure can be transferred to the supplier of chemical products. This is possible if the right incentives are in place: the payment terms of the supplier should be linked to increasing the effectiveness of the process and not to the quantity of sold chemical products. The supplier must use its knowledge to increase its profits by improving the process, managing the reuse of process water, and reduce specific micropollutants in the process.

ChL is a very flexible model enabling all involved to conclude agreements in line with their personal wishes, objectives, and means. But how is it different from already existing and proved business models like leasing, chemical management services (CMS), and outsourcing?

12.3
Differences between Chemical Leasing and Other Alternative Business Models for Chemicals

On a regular basis, managers claim that they already use ChL. When looking into the details, it most often is another business model, such as classical leasing, CMS, or outsourcing. All these models are service-based chemical procurement systems that show similarities with the business model ChL. However, basic differences can be isolated. The main differences are explained in the following.

12.3.1
Classical Leasing

Financial and operational leasing have limited resemblance with ChL. In ChL suppliers of chemical substances deliver the function of a chemical and not the chemical substance. As distinct from classical leasing, ChL is in essence a service and not a traditional product sale: the supplier is responsible for the substance during its whole life cycle. Another significant difference is the fact that in a classical leasing relationship the lessee specifies the material. In ChL, the customer of the supplier defines the end result and the supplier has the authority to choose the most suited chemical. Of course, when choosing the chemical substance, the supplier will have to respect the contractual arrangements on (for example) hazardousness of the chemical and suitability for the manufacturing process.

12.3.2
Chemical Management Services

CMS is a business model that is the closest to ChL. Parties share responsibility and knowledge. However, the same or very similar chemicals must still be procured, received, stored, quality-assured, distributed to point-of-use, and so on (Stoughton and Votta, 2003). The main difference between CMS and ChL is that the supplier within a CMS system usually is not paid on the basis of the function of the substances. Though the supplier may make occasional suggestions to improve process efficiency and chemical-handling processes in order to maintain customer loyalty, he essentially does not have the opportunity to choose the most suited chemical and/or industrial process achieving the required end result. With ChL, the end result achieved by the function is always the decisive parameter for payments and consequently the supplier of the function of the chemicals has more freedom. Thereby ChL implies in a lot of cases that liability is transferred from the consumer (user) to the supplier.

CMS hardly ever has quality control policies on the end product, or policies on the environmental impacts of the production processes. The opposite is true for ChL. ChL aims at closing the life cycle of chemicals. In practice, this is often unattainable, but the whole production process will be involved in trying to close the loop as much as possible. Each phase in the life cycle of the used chemicals is looked at. The only fixed parameter is the agreed quality of the end product of the customer.

12.3.3
Outsourcing

Outsourcing and ChL do have some similarities. Both business models aim at financial savings. But there are also differences. In ChL, both parties should benefit from the financial savings. Applying ChL implies a trust and transparency between the parties involved (supplier, consumer (user), etc.) in order to accept a fair sharing of the advantages between all parties. Another difference is that knowledge about chemicals and production processes are normally not shared in outsourcing. In the ChL model, both parties work closely together toward a more effective and efficient purchases, thereby sharing information and knowledge becomes an integral part of the cooperation. In the outsourcing model, payments can also be calculated on the quality and quantity of the end product. In ChL, this is always the case. Anyhow, good contractual arrangements are necessary in both models; however, the supplier of the chemical substances in a ChL model is motivated to reduce the use of chemicals by optimization of the production process. In ChL, the supplier and the customer are linked with respect to several aspects such as knowledge and responsibility for production processes. Contrary to outsourcing, in ChL, the pressure for innovation is with the supplier, who is profoundly involved in the whole production processes. The attention for environment and human health is also a characteristic of ChL. One of the driving factors is the protection thereof by good chemical management, usage, and development. This aspect is seldom found with outsourcing (Table 12.1).

12.4
Practical Implications of Chemical Leasing

On the basis of our study of the market and discussions with key actors in the chemical sector, we have found that ChL projects are mostly suited for medium and large-sized companies. The success of implementing ChL as a business model also depends on the availability of earlier experiences of multinationals or groups of enterprises in the same sector. The role of these front-runners is to act as a catalyst. This way, uncertainties linked to a new business model are more easily accepted.

Table 12.1 Significant differences between Chemical Leasing and other business models.

	Chemical leasing	Classical leasing	Chemical management services	Outsourcing
Payment for the function of the chemical	Yes	No	Yes/no	Yes/no
Explicit attention for environment and human health issues	Yes	No	No	No
Sharing of responsibility and knowledge between supplier and customer in relation to the use of the chemical	Yes	No	Yes	Yes/no
Optimization of the existing production process to improve efficiency	Yes	No	Yes/no	No
Changing the industrial process to achieve required end result	Yes	No	No	No
Shared financial savings	Yes	No	Yes	Yes

The suppliers, however, think that business models focusing on the product–service connection will provide no more than 20% of their turnover. They see ChL rather as a niche market, but still important enough to implement because of the additional value it creates and for the long-term relation with the customer. This conclusion is in line with the outcome of the nonlinear pricing approach within the ChL concept (Ehtamo et al., 2010).

On the side of the consumer (user) of the chemicals, it is clear that they will not apply the ChL concept to their core business processes. Indeed, core businesses are the reason for their existence and consequently companies want to retain full control. ChL is mainly suited for activities just outside the core business of the consumer (user) of chemical substances. Examples of such processes are cleaning, coating, printing, wastewater treatment, heating/cooling, lubrication, and so on. Activities further away from the core business are also suitable, but these are often outsourced or contracted. Experience shows that ChL can be envisaged as a suitable business model where proven technology is involved. The reason for this is that it is difficult to conclude solid contracts around new technologies, while many factors are still uncertain. Application of proven technologies from other processes is, however, also possible.

Every business change costs money. A change in business model could cost even more before it starts bringing in financial gains. Experience shows that the supplier of chemical substances should be prepared to invest before financial compensation can be concretized. This preinvestment by the supplier is required in order to find the right partners, to improve the supplier–customer relationship, and to create confidence between the partners. Such a preinvestment is acceptable if sufficient environmental benefit (e.g., reduction in energy use, waste generation, material use, water use) can be achieved, resulting in higher profits for the supplier. As a rule of thumb, it is required to achieve a total cost reduction of more than 20% due to an improved environmental performance. The preinvestment should be done by the supplier because the consumer (user) of chemicals – in particular for indirect use – generally sees little intrinsic value in the chemicals per se. Instead, the real value of a chemical resides in the function it performs. If the supplier can assure that these functions are delivered by managing certain aspects of chemical use and handling, then a move to ChL is possible.

It is not easy to get a good view on the potential costs and saving beforehand. Manufacturers frequently do not have an understanding of their chemical costs. Of course, no company willfully employs an inefficient chemical management system, but lack of attention is frequent. The reason for this may be that the use of chemicals is outside the core business and because the costs may be perceived to be relatively small. A quick scan of the process and the related (existing and future) financial costs could offer an assessment of the feasibility of the ChL project and provide parties with an overview of the investment needs/costs.

In order to optimize the financial compensation for all parties involved, it is advisable to focus on a process that is applied on a large scale by different manufacturers (consumer of chemicals). That way, the initial investment costs for the supplier and/or the customer could be reduced. In case the supplier prefinances the whole project, he should be allowed to use the same improved processes with other customers (users) to make his investment profitable. Again, this is a topic that has to be regulated in a solid well-balanced contract, with special attention to intellectual property, confidential business information, and competition requirements. These contractual arrangements should be adapted to the individual situation of the project and the parties.

Overall, the conclusion is that it is very difficult if not impossible to strive for a generic approach of ChL. The individual differences between companies or group of companies are too big. However, the flexibility of the ChL concept is an advantage. As mentioned before, ChL also supports sustainability and the protection of environment and human health. It is worthwhile to consider financial support or subsidies by official authorities for ChL or for part of such projects, like quick scans. This also enables policy makers to control the potential ecological and sustainable improvements of a proposed ChL project, which could be an advantage for securing sustainable choices in the execution of a ChL project.

12.4.1
Strengths and Opportunities for the Supplier

ChL offers several advantages for a supplier of chemical substances. On the basis of his involvement in the manufacturing process of his customer (user), the supplier can use and valorize his knowledge and expertise on chemicals. The valorization of knowledge leading toward reduced chemical consumption in the industrial processes of the customer yields greater margins for the supplier and cost savings for the customer (user) of chemicals. ChL provides the supplier with in-depth knowledge on the needs and strategic plans of the customer. Continuous improvement of the efficiency of the manufacturing process of the customer (user) becomes possible within ChL and is an objective of the supplier and the customer (user). The persistent drive toward improvement of the performance can be used as a concrete clue for innovation, research, and development efforts. Furthermore, the public image of the chemical industry will also benefit from the innovative, sustainable, and environmentally friendly focus and drive created by ChL.

Of course, there are some threats: important fluctuations in feedstock prices could undermine the pricing that was agreed upon. Administrative follow-up of the contractual stipulations and reporting efforts will increase. Last but not least, because of the close cooperation between supplier and customer (user), knowledge leakages are more probable, as well as problems with confidential business information. Many of these potential drawbacks can be prevented by solid and thorough agreements.

12.4.2
Strengths and Opportunities for the Customer

Concerning the customer (user), his administrative burden will decrease in the sense that the supplier will manage some legal obligations such as those imposed by Registration, Evaluation, and Authorization of Chemicals (REACH), waste legislation, or exploitation and environmental permits. Since legal obligations can be burdensome, this could clearly be a significant advantage for the customer, while not a big disadvantage for the supplier. Indeed, the latter can use his knowledge, expertise, and experience in this matter. On the other hand, owing to the displayed specialized information of the chemicals by the supplier, the knowledge of effective use of chemicals with the customer will increase. The supplier shall also support and advice the user on the safe and legal storage of chemicals, on the development and implementation of an adequate logistic management system, and on the collection of useful and correct information on the handling and use of chemical substances. Support from the supplier, preferably during the whole life cycle, will result in a safer and more adequate handling of chemical substances at the level of the customer (user). Parties could also agree that the supplier actively informs his customer on changes in regulations and consequently performs preparatory work. This seems logical since the supplier has the knowledge of the chemicals and the industrial processes at the level of the customer (user). Sharing information means also that the employees of the customer (user) will be better trained in dealing with chemicals. Consequently, safety standards will improve.

However, the user could experience this intensive collaboration as a disadvantage. It becomes increasingly difficult for him to change suppliers: the complexity of the relation in a ChL model makes it enduring, so the transition to another supplier becomes difficult. These elements should be tackled in the contract to avoid future problems.

12.5
Economic, Technical, and Juridical Aspects of Chemical Leasing

ChL is a business model that is simple in principle, but can become quite complex in practice. A lot depends on the industrial activity, the number of parties involved, and on the applicable law and regulations (frequently depending on the industry sector).

12.5.1
An Example

Let us analyze the packaging of frozen food with an imprint of the freshness date to make things clearer. Printing on packaging material for frozen food follows a specific method. And, since the date varies depending on the production moment and the material, this printing is done at the premises of the freezing. At this point, the process uses a lot of packaging material and creates waste polluted with ink. When ChL is implemented, the supplier of the ink will be paid per printed packaging instead of per volume ink sold. The supplier is thus interested reducing the consumption of ink and packaging material. Therefore, it is essential for the supplier of the ink to recognize and to implement the required improvement of the printing process at the level of the frozen food manufacturer in order to increase its efficiency. The frozen food manufacturer is interested since he will benefit from the optimization by sharing in the reduced costs and the reduction of waste.

This example illustrates a fundamental revision of the chemical supplier–customer relationship, one in which the incentives facing both parties are aligned in favor of reducing consumption of chemicals, water, and energy and minimizing waste streams. To summarize, the success of ChL for this example is related to the significant process ameliorations, whereby the volume of ink used and the waste generated were reduced. Not all industrial processes are as well suited for ChL as the one in the example. An overview of suitable processes, their presence in industry and the categories of substances involved are listed in Table 12.2.

12.5.2
Barriers to the Model

The ChL model is timely in many ways, responding as it does to a set of current trends and priorities in business management, including focus on core competencies, continuous improvement, reduction of environmental footprints, and suppliers as strategic resources (Reiskin *et al.*, 2000). However, ChL is a challenging business model and is generally not a management priority at the level of the consumer (user) of chemicals, in part because chemical purchases are generally a relatively small fraction of operating costs. This attitude is compounded by limited management awareness of the ChL model, as well as poor internal data systems and chemical cost awareness. All of these raise the barriers to serious consideration of ChL adoption.

Table 12.2 Selected examples of existing demonstration projects and experiences with Chemical Leasing.

Industrial process	Industrial sectors	Substances	Unit of payment
Purification, drying, and so on, of chemicals	Cosmetic industry, pharmaceutical industry, and petrochemical industry	Polar solvents, acids, and so on	€ kg^{-1} treated product or monomer
Cleaning of working areas and installations	Food industry, cosmetic industry, health care and cure, and hospitality business	Detergents	€ m^{-2} cleaned surface
Wastewater treatment	Broad spectrum of industries	Organic and inorganic compounds	€ m^{-3} treated water €/unit removed compound
Surface treatment: greasing/degreasing, powder coating, wet painting, galvanization, thermal zinc galvanization, electroplating	Automotive industry and metal industry	Polar solvents, paints, abrasives, acids, and lubricants	€/number of pieces € m^{-2}, € m^{-1}, € kg^{-1}
Use of agricultural chemicals	Agro sector	Pesticides and fertilizers	€ ha^{-1} disease-free farmland
Pre- and post treatment of textiles	Textile industry	Paints, fire retardants, antibacterial, and water-resistant products	€ m^{-2}, € m^{-1}
Printing	Food industry and newspaper industry	Ink and solvents	€ m^{-2}, €/standard page
Binding, glue	Construction and industrial labeling	Resins and solvents	€ m^{-2} banded surface, €/labeled piece

If ChL is considered as a new business model at the level of the consumer (user), internal proponents face a number of challenges. First, chemical management activities constitute a complex system. Transferring the management of this system to a supplier can be a daunting task because of its many linkages with other management and manufacturing systems such as procurement, material management, production engineering, and waste management. Secondly, as with any change process, implementing a ChL program is subject to individual and organizational resistance, system inertia, and risk aversion, especially when potential gains do not directly accrue to the parties essential to implementation. Thirdly, ChL creates increased interdependency between supplier and customer that requires high levels of supplier capability and customer confidence and trust in these abilities. In contrast to the traditional seller–buyer relationship, ChL requires long-term continuous and multifaceted interaction. Fourthly, ChL can evoke resistance from personnel, especially union personnel who may see ChL as a cause for staffing reductions and dislocation of part of the production. Fifthly, ChL implies a transfer of liability of certain elements of the industrial process from consumer (user) to supplier. Specific contractual arrangements between the parties should be defined in detail before implementing a ChL program.

With support from top management and a sound communication effort, a thoughtful implementation program can respond to all these challenges. While ChL does require considerable leadership from top management and receptivity across numerous staff functions, a roll-out of a ChL program may occur gradually by phasing in different stages of the life cycle and incorporating well-outlined industrial processes lying outside the core business.

Not all barriers to the ChL model are on the customer (user) side. For traditional chemical suppliers who are primarily chemical manufacturers, the prospect of reduced chemical sales under this new business model poses an obvious conflict. For these reasons, the suppliers who have interest in making a service transition are either suppliers of chemicals without any chemical manufacturing division or they have built effective firewalls between their manufacturing and service divisions. Finally, it should be noted that management of ChL contracts is far more complicated than that of supply contracts, as it typically requires coordination across multiple business units of the client firms and a strong information technology component.

12.5.3
Analysis of the Legal Requirements Impacting Chemical Leasing Projects

Since the ChL business model implies long-term commitments, it creates uncertainty with the customer (user). In relation to the continuation of supply, especially, it involves at least (but often more) two contracting parties and it impacts responsibilities of all involved. Seeking advice from knowledgeable legal and financial sources is a good investment.

12.5.3.1 The Importance of Contracts
Until now, there have been no standard contracts available for ChL. On the other hand, should we have such a generic model contract? Standards limit flexibility and ChL is a business model that is adaptable to the specific needs of the parties involved. The flexibility is certainly an advantage since a lot of variety exists between ChL projects. The definition of the cooperation consequently also varies considerably. Objectives, opportunities, bottlenecks, and possibilities are all to be taken into account. However, some important high-level features are omnipresent. These are discussed in the following paragraphs. The objective is to give one a first idea of the contractual arrangement that one would want to conclude. Individual appreciation and adaptation remains necessary and will provide one with sufficient legal certainty.

The practical application of ChL is quite complex. The number of parties involved, new cooperation methods, and shifts of responsibilities and liabilities could be an obstacle. In our contacts with industry, the legal aspects appeared to be a major concern. Good contracts between all parties involved are a condition *sine qua non* for success.

ChL thus demands a high level of legal expertise in civil and public law on the international as well as the national levels. For example, chemicals and waste are regulated or at least highly influenced by the international (European) rules, environmental permits are part of public law, as are subsidies. Although, in practice, these distinctions between the different legal fields are not important, legal advisors and lawyers are mostly trained and/or specialized in one field. Therefore, a legal advisor should be chosen carefully, taking into account the available knowledge in-house and the identified gaps.

Some elements that specifically should be included in a contract are the subject of the contract, the duration, the representation including the authority level of each party, as also product specifications, intellectual property, confidential business information liability rules, procedures for administration, reporting, and so on. In view of its importance, the method of calculation for payment of the supplier should also be included in the contract. Cooperation between enterprises does not work if the performance is not measured and followed up. Specifications of the end product, quality requirements, timing, contingency plans, and so on, should be agreed upon. Distinctive of the ChL concept are the ecological and sustainable elements incorporated in the model. Measurable and controllable criteria in relation to these aspects should also be described. The impact on the environment, the reduction of chemicals, the properties of the chemical substances used, and energy efficiency are only a few measurable suggestions. A specific item to be regulated is the cooperation between and the management of the employees of each contracting party. Much will depend on the extent, the organizational structure, the location of the work performance, the skills, the necessary competencies, and the required training. In such projects, a relation of trust between the supplier and the customer (user) remains paramount, even with a good contractual arrangement.

12.5.3.2 Competition Law and Chemical Leasing

As with all partnerships between economic actors, antitrust rules should be carefully followed when implementing the ChL model. Different legal regulations contain antitrust rules. The Treaty on the Functioning of the European Union describes the basic (and brief) provisions. Subsequently, a number of regulations have been adopted. Some of these rules regulate the general implementation of the Treaty provisions, thereby defining, among others, the investigative powers of the Commission. Other regulations deal either with particular types of conduct or with specific sectors. Regulation 1/2003 (EC, 2003) especially is worth mentioning. It deals with the relationship between Articles 101 and 102 of the TFEU and national competition laws. When Members States apply national competition law they shall also take into account the European rules. However, Member States are allowed to adopt and apply on their territory stricter national laws than the

European rules. The core principle is however similar: it is prohibited to hinder fair trade or the functioning of the common market by agreements, decisions, and/or concerted practices. On the other hand, agreements or decisions improving the production or distribution of goods or promoting economic or technical progress are tolerated on condition consumers receive a fair share of the resulting benefit. Since ChL can only be successful on the basis of a close cooperation between the different parties involved, a project-specific analysis of the possible antitrust risks is necessary before the implementation of the model. Dangerous information includes, for example, pricing, discounts, promotion policies, client preferences, production capacities, production volumes, sales volumes, import volumes, and market shares. It is important that all involved parties remain free to set their pricing and that information is only shared on a need-to-know basis. Another important criterion is the fact that the supplier of the chemicals should remain free to use his expert knowledge for other customers. With some creativity, solutions to these challenges have proved possible.

12.5.3.3 REACH and Chemical Leasing

After many years of discussions and counter fights by the industry plus negative comments from international trade partners, the European Chemical regulation "REACH" finally came into force in June 2007. The goals of REACH are ambitious: provide a high level of protection of human health and environment, enhance the competitiveness of the EU chemicals industry, and foster innovation. Responsibility for the management of the risks of substances should lie with the natural or legal persons that manufacture, import, place on the market, or use chemicals.

Users of chemicals having other activities as their core business usually do not have sufficient knowledge in-house to comply with the complex REACH rules. REACH is a challenge for them and sometimes even an insurmountable burden. ChL is a possible solution. REACH and ChL adhere to the same philosophy: transparency in the supply chain and communication. Cooperation between different actors in the supply chain and protection of the environment and human health are important.

The Added Value of Chemical Leasing for Compliance with the REACH Requirements
An important innovation brought by REACH, is that suppliers and users should prove that their chemical products are safe. Therefore, the precautionary principle should be followed (Article 1 (3) REACH). In the past, chemicals were regarded as safe until someone or something proved the opposite. Subsequently, the government had to prove the negative impact and the causal link before the chemical could be prohibited, or any other measures could be imposed. Now the burden of proof for safety of a chemical lies with the producer or user of a chemical.

Traditionally, the chemical industry knows the most about chemistry and chemical use. Working within the ChL model, this industry, that is, the supplier of chemicals, is motivated to attain the best result with the least of costs. His knowledge on chemicals ensures more effective and efficient use of substances and a safer production process, in line with the objective of REACH. In addition, if the supplier remains the owner of the chemical substances, several obligations from chemical legislation are now his responsibility. If the user still purchases the chemicals and consequently owns the substances, he could still rely on the knowledge and information of the supplier. In both cases, the compliance with chemical legislation becomes easier for the customer whose core business is not chemistry.

REACH-Data Collection   Users of chemicals have to collect information on the substances and preparations they use. This information is often difficult to find without sufficient knowledge of chemistry, despite the data included on labels and in permits. Knowledge on chemicals substances and on the management thereof is however the core business of the supplier of chemicals. In ChL the supplier has the relevant information that the user needs. And by managing the use of the chemicals and selling the function, the supplier becomes the user, as understood within the REACH legal framework, and thus he has to fulfill the related obligations instead of the customer.

Identified Use   A user is allowed to use a chemical substance on condition the considered use is registered. In case the registration did not happen, the user will have to notify his use to his supplier(s) (REACH, EC/1907/2006). After the registration of the use, conditions mentioned in the safety data sheet and in the exposure scenario (if obliged) have to be respected when working with the chemical. The responsibility of these obligations is attributed to the users. The same applies "*mutatis mutandis*" when no safety data sheet is required (Article 32 of REACH).

With ChL, the obligations on use and risk management become the responsibility of the supplier. After all, it is the supplier who uses the chemical to attain an agreed-upon result.

Authorizations and Restriction   REACH stipulates that some very noxious substances can only be used when they are authorized for an intended use. The objective is to replace these dangerous chemicals by less noxious substitutes. ChL enable providers of these noxious chemicals to look in a constructive way for solutions, while the customer (user) remains ensured of the desired process outcome. In case the supplier in a ChL project uses a chemical subject to restrictions, he will also be responsible for respecting the conditions of use as imposed on the basis of REACH.

12.5.3.4   Legal Aspects, a Bottleneck?

Experience shows that a breach of trust between contractual parties is in most cases based on price calculations (calculation measurements, calculation methods) and/or on the (lack of) possibilities to control the end product (Umweltbundesamt, 2010). Omitting arrangements on these aspects could become the stumbling stone of a ChL project. However, this should not be the case. Such situations are avoidable by contractual arrangements. The overview below shows the main points of attention, without being all inclusive. ChL has the advantage to be flexible and adaptable to the needs of the parties involved. It also gives parties the opportunity to shift responsibilities and activities to the most knowledgeable party, for example, on the compliance with the numerous complex environmental rules and obligations (Table 12.3).

12.6
Conclusions and Recommendations

A generic approach of industry for implementing ChL programs is neither desirable nor possible: industrial processes at the customer (user) of chemicals are too different and too complex. The growth of ChL in the chemical industry is driven by demonstration projects at plant level. However, realizing the environmental potential of the model and the related financial incentives depends critically on the scope of ChL programs. Successful implementation of ChL programs is strongly dependent on gain-sharing mechanisms between both partners, which strongly motivate environmental footprint reduction. ChL is a model that improves significantly several elements of the chemical life cycle and creates an industry whose focus is effective and optimal chemical, water, and energy use, providing alternatives for noxious and toxic chemicals and avoiding waste generation.

While the environmental benefit of ChL seems to be beyond dispute in case smart gain-sharing mechanisms are in place, financial incentives need further thorough consideration. As the critical reflection on quality assurance, proprietorship, allocation of risk, and profit sharing shows, several specific points of concern need to be answered. These answers can be given from field experiences.

Table 12.3 SWOT analysis of the main legal aspects of ChL.

Strengths	Weaknesses
Flexibility when concluding contracts	Long-term contracts
Sharing of information and duty to inform of the most knowledgeable party	Possible issues with intellectual property, also in case of innovations
Safer use of chemicals	Leaking of confidential business information caused by the necessary sharing of information
	Increased liability for the supplier of the chemical substances
Opportunities	**Threats**
Improved safety management, environmental management, and permit management, especially on company level	Complexity of the contracts between the parties involved
Professionalization of company management based on the need for (clear) quality standards and product specifications	Need for specific contracts per project, impossibility of standardization
	Administrative workload: follow-up contracts, audits, production controls, results and so on

It makes sense to start with government support to demonstration projects, to spread the successes of ChL (environmental and financial benefits) to the suppliers and consumers (users) of chemicals, and to link this support to the governmental ambition related to sustainable materials management.

References

Comethe (2008) COMETHE: Pour une Écologie Industrielle. Technicités du (juin 8 2008) no. 151, p. 14.

Cowi (2008) Promoting Innovative Business Models with Environmental Benefits, European Commission – DG Environment, Denmark.

EC (2003) Council Regulation (EC) No 1/2003 of 16 December 2002 on the implementation of the rules on competition laid down in articles 81 and 82 of the treaty. *Off. J.*, **L 1**, 1–25.

Ehtamo, H., Berg K., and Kitti M. (2010) An adjustment scheme for nonlinear pricing problem with two buyers; *Eur. J. Oper. Res.*; **201** (1) 259–266.

European Commission (2009) The Path Towards Sustainable Industrial Production, Community Research, July 2009, *ftp://ftp.cordis.europa.eu/pub/growth/docs/leaflet_sustainable_development.pdf*.

Geldermann, J., Daub, A., and Hesse, M. (2009) Chemical Leasing as a Model for Sustainable Development, Georg-August-Universität Göttingen, Göttingen.

Jakl, T. and Schwager, P. (2008) *Chemical Leasing Goes Global: Selling Services Instead of Barrels - a Win-Win Business Model for Environment and Industry*, Springer, Wien.

REACH (EC/1907/2006) *ec.europa.eu/enterprise/.../reach/index_en.htmEn cache - Pages similaires*.

Reiskin, E.D., White, A.L., Johnson, K., and Votta, T.J. (2000) Servicizing the chemical supply chain. *J. Ind. Ecol.*, **3** (2–3), 19–31.

Stoughton, V. and Votta, T. (2003) Implementing service-based chemical procurement: lessons and results. *J. Clean. Prod.*, **11**, 839–849.

Umweltbundesamt (2010) Chemikalienleasingals Modell zurnachhaltigenEntwicklungmitPrüfprozedurenunde Qualitätskriterienanhand von Pilotproject in Deutschland, Dessau-Rosslau, March 1, 2010, pp. 26–30, *http://www.reach-info.de/dokumente/Chemikalienleasing_Endbericht_UBA_FKZ_3707_67_407_Langfassung.pdf* (accessed 4 November 2010).

Vander Velpen, B., Hoppenbrouwers, M., and Geysen, D. (2010) Chemical Leasing: Mogelijkheden en Beperkingen in Vlaanderen. OVAM.

13
Sustainable Chemical Warehousing

Kenneth Sörensen, Gerrit K. Janssens, Mohamed Lasgaa, and Frank Witlox

13.1
Introduction

While chemicals contribute a great deal to our society, they also pose serious threats to human health and to the environment when improperly handled. In 2003, the European Commission proposed a regulatory framework for the Registration, Evaluation, and Authorization of Chemicals (REACH). The goal of REACH is to ensure sustainable development by contributing to the protection of human health and the environment through more accurate identification of chemical substances, while at the same time stimulating innovation and competitiveness in the chemical industry.

Although companies sometimes seem to be reducing their inventories at all costs, storing goods (whether they are finished goods, work-in-progress, or raw materials) remains an important step in the supply chain. Inventories for chemical products appear both as pipeline inventories and as warehouse inventories. Pipeline inventories include products on sea-borne logistics containers such as gas tank shipping, chemical tank shipping, and mineral oil shipping, while those by road include liquid bulk chemicals or even packed goods with chemicals. Distribution logistics offer warehouses to store the chemical products while they are not in transit to another destination or before processing. As many chemical products are hazardous to the environment or to the people, the effects of a poorly managed chemical warehouse can extend far beyond the boundaries of the warehouse itself, as has been demonstrated by several disasters.

Storing 10 000 m^3 of ammonia in a single location clearly presents an enormous environmental and safety hazard if not managed correctly, but smaller storage installations may pose a similar risk to human health and the environment. This chapter is intended for anyone with responsibility for storing chemical substances, regardless of the warehouse size. Proper and sustainable management should include measures to eliminate or reduce risk due to storage. This might be realized by good design of new facilities, but hazard identification, a quantification of risk, and mitigation strategies are also necessary. Control and documentation will specify emergency arrangements, training needs, and recommended audit processes.

Management Principles of Sustainable Industrial Chemistry: Theories, Concepts and Industrial Examples for Achieving Sustainable Chemical Products and Processes from a Non-Technological Viewpoint,
First Edition. Edited by Genserik L.L. Reniers, Kenneth Sörensen, and Karl Vrancken.
© 2013 Wiley-VCH Verlag GmbH & Co. KGaA. Published 2013 by Wiley-VCH Verlag GmbH & Co. KGaA.

13.2
Risk Management in the Chemical Warehouse

13.2.1
Hazard Identification

The first step in the risk management process involves identifying all potential hazards, as well as their potential impact on either the environment, the employees working in the warehouse, the emergency services in case of an incident, the general public off site, or any combination of the mentioned categories (Figure 13.1).

To be able to assess the hazards involved in storing chemicals, the risk manager needs to carefully study each of the chemicals stored and determine whether they are hazardous. A hazardous chemical can be categorized as a health hazard or as a physical hazard (OSHA – Occupational Safety & Health Administration).

For a chemical to be considered a health hazard there has to be statistically significant evidence, based on at least one study conducted in accordance with established scientific principles, that acute or chronic health effects may occur in employees exposed to the chemical. Chemicals covered by this definition include carcinogens; toxic or highly toxic agents; reproductive toxins; irritants; corrosives; sensitizers; hepatotoxins; nephrotoxins; neurotoxins; agents that act on the hematopoietic system; and agents that damage the lungs, skin, eyes, or mucous membranes.

A chemical is considered a physical hazard when there is scientifically valid evidence that it is a combustible liquid, a compressed gas, an explosive, flammable, an organic peroxide, an oxidizer, pyrophoric, unstable (reactive), or water-reactive.

The OSHA definition of hazardous chemicals includes not only generic chemicals but also paints, cleaning compounds, inks, dyes, and many other common substances. Chemical manufacturers and importers are required to determine if the chemicals they produce or repackage meet the definition of a hazardous chemical. A chemical mixture may be considered as a whole or by its ingredients to determine its hazards. It may be considered as a whole if it has been tested as a whole and a material safety data sheet (MSDS) has been issued accordingly. Otherwise the mixture must be evaluated by its components. If the mixture contains 1.0% or more of a hazardous chemical or 0.1% of an ingredient listed as a carcinogen or

Figure 13.1 A storage container for chemical products.

a suspected carcinogen, the whole mixture is assumed to have the same health and/or carcinogenic hazards as its components.

There are several sources of information available to identify potential hazards. The chemical container label includes basic identification, risks, and safety information. More detailed and important information regarding a specific chemical product is provided in an MSDS, which must contain at least:

- the product name and the names of hazardous components;
- the chemical and physical properties of the chemical;
- health hazard information;
- precautions for safe use of storage and handling the hazardous chemical;
- the manufacturer's or importer's contact information.

This very essential document should provide the employees with all necessary information to safely manage potential risks. Access to an MSDS can be provided in several ways including electronic databases or a hard copy. It is important that workers are familiar with the correct procedures on how to apply an MSDS.

Identifying all potential hazards is difficult since it involves thinking about consequences of often very unlikely events. The risk manager also needs to make a constant trade-off between the likelihood of a potential hazard and its probability of occurrence, to ensure that only relevant risks are listed. Only risks that are judged to be sufficiently likely to occur and have sufficient potential damage must be considered, lest the risk manager should lose focus and put too much effort into managing inconsiderable hazards.

For example, the risk manager may decide in this phase to not list spilling small quantities of a nontoxic chemical by careless handling of employees. Such spills may be very likely to occur but have very little consequence in terms of environmental or health damage. On the other hand, he may dismiss a nuclear strike on the chemical plant as a risk to be managed. Although such a terrorist attack could have enormous consequences, the risk may be deemed inconsequential.

Hazards from chemical products may affect different categories:

- damage to the environment
- material damage to property
- damage to the health of people in or around the warehouse (people working in the warehouse, emergency rescue workers, or the general public).

This chapter focuses on risks that are directly related to stored chemical products: environmental and chemical-related health hazards. Many other (health) hazards, such as employees hurting themselves by tripping, falling, or making wrong movements are not specific to the chemical industry, but should nevertheless be considered by the risk manager as an integral part of the risk management process.

Labeling of chemicals is a critical issue because it is the most visible hazard communication tool. The label is often the first source of information alerting users to the inherent hazards of a chemical and any instructions for its safe storage, handling, and use. All containers that contain chemicals must be labeled, irrespective of the size of the container. It is worth mentioning that most dangerous substances are (and should be) clearly labeled, for example, by the ADR (European Agreement concerning the International Carriage of Dangerous Goods by Road) regulatory regime. Figure 13.2 shows some symbols that are commonly used to label the chemicals of all different classes. Figure 13.3 shows the ADR dangerous goods classes together with their labels and descriptions.

When identifying potential hazards, some factors that must certainly be looked at are the following:

- The properties of the substances being stored: explosiveness, toxicity, flammability, poisonousness, and so on. Information regarding these properties should have been made available by the supplier of the substances. The supplier can make use of a classification system such as the CHIP (Chemicals (Hazard Information and Packaging for Supply)), using black or orange-yellow danger symbols, to indicate the danger statements.
- The actions that will be undertaken with each of the dangerous substances, especially those that may increase the hazard involved, such as transportation. Also, the circumstances of the regular (e.g., storage) and irregular use of the substance: in which environment the substance will be handled normally and in abnormal but predictable situations. For example, are there any ignition sources in the area in which an explosive substance is handled?
- The qualifications of the employees handling the dangerous substances. Employers are required to assess the risks to workers, which may arise owing to the presence of dangerous chemicals.
- The state of the equipment that is used to handle the substances. This also relates to equipment in warehouses such as fire alarm systems and automatic foam sprinkling.
- The effects of measures already taken to prevent dangerous situations from occurring.

Dangerous goods have their own well-defined standards of storage, which are set up by legislation to minimize the potential hazards pertaining to each specific substance. Clearly, these standards of storage should be carefully followed, and the list of potential hazards can serve as an input for the risk management process. However, additional hazards may stem from the storage of dangerous substances, and it is important to assess all potential hazards, even those that only arise as the result of a combination of several substances (Figure 13.4). Such an analysis allows the risk manager to ensure which hazardous products to segregate or to separate accordingly. (Guideline for Managing Risks with Chemicals in DET Workplaces (HLS-PR-006), Queensland Government - 2008).

- *Separation* is the storage of hazardous materials in different storage areas by at least fire-resistant walls and ceilings. Outside a warehouse, the hazardous products must be stored at a distance of at least 5–10 m depending on the combination of hazardous products stored.
- *Segregation* means storage in the same storage area, but the products of different classes are separated from each other by gaps or barriers or in cabinets. Products

Class 1 Explosives:
Substances containing a great amount of stored energy that can produce an explosion, a sudden expansion of the material after initiation, usually accompanied by the production of light, heat, sound, and pressure;
i.e, ammunition, fireworks, detonators.

Class 2 Gases:
Gases may be lighter or heavier than air. Heavier than air gases can collect in low-lying areas such as pits, depressions, and drains. Gases can be supplied as either compressed (e.g, aerosols), liquefied, refrigerated liquefied or gas in solutions. This class has three divisions:

- Division 2.1 – flammable gases, that is, butane, propane, acetylene, hydrogen, LPG

- Division 2.2 – nonflammable, nontoxic gases, that is, oxygen, nitrogen, compressed air

- Division 2.3 – toxic gases, that is, chlorine, ammonia, carbon monoxide

Class 3 Flammable liquids:
Flammable liquids produce vapor that can be ignited in air on contact with a suitable ignition source. By definition, these must have a flash point of less than or equal to 60.5 °C. Examples are petrol and alcohol, acetone, thinners, and kerosene.

Class 4 Flammable solids:
Substances liable to spontaneous combustion and substances that, in contact with water, emit flammable gases. This class has three divisions:

- Division 4.1 – flammable solids such as hexamine solid fuel tablets for camping stoves, self-reactive substances and desensitised explosives, magnesium, metal powders, sulfur, activated charcoal
- Division 4.2 – substances liable to spontaneous combustion under normal conditions such as phosphorus, which burns by itself when exposed to air, and sodium sulphide

- Division 4.3 – substances that emit flammable gases when they come into contact with water, that is, "Dangerous when wet" Examples are sodium, zinc particles, calcium carbide, alkali metals, and so on

Class 5.1 Oxidising substances:
Substances that, in themselves, are not necessarily combustible, but which, by yielding oxygen, may cause or contribute to the combustion of other materials. Examples are pool chlorine, sodium peroxide, potassium permanganate, and ammonium nitrate fertiliser.

Figure 13.2 ADR dangerous goods classes.

Class 5.2 Organic peroxides:
Organic chemicals containing the peroxy group (–OO). These are thermally unstable substances that may undergo heat generating, self-accelerating decomposition – which may be explosive, rapid, sensitive to impact or friction or react dangerously with other substances. Examples are hydrogen peroxide and methyl ethyl ketone peroxide (MEKP).

Class 6.1 Toxic substances:
Substances that are liable to cause death or injury if swallowed, inhaled, or absorbed through the skin. Examples are pesticides and poisons such as cyanide, paraquat, and arsenic compounds.

Class 6.2 Infectious substances:
Those substances known or reasonably expected to contain pathogens including bacteria, viruses, parasites and fungi, and clinical or medical waste.

Class 7 Radioactive material:
Substances that contain unstable (radioactive) atoms that give off [ionizing] radiation as they decay, that is, uranium, tritium, thorium.

Class 8 Corrosives:
Substances capable of causing the degradation and destruction of living tissue, steel, and other materials on contact. In the event of a leakage, corrosives can cause severe damage when in contact with living tissue or materially damage other property. Corrosive materials are either acids or bases/alkalis. Examples are nitric acid, hydrochloric acid, caustic soda, liquid chlorine, mercury and car batteries.

Class 9 Miscellaneous:
Comprises substances and articles that present a danger not offered by other classes, including asbestos, magnetic articles, molten bitumen, dry ice (solid carbon dioxide).

Figure 13.2 (Continued).

of similar classes may, in principle, be stored together in the same storage area. Exceptions to this are cases where specific storage regulations such as the regulations for explosives, organic peroxides, and flammable substances have to be observed. Segregated storage within a storage area may also be necessary, for example, because of special material properties possessed by certain individual products of the same storage class or by products of other storage classes, for which mixed storage in accordance with the mixed storage table is permitted.

In order to determine possible causes for hazards, the quality management literature has offered a structured problem-solving technique, which is called the *cause-and-effect diagram* (also called the *Ishikawa diagram*). This type of analysis serves to organize information after a brainstorming session, including the possible causes for a problem and how to verify them, besides selecting the most probable cause until a cause-and-effect relationship, which leads to a solution, is found. Chang and Lin (2006) investigated the causes of storage tank accidents. Such a cause-and-effect diagram is worked out in Figure 13.5. Perhaps surprisingly, the

Figure 13.3 Some symbols used to label dangerous substances.

most common cause of storage tank accidents was found by the authors to be lightning, maintenance, operational error, equipment failure, and sabotage.

13.2.2
Quantifying Risk: Probabilities and Consequences

Once all potential hazards have been identified, and having investigated how the problem occurred, further analysis is required to decide on appropriate action. In some cases the action may be obvious, but in others there may be questions regarding the likelihood of the occurrence and the magnitude of the consequences. Analysis of either hazard frequency or consequence may be sufficient to suggest an action. The risk manager needs to quantify the risks involved for each of them. To this end, the risk manager needs to determine two distinct values:

1) the probability of the hazard from occurring and
2) the consequences of the hazard.

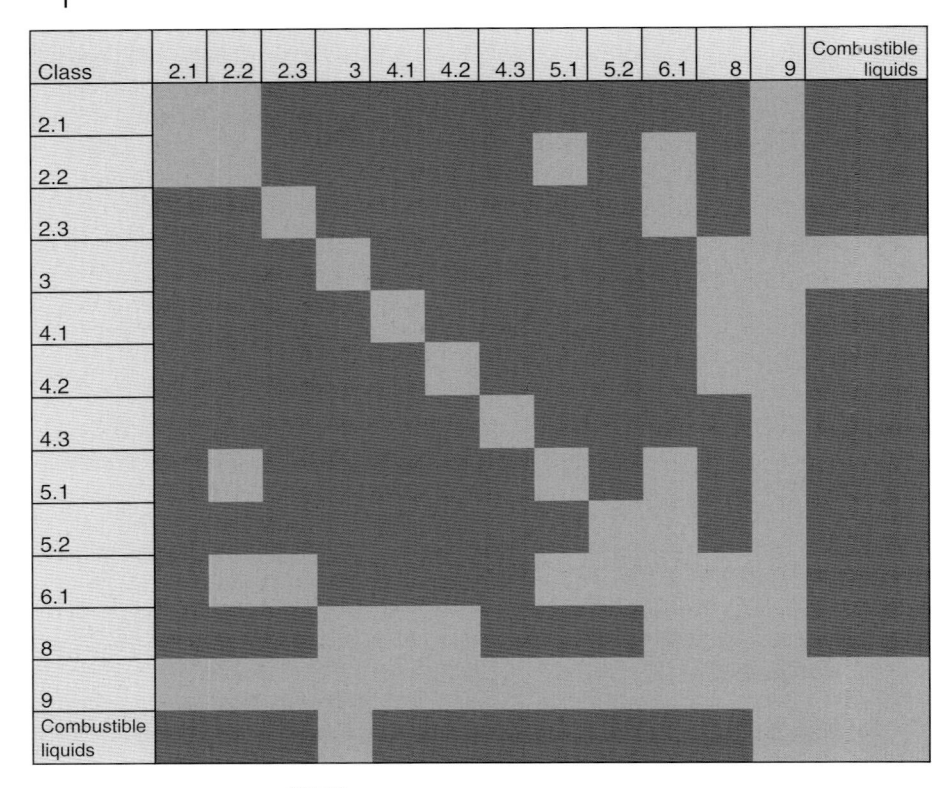

Class	2.1	2.2	2.3	3	4.1	4.2	4.3	5.1	5.2	6.1	8	9	Combustible liquids
2.1													
2.2													
2.3													
3													
4.1													
4.2													
4.3													
5.1													
5.2													
6.1													
8													
9													
Combustible liquids													

▫ Compatible in many cases with exceptions
▪ Likely incompatible, segragation or separation needed

Figure 13.4 Class compatibility matrix. (Guideline Safe Storage of Dangerous Goods and Dangerous Substances, Deutsche Gesellschaft für Technische Zusammenarbeit (GTZ) - 2002).

In industry, risk is generally defined as the multiplication of both values:

Risk = frequency × consequence

On the basis of this information, a *risk classification matrix* (similar to the one shown in Figure 13.6) can be constructed. This matrix reflects all identified risks categorized according to their likelihood of occurrence and the impact that these risks will have on the warehousing organization. As can be seen in the matrix, both probability and consequences are measured on an ordinal, three-point scale. Generally, risk probability and risk consequences will not be known with any kind of precision. For this reason, the matrix does not pretend exactness but rather provokes order-of-magnitude estimates.

The United States Department of Defense uses another risk classification matrix (Figure 13.7), in which consequences and probabilities are qualitatively expressed in words, making it easier to fill them in.

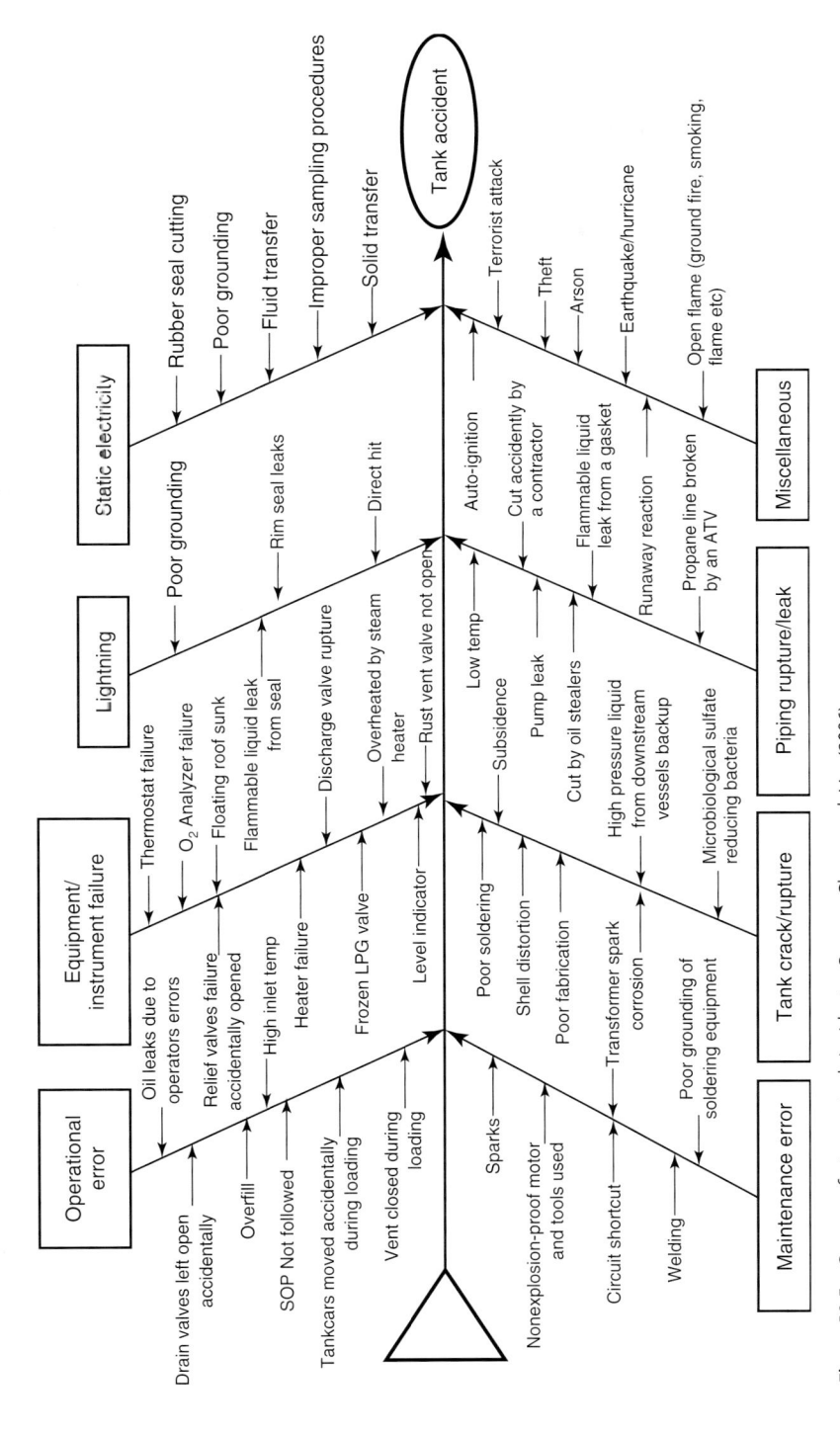

Figure 13.5 Causes of storage tank incidents. Source: Chang and Lin (2006).

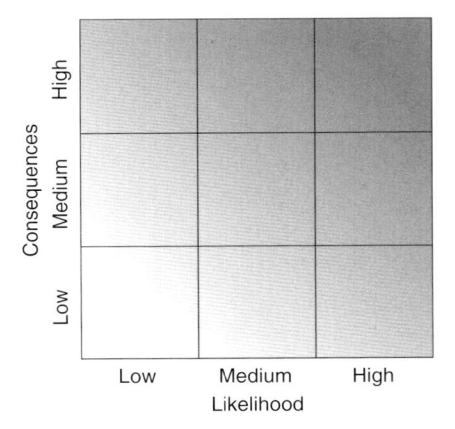

Figure 13.6 General risk classification matrix.

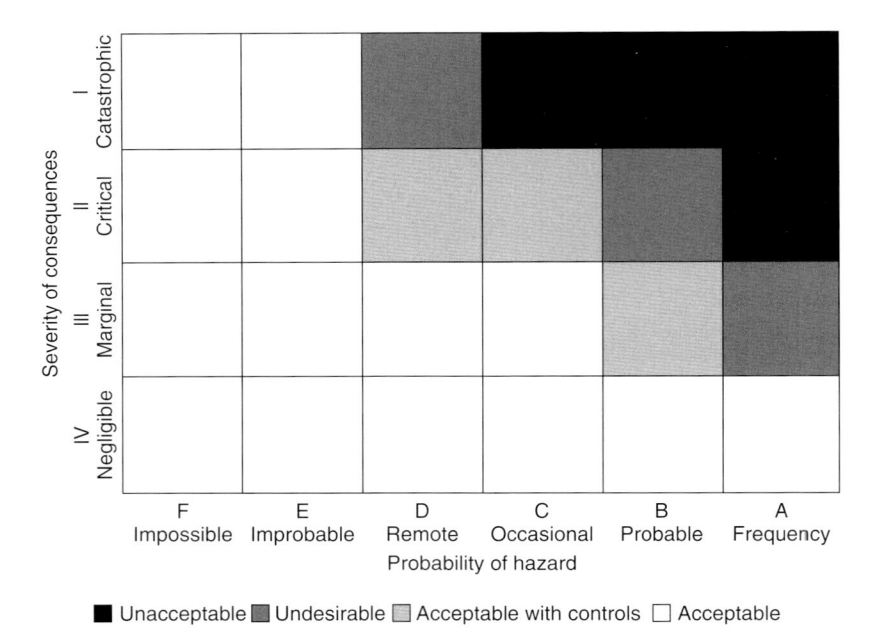

Figure 13.7 Risk matrix. Source: United States Department of Defense, 2006.

Probability combined with consequence gives a sense of the overall risk impact associated with a certain hazard. The more the hazard is situated toward the top right corner of the matrix, the higher it should be on the agenda.

Although some authors (e.g., Cox, 2008) argue that risk matrices do not necessarily provide useful information to prioritize risks or to distinguish between risks that can be disregarded and risks that cannot, most authors agree that using a risk matrix is preferable to using a less structured method.

Table 13.1 Risk factor calculation: examples.

Hazard	Probability	Consequence	Risk factor
Large earthquake	Low (0.1)	High (0.9)	High (0.81)
Electrical failure	Low (0.2)	Medium (0.3)	Medium (0.44)
Worker tripping	Medium (0.5)	Low (0.1)	Medium (0.55)

If a more quantitative analysis is desirable, the risk manager can make use of the so-called probability of failure (PF) and cost of failure (CF) measures and calculate from it a so-called risk factor (RF). Both measures should be expressed on a scale from zero to one. The risk factor is usually calculated as

$$RF = PF + CF - (PF)(CF)$$

As a rule of thumb, values of $RF < 0.30$ are considered "low risk" and values of $0.30 \leq RF < 0.70$ are considered "medium risk." Values of $RF > 0.70$ are "high risk." Table 13.1 shows some simple examples.

Several authors have proposed a system similar to the one mentioned here. Some authors (e.g., Kindinger and Darby, 2000) propose a system in which the hazards are ranked from high to low, in order to obtain a prioritization of risks. Broadly speaking, however, all risk identification systems combine a measure of the consequences of a hazard with a measure of the probability of that hazard turning into an overall measure of risk.

13.2.3
Mitigation Strategies

Once all potential hazards have been identified, and their corresponding risks quantified, the risk manager can start developing risk mitigation strategies, focusing in the first place on those risks that are considered high or even unacceptable.

Mitigation strategies can take on different forms, but most strategies can be categorized in one of the following categories.

- Minimize risk
- Transfer risk
- Accept risk.

13.2.3.1 Minimize Risk
The "minimize risk" category contains all risk mitigation strategies designed to either reduce the probability of a hazard from occurring or to reduce its consequences if it occurs. Several strategies in this category are put in place as a result of legal requirements, others as a deliberate common-sense strategy of the risk manager.

The location and design of storage tanks, for example, is an especially important factor in minimizing the risk of incidents. For example, storage tank design should ensure that no dangerous fluids or gases leak or evaporate. Storage tank location should ensure that the consequences of such a leak are minimal, for example, by ensuring that the leak can be quickly fixed and that all sides of the storage tank can be easily reached by the repair crew.

By their very nature, chemicals react with each other. Some reactions are violent, and produce noxious or otherwise dangerous substances. For this reason, several legal requirements usually dictate the distance that should be kept between storage tanks of incompatible chemicals. This not only applies to storage tanks but also to storage cabinets and to shelves. For this reason, chemical incompatibility matrices and tables have been developed.

Clearly, handling of dangerous substances should be done only by qualified personnel, whether they are employed by the company or by subcontractors. Access to these products and the warehouses they are contained in should be strictly monitored, so that only people who have sufficient knowledge about handling the product can access them. As an example, a container with a hazardous chemical might be well designed, but it offers no risk only when caps or lids are securely tightened.

To minimize the risk of explosion, adequate ventilation can often help to contain the buildup of dangerous concentrations of gases. Similarly, control of possible ignition sources is important (e.g., the use of cell phones is often forbidden in the vicinity of explosive fumes).

The substitution of a dangerous product with a less dangerous one is a final risk mitigation strategy that falls in this category; if the correct storage conditions cannot be met for particular dangerous substances, they should not be permitted on the site.

An often-used way to minimize risk by mitigating potential hazards is to reduce inventory levels, mainly cycle and safety stock, for products with a medium/high risk factor. Although strategy is mainly driven by a cost cutting objective, it has serious risk implications.

For this reason, the risk manager can combine his mitigation strategy with a cost savings program in cooperation with the operations/supply chain responsibles in order to convince the sales representatives. Traditional businesses still aim to have every Stock Keeping Unit (SKU) on stock for sale (push or supply methodology) with a "one-size-fits-all" distribution service for all SKUs (also for relative low margin SKUs) and its customer base. As a consequence, some customers are probably receiving a poorer service than they would like, while others are getting a better service than they require. It would be better to distinguish the standard of logistical service in line with customer requirements and related service costs (cost-to-serve). For the majority of the chemical industries, this would be more difficult to manage, but would offer benefits in improved efficiency, higher profits, and greater customer loyalty.

In addition, by use of a push model, incorrect SKUs and incorrect quantities are hold on stock, which is absolutely not cost efficient, and in our case it increases risk. Currently, front runners of the supply chain management in the chemical industry apply sales and operations planning (S&OP) processes to align sales forecasts

13.2 Risk Management in the Chemical Warehouse

Approach

A clear and agreed baseline for the model is important

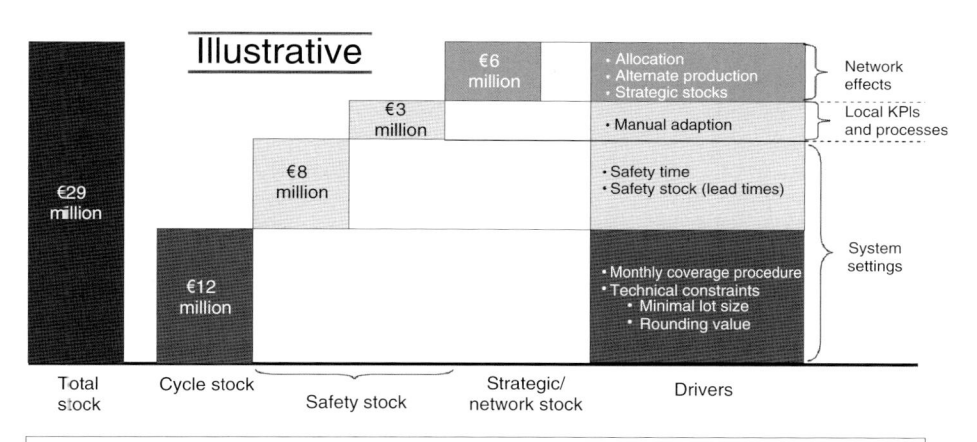

Illustrative

- Total stock: €29 million
- Cycle stock: €12 million
- Safety stock: €8 million
- Strategic/network stock: €3 million + €6 million

Drivers:
- Allocation, Alternate production, Strategic stocks → Network effects
- Manual adaption → Local KPIs and processes
- Safety time, Safety stock (lead times)
- Monthly coverage procedure, Technical constraints, Minimal lot size, Rounding value → System settings

With the settings the models best displays the site reality. We have a "gray" area where the coverage goal and manual planner activities drive which might be partially strategic.

Figure 13.8 Inventory costs (As-is situation) of a chemical company.

and customer demands (new to-be-produced and already allocated stock) to plan production capacities in order to have correct SKUs on stock. Operations, Finance, Sales, and Procurement have, that is, a monthly S&OP meeting to discuss what products and related quantities have to be produced against budget (demand-driven methodology).

As an example, we present the case of one of the largest multinationals in the chemical industry, which has implemented a Sustainable Inventory Management savings program for one of their strategic business units (Figures 13.8 and 13.9). The company also succeeded in reducing the cycle and safety stock of several hazardous products with medium/high risk factors and even increasing some service levels. In this case, the company was able to reparameterize its batch sizes on the SKU level.

13.2.3.2 Transfer Risk

The most common way to transfer a certain risk to another party is by taking an *insurance policy*. This is, however, not independent of the previous strategy. As insurance companies assess the risk involved in a certain hazard, they generally evaluate carefully how the company manages its risks, more specifically at the measures that are already in place to minimize the probability of the risk or its potential consequences. The more the company invests in minimizing the risk, the lower its premiums toward the insurance company will be. In many cases, risk

Results

Overview OF different optimization scenarios and inventory reduction potentials

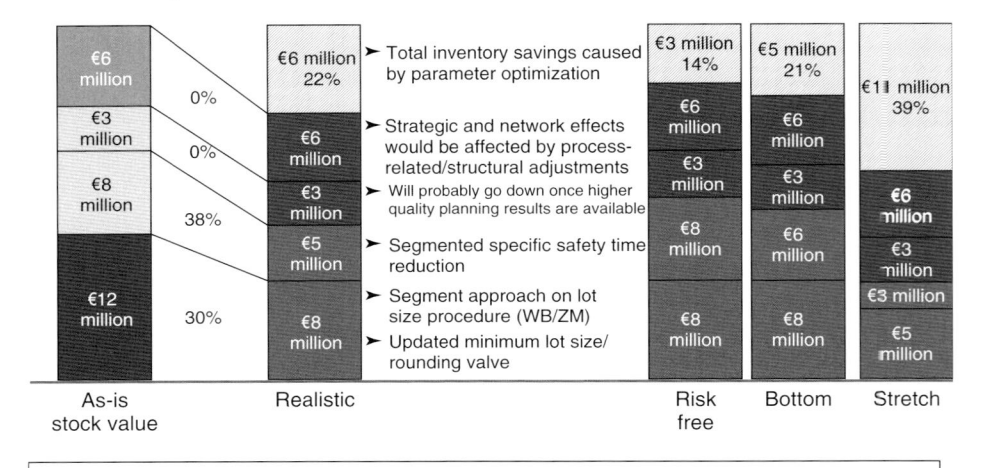

As-is stock value	Realistic		Risk free	Bottom	Stretch
€6 million / 0%	€6 million 22%	➤ Total inventory savings caused by parameter optimization	€3 million 14%	€5 million 21%	€11 million 39%
€3 million / 0%	€6 million	➤ Strategic and network effects would be affected by process-related/structural adjustments	€6 million	€6 million	
€8 million / 38%	€3 million	➤ Will probably go down once higher quality planning results are available	€3 million	€3 million	€6 million
	€5 million	➤ Segmented specific safety time reduction	€8 million	€6 million	€3 million
€12 million / 30%	€8 million	➤ Segment approach on lot size procedure (WB/ZM) ➤ Updated minimum lot size/rounding valve	€8 million	€8 million	€3 million €5 million

The site agreed to the realistic scenario

Figure 13.9 New inventory costs of a chemical company because of a savings program (Logichem 2009).

mitigation strategies are a condition for approval of the insurance policy, and the insurance company enforces its own list of risk minimization measures that need to be taken before the risk is insured.

Chemical leasing is a concept by which companies do not buy chemical products, but rather buy the *function* of these products. This may mean that, for example, the chemical product in question remains the property of the leasing company. This in turn may imply that the leasing company also carries the risk involved in the storage and handling of the product. Although chemical leasing contracts differ widely, it may be an interesting avenue to research whether risk can be transferred through chemical leasing. In general, it might be considered as an "extended product responsibility" that fits into the concept of sustainability. The EU REACH proposal aims at better identification of the properties of chemical substances by shifting the responsibilities of managing the risks and providing safety information to the producers and users of these substances. If a company does not have the use of chemical as its core business, it depends on transfer of knowledge from the producer. That producer might not be interested to transfer this knowledge. In a study by Ohl and Moser (2007), it is shown that the business model of chemical leasing might overcome this type of problem.

Vendor-managed inventory (VMI) is a well-established practice in many other industrial sectors, where it has been permitted as a more effective supply chain management tool for the mutual benefit of suppliers and customers (Cefic/Epca 2004). It relieves customers of the need to place orders in the conventional manner and puts the onus on suppliers to replenish customers' supplies within agreed limits. This gives the producer both visibility and control of the end-to-end supply chain, allowing it to manage inventory and transport capacity more efficiently. By doing this, the supplier can minimize the stock levels of some hazardous products.

From the customers' perspective, the main disadvantage of VMI is that it locks them into a single-sourcing arrangement with a particular supplier. Experience in other sectors, however, suggests that the risk associated with single sourcing can be more than outweighed by the cost and service benefits of VMI. Also, "vendor hubbing," a service provided by some logistics service providers, allows a customer to combine VMI with multiple sourcing from several suppliers.

13.2.3.3 Accept Risk

In some cases, either the probability or the consequence of a certain hazard is too small to warrant doing anything about it. In these cases, the risk can be accepted. The risk matrices in Figures 13.3 and 13.4 show the areas corresponding to acceptable risk.

Another factor that should be taken into consideration is the cost of the other strategies. For example, even though the risk of someone tripping over a certain step is deemed small and its consequence is minor, signposting the step with black and yellow tape might be an inexpensive measure that can be taken up regardless. On the other side of the spectrum, even though the consequences of a nuclear strike on the facility may be devastating, there is very little that a company can do about it except for taking out insurance. It is the cost of the insurance premium that should determine whether this is worth it, or whether the company should just trust that such a rare event will never occur during its lifetime.

13.2.4
Control and Documentation

It is crucial that the above-mentioned analysis, that is, risk identification and risk mitigation, is executed on a regular time basis and that the actions and their results are carefully recorded and documented. Risk control is therefore the final step in the risk management process. It is key in developing an organization memory for risk, as it allows risk managers to categorize various types of risks, the company's response to them, and the results of this response. Information systems projects have been the worst examples in business to control in terms of risk (mostly time, budget, and quality risk) as results are not reported against project proposals. As a result, a future project can never benefit from errors in forecasts made in the past.

Having a risk analysis database helps companies execute the previous steps in the risk management process more swiftly and with a higher level of confidence. It also makes it easier for new people in the risk management department to be trained, and avoid the same mistakes their predecessors had made.

The risk analysis database should contain records of each analyzed hazard, its probability and its consequences, the risk factor attributed to it, and the strategy that the company followed to mitigate it. Also, each risk should receive an assessment time frame that stipulates on which time scale the risk is reassessed. In this way, risks in the chemical warehouse can be managed systematically, minimizing the risk of oversight.

13.3 Conclusions

Adequate risk management is of utmost importance in a chemical warehouse. A sustainable warehousing policy therefore requires that the company puts in place formal risk management policies and procedures. These generally consist of four phases: risk identification, risk quantification, risk mitigation, and control and documentation.

References

Cefic/Epca (2004) Supply Chain Excellence in the European Chemical Industry.

Chang, J.I. and Lin, C.C. (2006) A study of storage tank accidents. *J. Loss Prev. Proc. Ind.*, **19** (1), 51–59.

Cox, L.A. Jr., (2008) What's wrong with risk matrices? *Risk Anal.*, **28** (2), 497–512.

Guideline for Managing Risks with Chemicals in DET Workplaces (HLS-PR-006). (2008).

Guideline Safe Storage of Dangerous Goods and Dangerous Substances. (2002) PN 1995.2501.5-001.00.

Kindinger, J.P. and Darby, J.L. (2000) Risk factor analysis-a new qualitative risk management tool. Proceedings of the Project Management Institute Annual Seminars and Symposium.

Logichem (2009) Supply Chain Risk Management in Chemical Industry.

Ohl, C. and Moser, F. (2007) Chemical leasing business models – a contribution to the effective risk management of chemical substances. *Risk Anal.*, **27** (4), 999–1007.

United States Department of Defense (2006) Risk Management Guide for DoD Acquisition, August 2006.

Part V
Sustainable Chemistry in a Societal Context

14
A Transition Perspective on Sustainable Chemistry: the Need for Smart Governance?

Derk A Loorbach

14.1
Introduction

This chapter positions the search for sustainability in the chemical industry and its surroundings in a transition context. We frame efforts to develop alternative technologies, processes, and products based on renewable resources and efficient use of (renewable) energy as part of a longer term process in which the dominant culture, structure, and practices in the chemical industry are reinvented. We distinguish between different emerging pathways that increasingly start to compete; on the one hand, efforts to improve the performance of the existing systems (or regimes) through efficiency increases and replacement of fossil by bio-based resources and, on the other, attempts to move to fundamentally different types of chemical industry based on second- and third-generation biomass technologies. From a transition perspective, we seek to understand the dynamics in and between these pathways to develop thoughts on the need for and possibility of governance strategies that might increase the chances of an accelerated sustainability transition.

Broader landscape changes are increasingly driving society toward more efficient use of renewable resources: consumption and production systems are fundamentally challenged, dominant economic paradigms are increasingly put under stress and global governance processes seek to (incrementally) steer industry and society toward internalizing external costs. Examples are the European Union policies on registration of chemicals (Registration, Evaluation, and Authorization of Chemicals, REACH), on electronic materials (Waste Electrical and Electronic Equipment Directive, WEEE) and their resource strategy.[1] External pressures such as these provide incentives at the level of the chemical industry to include sustainability in their strategies, moving toward more efficient use of energy, using alternative resources, and rethinking lifecycles of products and materials (Whitesides and Ismagilov, 1999). Simultaneously, we see emerging

---

1) http://ec.europa.eu/resource-efficient-europe/

niche innovations that develop on a different basis starting from the concept of green chemistry (Kirchhoff, 2005). Here, we find the quest for a new type of chemical industry fully based on closed cycles, renewables, and producing societal value.

In this dynamic perspective, there is an inherent struggle between those structures and industries with established positions that seek to sustain themselves by moving toward sustainability and the alternative models in niches that do not carry the burden of a historically grown regime but also lack the position and power to quickly scale up and take over. These struggles and the associated tensions are visible in many industries in the post WWII era, such as the steel industry (Rynikiewicz, 2008). These struggles, however, are not necessarily fought between insiders and outsiders; there are many examples of established industries that did develop radical innovations, which they then used to improve their positions. It seems that while large established companies might have the financial and political power to develop those alternative technologies and solutions that might open up pathways toward a sustainable chemistry, they are unable or unwilling to also transform the structure of chemical production systems at large. In other words, innovation, radical or incremental, is mainly approached as a means to sustain individual business.

The premise, however, is that the shift toward a sustainable chemistry does presuppose a more radical transformation in culture, structure, and practices in the chemical production systems. As regimes are inherently unable to make a radical systemic change happen by themselves, the question is how the interaction between regime improvement and radical innovation from niches can be governed in such a way that the overall sustainability transition is accelerated. This dilemma of, on the one hand, hanging on to existing assumptions about availability of resources and consumption levels and, on the other, the need to make a radical shift is illustrated in the vision of the Dutch organization for chemical industry. This vision expresses the ambition to transition to a renewables or bio-based chemical industry in 2080.[2]

Given all the evidence on impacts of climate change, resource availability and prices, and increasing consumption levels in different parts of the world, this seems a rather incremental and too long a time horizon. We therefore raise the question, in this chapter, of how to understand the dynamics currently going on in the area of sustainable chemistry and explore the conceptual possibilities and contours of an accelerated transition. This raises the issue of transition governance: is it possible or feasible to develop strategies that enhance the chances of emerging niches and accelerate the momentum within the existing regimes? And if so, what would be the desirable results in terms of the direction of this transition? And how could such a strategy take shape and be implemented? To explore these last questions we will further develop on the ideas of transition management.

2) http://www.vnci.nl/actualiteit/nieuwsbrief/nieuwsbrief-artikelen/11-11-15/_Transitie_naar_biobased_rond_2080_voltooid_.aspx

14.2
A Transitions Perspective on Chemical Industry

Transitions are major, nonlinear changes in societal cultures, structures, and practices (Grin et al., 2010) that arise from the coevolution between economy, society, and ecology. Transitions can be viewed as a shift from one dynamic equilibrium (e.g., a fossil-based centralized energy system) to another (a renewable-energy-based decentralized system). They are the result of interacting developments at different levels of scale that, under specific conditions, might over time fundamentally alter dominant practices, paradigms, and structures. Usually, transitions take a very long predevelopment phase in which there is a gradual buildup of pressure on a dominant regime, which may be understood as the dominant structure, culture, and practices in a societal system. This pressure stems from an internal dysfunction of the regime, increasing competition of alternatives, or a changing external context (e.g., a financial crisis). When these pressures start to reinforce each other, a relatively rapid systemic change might occur. In transitions research, transitions are visualized as processes of multilevel (Geels, 2002), multiphase (Rotmans, Kemp, and van Asselt, 2001) changes (Figure 14.1).

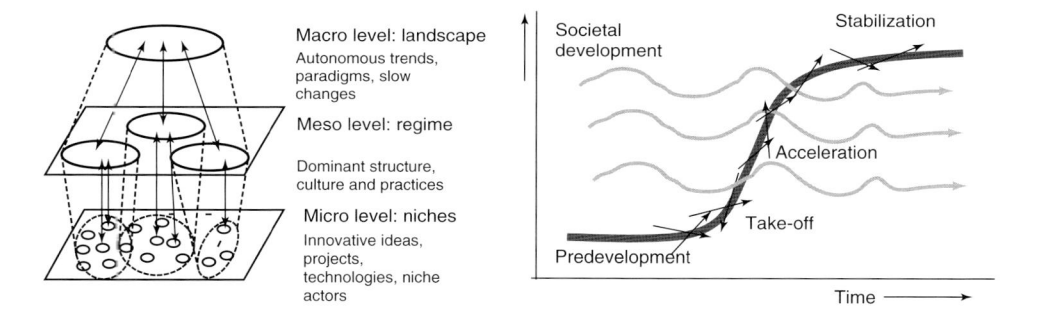

Figure 14.1 Multilevel and multiphase concepts of transitions.

In the context of sustainability, transitions seem inevitable from the perspective of limited resources, ecological thresholds (Rockstrom et al., 2009), and changing economic and demographic landscapes. They do, however, also provide possibilities for innovation, green economic growth, and new business. According to Keijzers (2002), sustainability efforts change the internal governance of the firm as new environmental preservation demands come forward, and the external governance of businesses as new initiatives need to be employed to ensure adequate and sustainable social, technological, and infrastructural conditions for production. Taking a transition perspective, the historical developments in terms of thinking about environmental and social issues (e.g., the limits to growth, energy efficiency, climate change debates, environmental policies, etc.) have coevolved with business strategies that adapted to this.

Current transitions that are deemed likely and possible in sectors such as food provision, energy, built environment, and mobility (Grin *et al.*, 2010; Ernstson *et al.*, 2010; Cherp, Jewell, and Goldthau, 2011; Clark, Crutzen, and En Schellnhuber, 2005; Nobel Laureate Symposium, 2011) will arguably also fundamentally impact existing industries. This perspective can be interpreted in multiple ways. Possible transitions might threaten existing industries (strategies) and require crisis management and adaptive strategies (or, in some cases, existing industries may be a factor in slowing down transitions), or industries might proactively try to anticipate possible transitions and play a role in guiding these toward more desired situations.

The current industrial systems have emerged over the past century or so as part of the historical societal transitions toward a high-energy, high-carbon welfare state. In coevolution with demographic changes, technological progress and economic growth, industrial production systems have developed exponentially. Especially since the late 1950s in the European Union and the United States, the chemical industry has grown and fueled economic and consumption growth based on fossil resources. It has developed in a context where oil especially has been widely available at relatively low prices and detrimental effects were not so obvious. It has also developed in an increasingly globalized world, leading to a global chemical industrial complex based often on cost and (fossil) resource efficiency.

The chemical industry originated in the nineteenth century, gradually developing into a global industry after WWII in which the important German industrial plants were largely destroyed. Around WWII also, the shift took place from a coal-based to a petro-based chemical industry producing plastics, fibers, and so on. This led to an enormous post-war boom of the chemical industry, driving exponential growth in production and consumption. In the 1960s and 1970s with publications such as Rachel Carson's Silent Spring (1963) and Limits to Growth (Meadows *et al.*, 1972), a new debate emerged drawing attention to the negative environmental impacts of chemical production as well as the use of fossil resources. Both environmental impacts as well as the availability and the instability of oil supplies and prices (as became evident during the 1973 oil crisis) triggered policies to deal with these effects and increase the (energy) efficiency in the chemical industry.

During the 1980s and 1990s, the then established bulk chemical regime of large-scale plants, oil-based production, and increasing efficiency in energy and resource use was further optimized. This optimization took place in three dimensions: in the production processes, in the (re)location of production plants for bulk chemicals (to the Middle East and Asia), and in the environmental performance of the sector. In the context of globalization and shifting economic balances, chemical production has moved across the globe, searching for minimal production costs as well as minimal environmental costs. Figure 14.2 well illustrates how around WWII chemical production started to accelerate, largely based on the use of fossil resources (oil and gas). The figure represents the amounts of sulfur produced as a by-product of fossil resources (which is cheaper than through mining) and thus indirectly how the use of fossil resources exploded over the past decades.

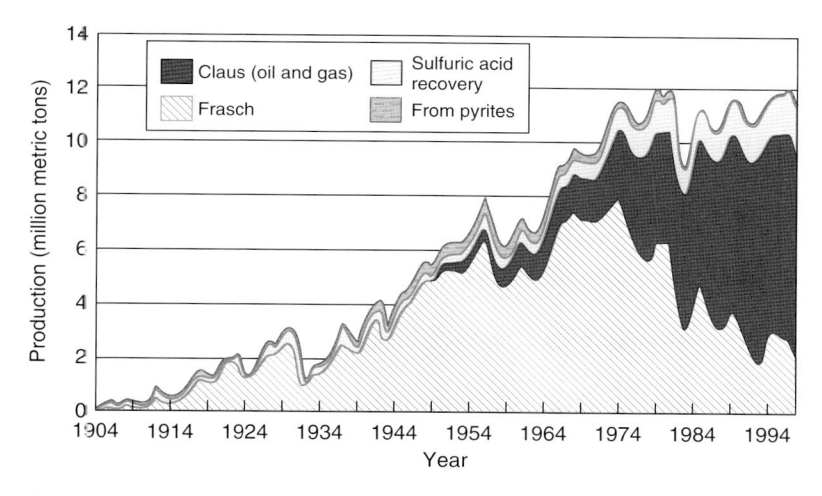

Figure 14.2 Historical sulfur production (1904–1994) in the United States. Source: U.S. Bureau of Mines, 1927–1996 and U.S. Geological Survey, 1901–1927, 1997–2000.

Historically, this can be interpreted as a transition; from relatively low levels of production and technology, to high-tech mass production of bulk chemicals in the 1990s. Especially because of technological improvements and environmental policies, this historic transition stabilized in terms of growth increases and entered a period of optimization in Europe and the United States. Especially with regard to resource and energy use as well as reduction of emissions, chemical industries improved significantly both in bulk chemical and in fine chemical production. This can be illustrated by the reduction of CO_2 emissions in the European chemical industry as well as the decoupling between energy use and production output (Figures 14.3 and 14.4).

Over the past two decades or so, we argue that a new type of dynamics is emerging. From a transition perspective, existing chemical industries have developed incumbent regimes: dominant cultures, structures, and practices that are increasingly inflexible and locked-in: based on fossil fuels, large-scale centralized production in a context of economic and demographic growth, cost reduction and efficiency, and with linear cradle-to-grave production processes. A number of landscape trends are increasingly putting pressure on this regime: resource prices and availability; environmental policies on resources, waste, and emissions; and stagnating economic growth in major parts of the world. Simultaneously, new niches have emerged around sustainable chemical production and concepts such as industrial ecology (Baas, 2008; Ayres, 2004; Morioka, Saito, and Yabar, 2006) and cradle to cradle. These niches are still relatively underdeveloped and small scale, but simultaneously seek to fundamentally address the problems inherent in the dominant regime. In the next section, we reflect on the current dynamics around the chemical industry from this transition perspective.

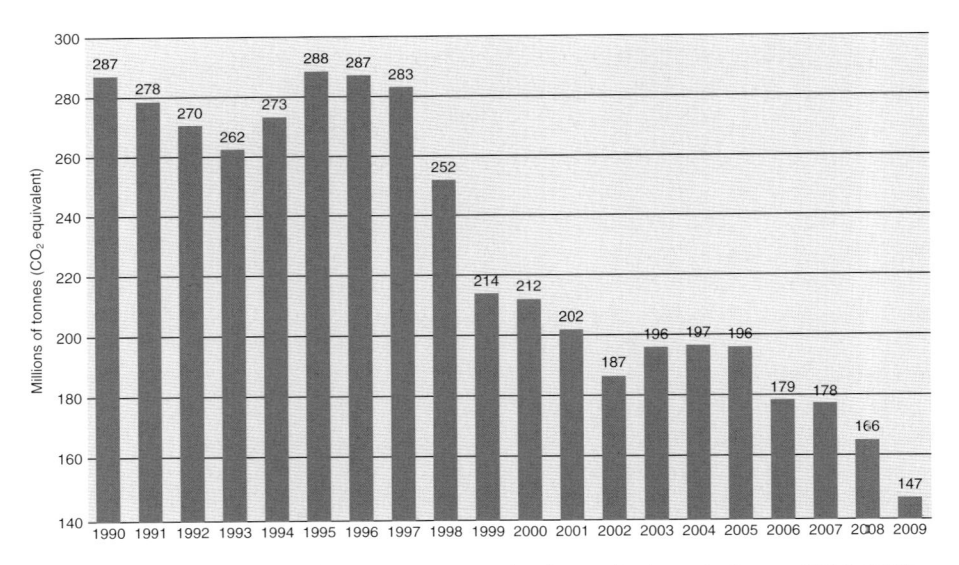

Figure 14.3 Reduction of CO_2 emissions in the chemical industry in Europe (1990–2009). Source: European Environment Agency (EEA) and CeficChemdata International.

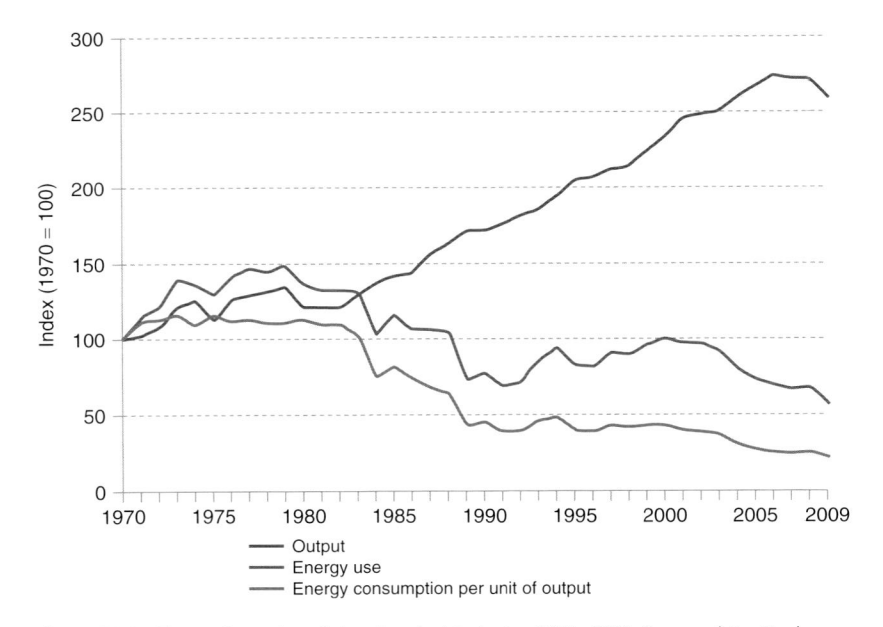

Figure 14.4 Energy intensity of the chemical industry 1970–2009 Source: http://onl.ne.businesslink.gov.uk/Environment_Efficiency_images/EE2870_popup1.jpg (accessed 13 February 2012).

14.3
A Tale of Two Pathways

Although past and future transitions are expected to share a generic dynamic pattern, namely, transformation due to increasing macro pressures and micro alternatives that break through, the current context appears different. The historical development of the now dominant chemical industrial regime has been largely driven by technological progress, economic and demographic growth, and the broader process of modernization (Grin *et al.*, 2010). While obviously these factors still play a major role in driving change in production and consumption in many parts of the world, the notion of sustainability transitions might also imply a partial breakdown of the existing structures, interests, and routines. Historical transitions certainly did not take place without resistance, breakdown, and backlashes, but it can be argued that the competition between alternative pathways and the unequal distribution of (financial, regulatory, and political) power between these different pathways requires a more encompassing analysis as well as governance strategies to deal with (Geels and Schot, 2007). It also implies an inherently normative and ambiguous process of change in which "sustainability" values will be continuously debated, negotiated, and redefined. In that sense, perhaps, the current resource and energy transitions have a more contested and normative character as opposed to historical transitions: it is not only about progress and increasing welfare but also about equity, justice, access, ecology, and so on.

From this perspective, there are at least two ideal typical pathways that seem to emerge: one building on and improving the existing and the other toward a fully green-resources-based chemical industry (organic chemistry). In the chemical industrial regimes themselves, dynamics started to shift from strategies focusing on improvement to strategies focusing on more fundamental adaptations to the changing context. Plants are increasingly cleaner, green resources are increasingly used, and new technologies such as biomaterials and carbon capture and storage are aggressively pursued to prolong the lifetime of centralized, fossil-based production (King and Lenox, 2000). The organization of the chemical industry in the Netherlands, VNCI, presented its vision in 2011. This vision illustrates the growing awareness within the regime that a shift toward sustainability is inevitable, but projects this shift along a very long time frame (2080). However understandable, this clearly is the perspective from the inside out: reasoning from the existing infrastructures, investments, technological possibilities, and vested interests.

Following the transition perspective, it is however likely that this pathway will be increasingly put under pressure by the mentioned landscape trends. Developments in technology, energy, and resources will no doubt continue to decrease the possibilities for maintaining existing operations and increase the space for radical alternatives. There is an increase in investments and production of green-resource-based chemistry and the broader context of "clean tech," which is growing in countries such as the Netherlands (Roland Berger, 2009) and projected to accelerate (Figure 14.5). From a transition perspective, however, the historical development of the fossil-based chemical regime is by necessity only able to develop incrementally.

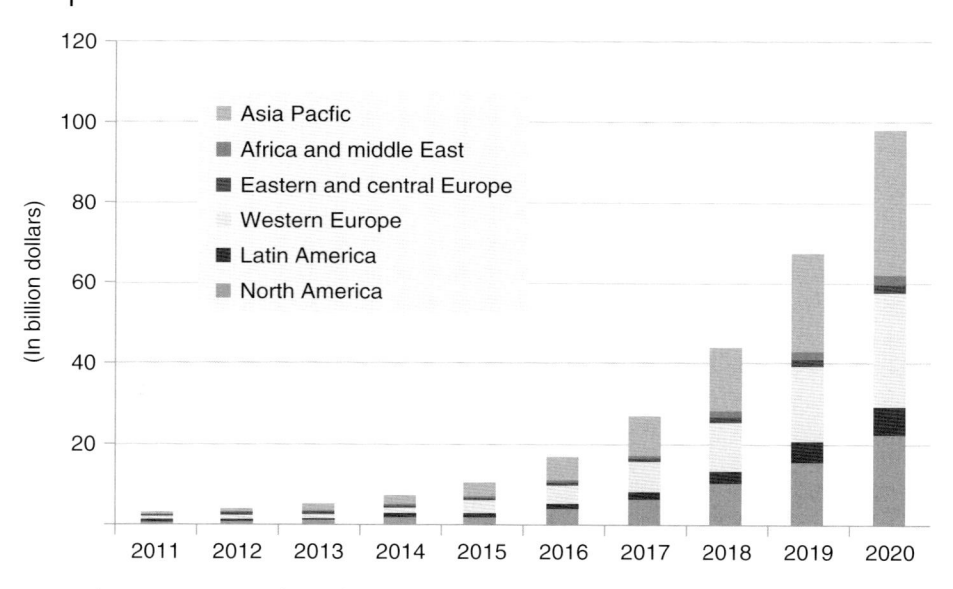

Figure 14.5 Green chemical market by region, Worlds Markets: 2011–2020. Pike Research

Because of the lock-in in terms of embedded routines, sunk costs, institutional and regulatory schemes, and material infrastructures, the radical innovations associated with sustainable chemistry are automatically threatening the existing regime. The combination of increasing competition from radical alternatives with pressure from landscape trends and performance issues within the regime will inevitably lead to nonlinear and unpredictable change.

Our argument is that we are witnessing the first signs of a transition in the chemical industry; one might even claim that in some cases we are already approaching the takeoff phase of such a transition. However, how this future transition might unfold is highly uncertain and unpredictable. A transition to a sustainable chemical industrial system, which may be defined more precisely than biomass-based and emission-free, is only one of the possible pathways and perhaps even a very unlikely one. Especially, the political and economic power of existing industries (both chemical industries as well as fossil energy production) will be no doubt used to slow down the transition or steer it into an enhanced optimization pathway.

The increasing tensions between the optimization and transition pathway might lead us to very different energy futures. As the competition between these different pathways increases, we move closer to a takeoff or reconfiguration phase, and the necessity for transition-based strategies increases. One of the major challenges in this context is to create the conditions under which more desirable pathways are more likely to be followed. Such conditions could be technological, financial, regulatory, but could as well include a broad consensus on direction, strong networks, or institutional change. Given this, there are a large number of challenges for transition governance in both understanding and anticipating possible future dynamics, especially regarding the possibilities of different types of pathways associated with the acceleration of transitions and/or breakdown or lock-in scenarios.

14.4
Critical Issues in the Transition Management to Sustainable Chemistry

Given the described nature of transitions, their inherent complexity, uncertainty and long-term horizon, choices on investments, institutional change, technological innovation, and so on are by definition political and disputed. In case of the transition to green or sustainable chemistry, this is no different. A real takeoff of sustainable chemistry will require structural changes in the existing regime, which is sustained by regulatory, technological, and financial as well as behavioral and mental structures. Especially the last aspect is found to be increasingly important in transition studies in both consumers (Shove and Walker, 2010) and in institutions (Walker *et al.*, 2009; van den Bergh and Stagl, 2006). Existing convictions and assumptions can be quite persistent when it comes to radical innovations: the promise and science of radical innovations is systematically underestimated and denied by incumbents. Transitions therefore are, the closer they come to takeoff, a battle of beliefs as much as that they are power struggles.

What is at a certain point in time defined as sustainable pathways, accordingly is inherently contested and ambiguous, and subject to change over time (Rotmans and Weaver, 2006). Understanding the role of agency and governance in this context is crucial, but so far lacking in a more systematic way. Much of the literature within transition studies, as well as in the fields of governance or political and policy sciences, focuses on partial processes, formalized policies, and articulated innovation strategies. This implies development of transition governance and management as a meta-governance approach. The picture that emerges from the brief transition sketch in the previous section is very clear. We cannot expect a smooth transition to a sustainable chemistry and a bio-based economy overnight. Actors related to the industry, be it incumbents or niche actors, are competing over financial resources, legitimacy, and political power. From a societal perspective it seems necessary to systematically deal with these dynamics so as to provide the conditions for a desired transition.

On the basis of this, there are at least three critical issues relating to the structures of the existing regime that need to be addressed in order to create space for a sustainable chemistry transition:

- **Infrastructure** A consensus is emerging that a cleaner, yet affordable and sustainable chemistry in the future might be fundamentally different from the fossil-based, centralized, and liberalized bulk chemistry we have nowadays. Be it in a centralized or a decentralized way, each of the bio-based options, more advanced than the first-generation biofuels, will require innovations and changes in the existing infrastructural and transport systems or infrasystems (Frantzeskaki and Loorbach, 2010). In the coming decade, the clash between existing and new visions and paradigms will shift slowly toward more tangible problems such as the investments, regulation, and use of infrasystems. Over the past years, sustainable alternatives have developed on their own both within the context of optimizing existing chemistry (companies such as BASF and DSM are successfully replacing fossil resources by organic ones on a modest scale) and within radical niches. With this, the increased sense of urgency and the mere necessity for change are setting the stage for a battle over plants, pipes, and production. We now enter a time in which choices need to be made and clever strategies developed to ease the transition toward more socially, economically, and ecologically benign chemical industries. Key questions that will emerge are related to the way in which existing social and technological structures can be adapted in an incremental yet relatively fast way overcoming the barriers of incumbent actors and sunk costs and without limiting the flexibility needed to change with societal contexts.

- **Institutions** Transitions require institutions that are reflexive, intermediating, coordinating, and adaptive. It implies institutions that are able to facilitate and guide emerging niches and niche regimes as well as reorienting the trajectories of existing regimes in light of possible transition pathways. As the dominant institutions in the chemical industrial system, such as trade organizations, regulatory bodies and regulations, scientific research and economic conditions, are part of the regime, they are inherently focused on improving the existing and tied to global energy and resource issues, georegional industrial systems, or specific markets. The inherent uncertainties combined with the need for fundamental reform imply novel ways in dealing with accountability, with output/measurement and with diversities of options, goals, and societal objectives. Governance of the transition to a sustainable chemistry thus requires institutional reform as well as establishment of novel governance institutions.

  A key problem in terms of institutional innovation is the problem of control and coordination. The resource and energy problem itself is a multilevel one. On a global scale, issues such as ecological impact and climatic change, availability and distribution, prices, and so on call for a debate and governance. On increasingly lower levels of scale, these global issues manifest themselves in different ways, ranging, for example, from regions without access to cheap energy and resources, to regions in which the impacts of harvesting fossil resources have grave social or ecological impacts. There is thus also a need for institutions at lower levels of scale that are able to mediate between global issues and local impacts as well as between global solutions and local implementation. In addition, there is a need for intermediating institutions that are able to connect different levels of scale in terms of overall direction, exchange of knowledge, and scaling up of successful innovations.

- **Instruments** Such examples point to the need for institutional change but as well call for a change in practices and instruments facilitating the transition. The current policy instruments are based on the idea of centralized production, security of supply, and safety of production. They are thus intrinsically linked to and part of sustaining and improving chemical industrial production. An important innovation challenge in terms of (policy) instruments to facilitate sustainability transitions thus lies in developing both general and context-specific instruments for transition governance.

14.5
Governance Strategies for a Transition to a Sustainable Chemistry

Transition management is a governance approach developed over the past decade to deal with complex long-term societal transitions so as to enhance the possibilities of sustainability transitions (Loorbach, 2007). The approach provides a way to articulate and structure what such frontrunners are doing and provides a framework to attempt to translate their approach into a more general and transferable methodology. It uses the transition perspective on nonlinear change in dominant regimes to analyze and accordingly try to influence the speed and direction of the transition. By distinguishing different types of governance through which changes in cultures, structures, and practices occur, the transition management perspective offers a framework to explain the possibilities of having influence and the various roles of firms in the context of transitions. Furthermore, transition management provides a framework for more specifically selecting partners, instruments, and actions related to organizations' and individuals' own goals and ambitions.

Transition management is based on a number of specific starting points, derived from literature on complex adaptive systems (Rotmans and Loorbach, 2009):

- *Approach the chemical production system as a complex adaptive system in its environment.* The transitions approach offers a systemic way to analyze debate and influence changes in the energy domain.
- *Dealing with lock-in.* The lock-in of the present chemical production systems is more than a technological lock-in and the direction of development is shaped by more than innovation or economy alone.
- *Dealing with uncertainty.* Some forms of uncertainty can be reduced by doing research (such as integrated systems analysis), some aspects are inherently uncertain (for instance, what we will learn in the future, or system responses after thresholds crossing).
- *Approach the transition as a multiactor problem-solving process.* Transitions are defined as broad societal processes of transformation and result from the actions of numerous actors.
- *Stimulate new combinations.* New combinations of knowledge (e.g., multidisciplinary knowledge), stakeholders, technologies, policy instruments, and so on, might trigger innovation and set off new dynamics.
- *Be reflexive in the management approach.* Every intervention is based on an incomplete model of the world. Each intervention will also produce unintended side effects and adverse boomerang effects that can partially be anticipated, and partially need to be responded to.
- *Creating space for agents to build up alternative regimes is crucial for innovation.* Stimulating emergence and divergence is crucial for innovation.

These basic principles have been translated into a process approach and have so far predominantly been initiated by governmental organizations[3] (Loorbach

---
3) See, for example, www.creativeenergy.nl, www.tplz.nl, www.plan-c.eu, www.duwobo.be.

and Rotmans, 2010). It is built up as a cyclical process of development phases at various scale levels (Loorbach, 2007). The core idea is that four different types of governance activities can be distinguished when observing actor behavior in the context of societal transitions: strategic, tactical, operational, and reflexive. In short, these different types can be described as follows:

- *Strategic*: activities at the level of a societal system that take into account a long time horizon, relate to structuring a complex societal problem and creating alternative futures often through opinion making, visioning, and politics.
- *Tactical*: activities at the level of subsystems that relate to the buildup and breakdown of system structures (institutions, regulation, physical infrastructures, financial infrastructures, and so on), often through negotiation, collaboration, lobbying, and so on.
- *Operational*: activities that relate to short-term and everyday decisions and action. At this level, actors either recreate system structures or they choose to restructure or change them.
- *Reflexive*: activities that relate to evaluation of the existing situation at the various levels and their interrelation of misfit. Through debate, structured evaluation, assessment, and research, societal issues are continuously structured, reframed, and dealt with.

This governance typology has been used to develop specific "systemic instruments" and process strategies (see also Loorbach, 2010). At the core, these instruments are based on the notion of "selective participation": bringing together frontrunners with different backgrounds in another environment (a "transition arena") in which they develop collectively a new way to understand and talk about a complex environment and their role within it. These instruments and the process strategy in which they are embedded are captured in the so-called transition management cycle (Figure 14.2), which consists of the following components[4]: (i) structure the problem in question and establish and organize the transition arena; (ii) develop a transition agenda, a vision of sustainability development, and derive the necessary transition paths; (iii) establish and carry out transition experiments and mobilize the resulting transition networks; and (iv) monitor, evaluate, and learn lessons from the transition experiments and, based on these, make adjustments in the vision, agenda, and coalitions. According to our experiences so far, there is no fixed sequence of the steps in transition management and the steps can differ in weight per cycle. In practice, the transition management activities are carried out partially and completely in sequence, in parallel, and in a random sequence.

From this perspective, it seems that at the operational and tactical level, a lot of activities are taking place. On the basis of the increasing number of investments (Figure 14.5), patents (Nameroff, Garant, and Albert, 2004), and plants, there are many experiments and niches emerging. Around this, there are quite some tactical activities in terms of developing research agendas, networks, platforms, and so on.

---

4) For extensive description of these activities see: (Rotmans and Loorbach, 2009; Grin et al., 2010; Loorbach, 2010).

It seems that by far the biggest challenge lies in developing a more strategic understanding of the interactions between emerging structures and practices around sustainable chemistry and the incumbent regime as well as a sense of direction in terms of a sustainable chemistry sustainability transition. It is in the context of rhetoric around a bio-based economy that there is perhaps no clear idea on the nature of this challenge as well as the basic principles on which this bio-based economy should be built.

The transition to a bio-based economy and sustainable chemistry certainly seems to be one of the vital components of a sustainable society. One could even claim that this is the "mother of transitions" in that it presupposed transitions in the domains of energy, agriculture, and consumption. The other way around, a bio-based transition could help accelerate the others. Without actually defining the desired future trajectory, there are at least a number of crucial issues that are so far barely addressed in coherence at a systems level but certainly need to be a part of the transition. These are as follows:

- *Resource sustainability*: there are multiple efforts to develop criteria for sustainable biomass, especially since the competition between biomass and food led to sharp price increases and biomass production led to dramatic land-use changes. A perhaps more fundamental issue is whether the large quantities of biomass that are now produced and the infrastructure to trade and transport these are actually helping to sustain the existing chemical industry or opening up the transition pathway. The emerging international biomass market seems to primarily support production and trade of cofired biomass, often wood pellets for coal-fired power plants. The more specialized biomass streams that are used in chemistry can potentially benefit from this development, but it could also be that these relatively small flows are further marginalized and require completely different production, trading, and transport systems. A poignant question thus is how to develop large-scale global sustainable biomass production, trade, and transport in such a way that it accommodates the transition to a bio-based economy.
- *Circular resource economy*: one of the key arguments for a shift toward sustainable chemistry is that it would have an inherently lower (or even positive) impact. Besides environmental issues, this should also include social as well as economic aspects. The question is therefore not only whether the bioresources used in sustainable chemistry can become cradle to cradle in the sense that they can be either fully reused or composted, but also whether they add socioeconomic value throughout the whole chain. In other words, could sustainable chemistry provide a route to economic and social welfare in producing countries and lead to an inclusion of externalities in developed countries? This bigger issue of an inclusive and inherently sustainable circular economy has a direct link to the transition to sustainable chemistry.
- *Consumption levels*: as the current production and consumption levels are based on the principle of ever growing and faster consumption cycles, a sustainable-chemistry-based production–consumption system should be based as much as

possible on a long user cycle and perhaps even a nongrowth economy. Besides a fundamental societal debate, probably outside the scope of the green chemical industry, it also calls for reflection within the context of sustainable chemistry. Can sustainable chemistry be sustainable when based on the idea of ever growing consumption?

These three issues are by far the only ones but they are critically important for opening up and defining the sustainability space within which a sustainable chemistry transition needs to unfold. From a transition management perspective, this underlines the need for a more strategic transition-arena-type debate: involving frontrunners from various societal domains related to the challenge of sustainable chemistry to develop a new and shared discourse around this.

By necessity, such a transition arena would not only include engineers and business but explicitly also policy, science, and NGOs. In addition, such a transition arena should include actors from different areas such as energy, resources, food, agriculture, and perhaps also domains such as consumption. The function of such a transition arena would be to develop a shared and integrating understanding and narrative about the type of transition as well as the desired direction. However, it would also provide with this new discourse the context within which the current experiments and debates can be framed. Rather than only developing the technology, transitions are about developing new societal systems. And it is through learning and experimenting that such system innovations are driven.

14.6
Conclusions and Reflections

This chapter provides a preliminary and brief exploration of a transition perspective on the challenge and changes around sustainable chemistry. Rather than being based on an encompassing analysis, we have tried to formulate a number of hopefully interesting perspectives and challenges for both governance and research in this context. We assessed that a transition toward sustainable chemistry will either go very slow and incremental (and perhaps never get there) or will necessarily require a breakdown of existing regimes. In this case, the transition will bring forth uncertainties, power struggles, and other negative side effects.

This transition scenario is, however, both likely, based on the multilevel dynamics described, as well as desired from the perspective of the urgency to move to a sustainable society. A transition *per se*, however, is no guarantee for a sustainable chemistry. As we have tentatively pointed out, there are quite a number of critical sustainability issues that need to be taken up much more systematically. Our argument therefore is that a transition to a sustainable chemistry is not so much a technological challenge as that it is an institutional, economic, and political challenge.

This also implies that much more attention is required in these issues to understand the nature of change as well as the possibilities to accelerate and guide a sustainability transition. The transition management approach might provide some ideas there, but these obviously need to be translated and applied to the specific contexts. The suggestion here is to broaden the context in which sustainable chemistry is framed and to invest in structured dialogs and experimentation with frontrunners from other backgrounds and sectors. Such an approach might help overcome the possibilities of a lock-in into suboptimal pathways as well as to help prevent an encapsulation by the existing regime.

As mentioned, this chapter only made a first attempt and many issues are open, of which these are the most obvious:

- What are the multilevel dynamics at a global, continental, regional, and local scale?
- What are the existing and competing governance regimes (including regulatory, financial, and political institutions)?
- What are the main barriers and drives for global as well as regional transitions to sustainable chemistry?
- What are the possibilities for developing transition governance structures across different levels of scale and different sectors?
- How can more space for experimentation be created and successful alternatives be scaled up and diffused?

References

Ayres, R.U. (2004) On the life cycle metaphor: where ecology and economics diverge. *Ecol. Econ.*, **48** (4), 425–438.

Baas, L. (2008) Industrial symbiosis in the Rotterdam harbour and industry complex: reflections on the interconnection of the techno-sphere with the social system. *Bus. Strateg. Environ.*, **17** (5), 330–340.

van den Bergh, J.C.J.M. and Stagl, S. (2006) Co-evolution of economic behaviour and institutions: towards a theory of institutional change. *J. Evol. Econ.*, **13** (3), 289–317.

Carson, R. (1963) *Silent Spring*, London: Hamish Hamilton.

Cherp, A., Jewell, J., and Goldthau, A. (2011) Governing global energy: systems, transitions, complexity. *Global Policy*, **2** (1), 75–88.

Clark, W.C., Crutzen, P.J., and En Schellnhuber, H.J. (2005) Science for Global Sustainability: Toward a New Paradigm, CID Working Paper No. 120, Harvard University, Cambridge, MA.

Ernstson, H., van der Leeuw, S., Redman, C., Meffert, D., Davis, G., Alfsen, C., and Elmqvist, T. (2010) Urban transitions: on urban resilience and human-dominated ecosystems. *AMBIO: A J. Hum. Environ.*, **39** (8), 531–545.

Frantzeskaki, N. and Loorbach, D. (2010) Towards governing infrasystem transitions: reinforcing lock-in or facilitating change? *Technol. Forecast. Soc. Chang.*, **77** (8), 1292–1301.

Geels, F.W. (2002) Understanding the Dynamics of Technological Transitions: A Co-Evolutionary and Socio-Technical Analysis, Universiteit Twente.

Geels, F.W. and Schot, J. (2007) Typology of sociotechnical transition pathways. *Res. Policy*, **36** (3), 399–417.

Grin, J., Rotmans, J., Schot, J., With, I.C., Loorbach, D., and Geels, F.W. (2010) *Transitions to Sustainable Development; New Directions in the Study of Long Term Transformative Change*, Routledge, New York.

Keijzers, G. (2002) The transition to the sustainable enterprise. *J. Clean. Prod.*, **10** (4), 349–359.

King, A. and Lenox, M. (2000) Industry self-regulation without sanctions: the chemical Industry's responsible care program. *Acad. Manag. J.*, **43** (4), 698–716.

Kirchhoff, M.M. (2005) Promoting sustainability through green chemistry. *Resour. Conserv. Recycl.*, **44** (3), 237–243.

Loorbach, D. (2007) *Transition Management: new Mode of Governance for Sustainable Development*, International Books, Utrecht.

Loorbach, D. (2010) Transition management for sustainable development: a prescriptive, complexity-based governance framework. *Governance*, **23** (1), 161–183.

Loorbach, D. and Rotmans, J. (2010) The practice of transition management: examples and lessons from four distinct cases. *Futures*, **42**, 237–246.

Meadows, D., Meadows, D., Randers, J., and Behrens, W. (1972) The Limits to Growth: A Global Challenge. Club of Rome, New York.

Morioka, T., Saito, O., and Yabar, H. (2006) The pathway to a sustainable industrial society. *Sustain. Sci.*, **1**, 65–82.

Nameroff, T.J., Garant, R.J., and Albert, M.B. (2004) Adoption of green chemistry: an analysis based on US patents. *Res. Policy*, **33** (6–7), 959–974.

Nobel Laureate Symposium (2011) The Stockholm Memorandum, tipping the scales towards sustainability. 3rd Nobel Laureate Symposium on Global Sustainability, Stockholm, Sweden 2011, Resilience Centre.

Rockstrom, J., Steffen, W., Noone, K., Persson, A., Chapin, F.S., Lambin, E.F., Lenton, T.M., Scheffer, M., Folke, C., Schellnhuber, H.J., Nykvist, B., De Wit, C.A., Hughes, T., Van, D.L., Rodhe, H., Sorlin, S., Snyder, P.K., Costanza, R., Svedin, U., Falkenmark, M., Karlberg, L., Corell, R.W., Fabry, V.J., Hansen, J., Walker, B., Liverman, D., Richardson, K., Crutzen, P., and Foley, J.A. (2009) A safe operating space for humanity. *Nature*, **461** (7263), 472–475.

Roland Berger (2009) Clean Economy, Living Planet; Building the Dutch Clean Energy Technology Industry.

Rotmans, J., Kemp, R., and van Asselt, M. (2001) More evolution than revolution: transition management in public policy. *Foresight*, **03** (01), 15–31.

Rotmans, J. and Loorbach, D. (2009) Complexity and transition management. *J. Ind. Ecol.*, **13** (2), 184–196.

Rynikiewicz, C. (2008) The climate change challenge and transitions for radical changes in the European steel industry. *J. Clean. Prod.*, **16** (7), 781–789.

Shove, E. and Walker, G. (2010) Governing transitions in the sustainability of everyday life. *Res. Policy*, **39** (4), 471–476.

Walker, B., Barrett, S., Polasky, S., Galaz, V., Folke, C., Engström, G., Ackerman, F., Arrow, K., Carpenter, S., Chopra, K., Daily, G., Ehrlich, P., Hughes, T., Kautsky, N., Levin, S., Mäler, K., Shogren, J., Vincent, J., Xepapadeas, T., and de Zeeuw, A. (2009) Looming global-scale failures and missing institutions. *Science*, **325** (5946), 1345–1346.

Weaver, P. and Rotmans, J. (2006) Integrated sustainability assessment. *Int. J. Innovation Sustainable Dev.*, **1**, 284–303.

Whitesides, G.M. and Ismagilov, R.F. (1999) Complexity in chemistry. *Science*, **284** (5411), 89–92.

15
The Flemish Chemical Industry Transition toward Sustainability: the "FISCH" Experience

Luc Van Ginneken and Frans Dieryck

15.1
Introduction

15.1.1
Societal Chemistry

Chemistry is ubiquitous. It plays a fundamental role in almost every aspect of modern society. The products of the chemical industry contribute to the quality of our lives, particularly in areas as diverse as food, health, agriculture, clothing, construction, transport, communication, mobility, and leisure. As the enormous populations in China, India, and the emerging nations have started demanding Western standards of living, the demand for chemical products and consumer goods has surged enormously, thereby putting substantial pressure on available natural resources worldwide (SusChem, 2005).

However, to achieve prosperous and peaceful global development for the increasing world population, thereby safeguarding food and water supplies, efficient use of the available resources and improvement of healthcare systems are indispensable prerequisites. These objectives are best pursued through a balanced effort, taking into account economics, environmental protection, and quality of life. Major contributions from the chemical industry supported by new research and subsequent innovations are necessary to meet these challenges. In this way, chemistry plays a clear role in providing technological solutions to the challenges faced by society currently and will be at the center of stimulating the worldwide economy – providing new opportunities and wealth creation beneficial to all citizens (Eder and Sotoudeh, 2000; SusChem, 2005).

15.1.2
The Belgian and Flemish Chemical and Life Sciences Industry in a Global Context

Besides its important societal role, the chemical and chemistry-using industries are also of vital importance to the economy and employment in Europe. Belgium, located in Western Europe, is one of the largest chemical producing countries, despite being a small country. The importance of the chemical and life sciences industry (expressed as the ratio of turnover relative to the gross domestic product of the country) is about 2.5 times higher in Belgium, as compared to all other considered European countries. Only Ireland scores better with a ratio of 3.5 (essenscia, 2008). In addition, looking at the turnover within the European chemical and life sciences industry, Belgium ranks after the four biggest countries of the European Union (i.e., Germany, France, the United Kingdom, and Italy) but has a size comparable to countries such as Spain and the Netherlands. In terms of turnover per capita, only Ireland performs better than Belgium, which scores 1.5 times higher than the Netherlands and more than two times higher than Germany, France, and the United States. While Belgium only represents 2.1% of the total EU-27 population and 2.7% of the European GDP, the Belgian chemical and life sciences industry also covers about 6% of the total European turnover in this sector and accounts for not less than 14% of the European Union's total extra-European exports of chemicals and life sciences products (essenscia, 2008).

Belgium is a federal state consisting of three regions: Flanders in the north, Wallonia in the south, and Brussels-capital in the center. With a turnover of €35.8 billion, the chemical and life sciences industry in the Flemish region represents over 70% of total turnover of the chemical and life sciences sector in Belgium (essenscia, 2010). These industries also account for 24% of the regional industrial contribution and for one third of the export, and they employ directly and indirectly more than 160 000 people. More than 3500 companies, ranging from large multinationals to a multitude of small and medium enterprises (SMEs), rely on chemistry for their businesses, making the sector of the chemical industry and life sciences the second largest industry in Flanders (essenscia, 2010).

Currently, Flanders is recognized as hosting one of the highest chemical industry concentrations not only in Europe but in the world, its petrochemical cluster (located in the Antwerp harbor region) being the largest and most diversified one in Europe.

15.1.3
The Challenge of Sustainable Development for the Chemical Industry in Flanders

The (Flemish) chemical and chemistry-using industries, however, are facing major challenges (Eder and Sotoudeh, 2000; Verbeeck, Debackere, and Wouters, 2004; SusChem, 2005): their ecological footprint is large and their manufacturing processes have to become cleaner and more energy efficient, they have to use less fossil raw materials or renewable alternatives, and they have to produce more harmless and less dangerous intermediates and end products. Increased globalization has led to increased competition; emerging economies and countries, such as the Middle East and China, leverage on cheap labor and bulk production, thereby strongly increasing their competitive edge. At the same time, our chemical industries still struggle with their public image, largely due to concerns over adverse environmental impact; according to a Cefic-led (pan-European) survey, the image of the chemical industry in 2010 was still at the same level as in 2008, but the sector was ranked below the overall average of eight benchmark industries in terms of having a favorable image (Cefic, 2010). These image issues slow down the chemical industries' future growth and potential investments in Flanders (as the influence of environmental pressure groups combined with their public acceptability effectively constrain governments to introduce legislation, forcing industry to adopt more efficient and sustainable manufacturing technologies). The industry is also facing a shortage of human capital (i.e., skilled and well-trained workers), despite the fact that the chemical sector is one of the most financially rewarding sectors in Flanders. Finally, the outlook of the global economy is gloomy as a consequence of the recent financial crisis, and this phenomenon will badly deteriorate the functioning of the Flemish chemical sector in the near future.

Considering the major economic role of the Flemish chemical and chemistry-using industries, all stakeholders involved will have to join forces to tackle these challenges and to build the foundation for a sustainable and prosperous chemical sector in Flanders. Sustainable development is nowadays regarded by governments, industry and the public as a necessary tool to achieve the desired combination of environmental, economic, and societal objectives (the so-called Triple-P philosophy). The big challenge herein is the development of new products, processes, and services offering all benefits of sustainable development. The latter requires an alternative approach minimizing the processes' materials and energy input, and thus operating at maximum efficiency, combined with minimized or no emission of harmful chemicals in the environment. Moreover, the use of renewable resources must be stimulated and combined with improved durability and recyclability of products (taking into account "cradle-to-cradle" principles). All the latter goals, however, must be achieved in a way still providing economic benefit to the chemical manufacturers.

"Sustainable chemistry" (also often referred to as "green chemistry") is the worldwide term used to describe this relatively new approach to develop more eco-friendly and eco-efficient sustainable chemical products and processes. The term has been defined at the end of the twentieth century by pioneers Paul Anastas and John Warner as follows: *"a new way of looking at chemicals and their manufacturing processes to minimize any negative environmental effects"* (Anastas and Warner, 1998). In designing and redesigning their chemical processes and chemical products, the Flemish chemical and chemistry-using industry thus will have to transform into an increasingly sustainable industry by applying more intensively the principles of sustainable chemistry to ensure not only efficiency but also human, environmental, and economic compatibility.

Because of its geography and demography, the presence of one of the largest chemistry clusters in Europe, the huge industrial activity in downstream use of chemicals, and a vibrant, emerging biotechnological industry, Flanders may be considered as a unique laboratory to implement the complex interdisciplinary, intersectoral, and value-chain life-cycle concepts of sustainable chemistry in its industry.

In this section, we briefly elaborate on the first steps taken by the Flemish chemical and chemistry-using industries in their transition toward (more) sustainability.

15.2
Transition of the Chemical Industry in Flanders: the "FISCH" Initiative

15.2.1
Setting the Scene: the "FISCH" Feasibility Study

Late 2007, the sector organization *essenscia flanders* (i.e., the Flemish industry federation for chemistry and life sciences that represents and defends the sector's interests at the Belgian and European level) started discussing with some of its stakeholders the type of sustainable chemistry initiative that was needed to support sustainable development of the chemical and chemistry-using industries in Flanders. In order for Flanders to establish a powerful chemical sector in the years to come, the stakeholders involved felt that excellent academic research (leading to innovative breakthroughs) and a powerful and vivid business community was needed. However, none of the players involved (industry, academia, research institutes, regulatory bodies, facilitators, societal stakeholders) solely possessed all necessary knowledge, technologies, and processes to achieve these objectives. Therefore, multidisciplinary and precompetitive collaborative partnerships were perceived as being essential to realize sustainable development and to maximize the valorization of the Flemish knowledge and know-how across the entire value chain. essenscia flanders and its partners were convinced that this would require a lot of research, development, and innovation (R&D&I) in a model of "open innovation", thereby bridging the gap between science and innovative business activity.

At that time, it was envisioned that Flanders needed a (strategic) platform to anchor world-leading innovation in sustainable chemistry to become a powerhouse of wealth creation and to boost the quality of life, thereby addressing the chemical sector's major needs and challenges (Figure 15.1). In the course of 2008, a feasibility study that aimed to investigate how the envisioned Flemish strategic platform on sustainable chemistry should look like and how it could address the major challenges of the Flemish chemical sector, was initiated by essenscia flanders and VITO (the Flemish Institute for Technological Research). Financial support for this one-year study (titled *"Flanders strategic Initiative for Sustainable CHemistry (FISCH) – a strategic initiative for sustainable development by clean chemistry, life sciences, and their manufacturing industries"*) was provided by IWT (the Flemish Agency for Innovation by Science and Technology), and also by essenscia flanders and 20 of its large enterprise (LE) members. In December 2008, a dedicated core team (under the leadership of essenscia flanders and with contract researchers from VITO; the SME Value-for-Technology; and the Universities of Ghent, Leuven, and Antwerp) began developing the outline of a strategic initiative for sustainable chemistry in Flanders. The general objectives of the feasibility study were to achieve the following:

- Identify the knowledge, innovation, and research and development needs from a socioeconomic perspective for a long-term new start of the chemistry-using industries
- Create broad support for this initiative across knowledge centers, university associations, industry, governments, and society

- How to reduce the use of fossil raw materials and energy input?
- How to stimulate the use of renewable resources and how to combine this with improved durability and recyclability of products?
- How to reduce the emission of harmful chemicals in the environment?
- How to increase competitiveness of the Flemish chemical industry?
- How to increase the number of students for sustainable-chemistry-related studies?
- How to present chemical sciences and technologies to the public as being an essential and relevant part of modern society?
- How to communicate the benefits and risks associated with novel technologies to the public?
- How to comply with regulations and standards?
- How to develop new businesses with sustainable products and processes?
- How to stimulate open innovation?
- How to use and develop measuring methods for sustainable development of chemical processes and products?
- How to organize innovation infrastructure (incubators, pilot plants, application and innovation laboratories, and research infrastructure) in an open collaborative model?
- How to develop and apply new service and business models for sustainable chemical products?

Figure 15.1 Major needs and challenges of the Flemish chemical sector.

- Put the acquired information and inventories in the perspective of sustainable development and related future solutions for the chemistry-using industries
- Give special focus to SMEs, for which the study wanted to create a forum to solve the sustainability questions and to facilitate the sustainability projects.

Representatives from small businesses, multinational companies, academic research groups, government departments, and environmental groups (which were already involved in "critical discussions" with industry on these issues) were consulted by the core team conducting the study using two questionnaires, six full-day workshops, and face-to-face interviews. On the basis of this input, supplemented with views and data from earlier studies and (policy) papers, preparative work of the core team, bimonthly decision-making meetings of a steering team, and quarterly governance meetings of a governance board, a blueprint outlining goals and recommendations for the Flemish chemical industry's transition to more sustainability was developed.

15.2.2
Outcome of the Study – Goals and Overall Setup of "FISCH"

15.2.2.1 Vision, Mission, and Setup of FISCH
The FISCH study obtained massive support from its stakeholders. More than 370 individuals from over 185 chemistry-related organizations actively participated

in the feasibility study, including 71 SMEs (74 individuals) and 31 non-profit SME-supporting organizations (71 individuals).

In early 2010, the final results of the FISCH study were translated into a business plan that set the scene for a world-class Flemish platform for sustainable chemistry (the so-called FISCH platform). In this business plan, the vision of FISCH was summarized as follows: *"To maintain and to guarantee the world-class competitive position and image of our region's chemical industry by sustainable chemical innovation."* To help realize its vision, FISCH defined the following mission statement: *"FISCH is an open knowledge organization enabling and stimulating innovation, education, entrepreneurship, and training in sustainable chemistry; FISCH promotes the sharing of knowledge and resources among industry, academia, knowledge centers, and other stakeholders. This will lead to new innovative breakthrough and collaborative programs, followed by valorisation trajectories creating new products and more sustainable processes; FISCH catalyzes the transition of the chemical industry into an innovation hotbed, supporting Flanders in anchoring world-leading innovation in sustainable chemistry, creating new and developing existing SMEs, contributing to a higher welfare and better quality of life in Flanders."*

To achieve the ambitious goals of FISCH, the business plan recommended that a structure consisting of four elementary building blocks be set up (Figure 15.2): (i) a Strategic Innovation Agenda (SIA), (ii) an Open innovation Infrastructure Cluster (OIC), (iii) a Sustainable Chemistry Knowledge Center (SCKC), and (iv) a supporting FISCH pontoon.

Strategic Innovation Agenda (SIA) The FISCH SIA deals with the Flemish research needs that must be addressed in the short term to realize the full potential of R&D&I in the field of sustainable chemistry, as such establishing a sustainable future for the Flemish chemical industry. Outlining the FISCH SIA started with identifying key areas of research, which were based on the research themes identified in the implementation action plan (IAP) of the European Technology

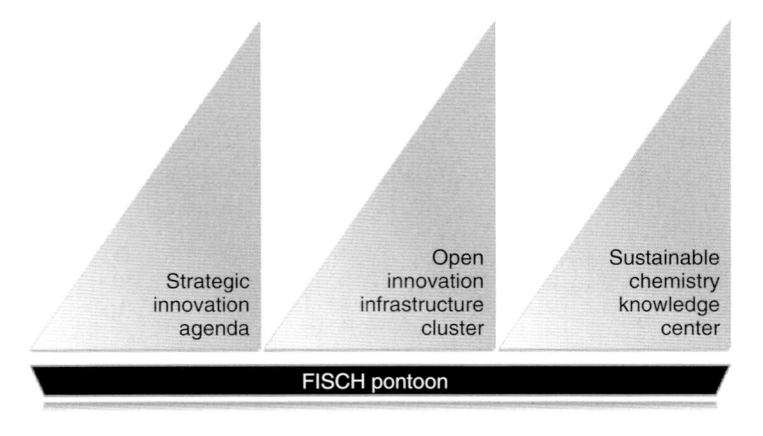

Figure 15.2 Overall setup and structure of the FISCH platform.

Platform "SusChem" (SusChem, 2006). More than 30 different research topics, considered to be important in a Flemish context, were withheld by the core team. These topics were both technical and horizontal in nature. The technical topics covered the overall research themes "industrial biotechnology", "reaction and process design", and "materials technology", which were then subdivided into more concrete research topics. The more horizontal topics covered (among others) public dialog, training, marketing strategies, and stimulation of innovation.

On the basis of stakeholder response gained from a written questionnaire, a shortlist of nine innovation breakthrough topics for Flanders, further clustered in three breakthrough domains for sustainable chemistry in Flanders, were extracted (Figure 15.3). It is expected that R&D&I in these three breakthrough domains for sustainable chemistry will offer "blue ocean" opportunities in Flanders, which create eco-innovative environments where large entrepreneurial opportunities are expected (Kim and Mauborgne, 2005). A clear interdisciplinary approach within the three breakthrough areas is needed to successfully tackle the challenges ahead. This approach will be supported by cross-cutting actions in areas such as education and training, stakeholder dialog including public engagement activities, and measures to create a Flanders that has the appropriate regulatory and financial environment for innovation and investment (i.e., for Flanders to become the preferred location for innovation activities).

For each of the nine initial FISCH innovation breakthrough topics, FISCH will set up so-called innovation programs. A FISCH innovation program is a long-term and a very performant open collaborative innovation consortium – involving all necessary stakeholders (LEs, SMEs, knowledge centers, academia, and service providers) – that, through the generation and execution of innovation projects, will

Alternatives for fossils
- Biomass conversion
- Micro–algae
- Valorization of side and waste streams

Process intensification
- Separation technologies
- Green solvents
- Micro–technology
- Catalysis and alternative energy input

Sustainable chemical products and processes
- Multi–factorial performance increase methodologies
- Sustainable chemical products

Figure 15.3 The Strategic Innovation Agenda (SIA) of the FISCH platform.

strive to evolve their activities into entrepreneurship (i.e., job creation), transfer of knowledge, training, and education. Through these innovation programs, FISCH aims to accelerate the creation of new value chains in Flanders. In addition, they must create added value for existing and emerging SMEs. A FISCH program will define and manage roadmaps and transition scenarios; inventorize existing collaborative projects and available infrastructures; define, initiate, and execute collaborative projects related to the consortium topic; define infrastructure needs for research, pilot plants, and application labs; strengthen existing and initiate missing networks considered to be of importance; and define training and education needs, and so on. For these programs, FISCH will cofinance their projects and ventures.

The FISCH SIA programs will be periodically evaluated/benchmarked by the FISCH platform for their international excellence, and phased out if expectations are not met. Other FISCH programs (if promising) can emerge.

Open Innovation Infrastructure Cluster (OIC) Highly specialized, state-of-the-art and usually expensive research infrastructure and equipment are necessary to execute innovation projects on sustainable chemistry. To facilitate companies (especially SMEs) and research centers to conceptually design, develop, test, and/or scale up sustainable chemistry technologies, products, and/or processes, cutting-edge capabilities (available at particular companies, research centers or academia) need to be clustered and open access to these different infrastructure capabilities provided. As such, these capabilities can provide companies and research institutes with low-cost access to state-of-the-art equipment to test and develop their new or improved products and/or processes. To this end, an inventory of key capabilities available at a range of companies, universities, and research centers across Flanders will be made. From this inventory, unique infrastructure capabilities will be selected and the conditions under which other companies/research institutes can use this equipment in an "open access" model will be discussed. It is envisioned that such open innovation infrastructure clusters will develop into internationally leading capabilities to assure in-sourcing of new users from abroad. As a consequence, Flanders will have a unique network of open innovation infrastructure clusters within driving distance. Funding for these infrastructure clusters will come from companies and will be matched with public funding to assure internationally competitive innovation productivity to drive the Lisbon targets. These clusters/capabilities will be integrated in a single (virtual) network, the so-called FISCH OIC, that will be managed as a single entity. This cluster will be advertised as a brand of the FISCH platform.

By providing open access to pilot installations and application laboratories, the challenging (and problematic) issue of prolongation of the innovation pathway for SMEs can be partly dealt with. Indeed, by making available to SMEs the existing state-of-the-art infrastructure present at LEs and research institutes, these SMEs can *"stand on the shoulders of giants"*. In return, the LEs can benefit from the often highly specialized knowledge of these SMEs.

Sustainable Chemistry Knowledge Center (SCKC)   In accordance with its three basic principles (i.e., People, Planet, Profit), sustainable chemistry implies a horizontal approach, including various aspects that influence the sustainability of products, processes, and businesses. This concerns (among others) criteria, measuring methods, business models, transition management, and societal expectations for sustainable chemical products, processes, and businesses. The SCKC of FISCH will cluster and align all existing efforts and know-how in Flanders to leverage R&D and to disseminate the results to the diverse stakeholders, so that they can be used in their businesses, research, education, and consulting activities. An SME helpdesk (named *SUSCHEManswers*) acts as a front office for all questions that SMEs might have with regard to all their sustainability issues: sustainable innovation (including cooperative projects, IPR issues, access to risk capital), sustainability measuring methods, transition methods, and societal expectations, and so on.

The SCKC will be organized around the following key themes: (i) metrics and indicators for sustainable development, (ii) transition to sustainable development, and (iii) meeting societal expectations. The main activities/responsibilities of the SCKC will be to perform the following:

- Provide the FISCH platform with advice on sustainability
- Develop, disseminate, and valorize new know-how on sustainable chemistry through nontechnological research projects
- Manage, optimize, standardize, and implement the use of metrics and indicators for sustainable development in the chemical value chains
- Follow up and advocate existing and emerging product safety laws (Registration, Evaluation, and Authorization of Chemicals (REACH), CLP, IPPC, etc.)
- Define and follow up on voluntary industry agreements (Responsible Care®, Environmental Policy Agreements, covenants, codes of conduct, codes of good practices, sustainability reports, etc.)
- Define, develop, and manage the transition to sustainable development
- Define, promote, and implement alternative business models, organizational models, and all other activities (production, transportation, services, recycling, etc.) of the chemical and chemistry-using organizations.

FISCH Pontoon   Basically, the role of the FISCH pontoon is internally oriented (coordination, and enabling and supporting of the SIA, OIC, and SCKC) as well as externally oriented (single point of contact, communicating, informing, advising, promoting, and networking). More specifically, the FISCH pontoon will commit to the following roles and responsibilities: (i) general management of the FISCH platform, (ii) PR and communication, stakeholder relations, public acceptance, and networking, (iii) organization of an SME helpdesk, (iv) supporting training and education, and (v) managing FISCH consortia and their associated projects.

15.2.2.2 FISCH in a Flemish and European Context

In the past couple of years, Flemish research centers, academia, and their associations, as well as the Flemish chemical industry have launched some smaller initiatives that (partially) meet in one way or another the objectives of sustainable chemistry. Most of these initiatives, such as projects on alternative energy, cleantech, smart materials, biomass conversion, and so on, are geared toward some smaller aspects of sustainable chemistry. However, none of these initiatives start from a holistic view on sustainable chemistry, and hence miss the power to exploit and support the potential of sustainable chemistry to its fullest extent. To really influence the mind-set of the different stakeholders (which are quite diverse in nature) and to create new opportunities for R&D, investments, and businesses, the powerful setup offered by FISCH is required. For an efficient use of (skilled) resources, FISCH will be structurally embedded in the Flemish R&D landscape, thereby consolidating the present Flemish initiatives and using as much as possible the existing financial resources and fiscal incentives. This will, at the same time, create the necessary critical mass and give a clear view on the efforts, costs, and results, the latter being more accessible to a wide range of stakeholders.

In the past, Flanders did not have a specific funding program for chemistry, nor did it have a dedicated platform on sustainable chemistry. The European Technology Platform "SusChem," initiated jointly in mid-2004 by the industry associations Cefic and EuropaBio to help and foster European research in chemistry, chemical engineering, and industrial biotechnology, aims to establish a national platform in each of its EU member countries. The purpose of these SusChem national platforms is to be a focal point for chemical and biotechnology R&D&I in each country. They also provide a single voice to promote relevant chemical ideas to the federal agencies and to facilitate cooperation. These national platforms also facilitate communication and strengthen the role of industry in R&D policy, by defining strategic areas in the field of chemistry and industrial biotechnology to be included in the current R&D&I plans at the national and regional levels. In addition, it is also envisioned that these platforms can establish better cooperation between different funding organizations at the national level to optimize the budget spending (Chynoweth, 2008).

So far, 13 national platforms exist. In September 2010, FISCH was officially accepted by SusChem to act as their Flemish/Belgian platform. The FISCH pontoon will host the "SusChem-Flanders" platform, which will provide FISCH with international credibility. The vision, the SIA, and the business plan developed within the FISCH study will be adopted for the "SusChem-Flanders" platform. The Flemish platform envisions defining the regional strategies needed to address societal needs, boosting the competitiveness of the Flemish chemical and biotechnology sectors, exploiting research competences and skills in transnational research programs and stimulating future regional (and national) R&D programs to be aligned with those at the EU level. Because the platform will work very closely in the cluster of regional SusChem platforms, one of the main goals of SusChem-Flanders is to build a network for sustainable chemistry research across Europe.

15.2.2.3 Added Value of "FISCH" and Spillover Effects

To estimate the impact of FISCH, a macroeconomic estimation exercise was performed using a "business as usual (BAU)" scenario (without the implementation

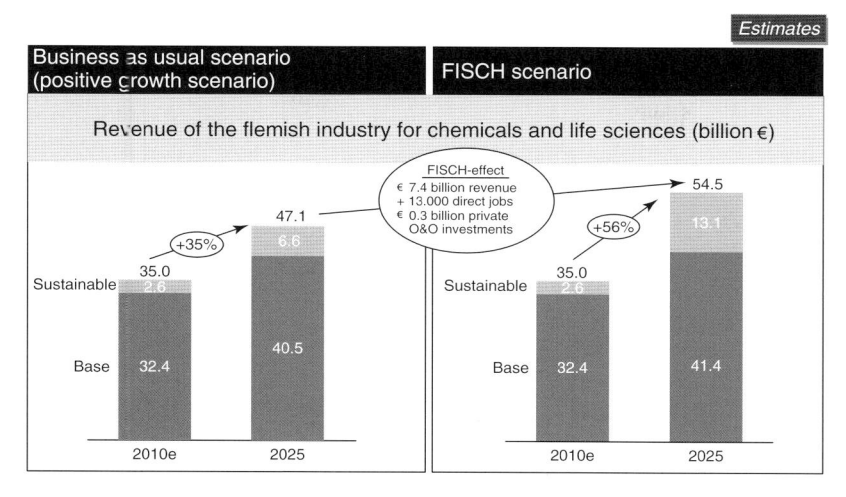

Figure 15.4 Estimated economic impact of FISCH in Flanders. Executed by essenscia and Arthur D. Little.

of the FISCH platform) and comparing it to the "FISCH" scenario (that takes into account that sustainable chemistry will be implemented in Flanders through FISCH). The turnover in the FISCH scenario is estimated to be €54.5 billion in 2025. This implies an added value of €7.4 billion as compared to the BAU scenario (Figure 15.4). In addition, the exercise forecasts an additional 13 000 jobs and an additional €0.3 billion of R&D spending in the case of the FISCH scenario.

Apart from the economic value added, FISCH will also have positive spillover effects: it will induce more open innovation, increase discussion with stakeholders, induce new and growing innovative SMEs and services, anchor R&D of international companies in Flanders, cluster state-of-the-art innovation infrastructure/equipment and create new business models that can be applied in other industrial sectors as well, and it will make the chemical and chemistry-using industries more attractive to researchers and youngsters. More Flemish products, processes and knowledge (business models and patents) will find their way to the strong growth markets in the world. It is also envisioned that by applying Flemish chemistry innovations, the EU climate goals and the Lisbon innovation targets (1% of the GDP contributed by the public and 2% contributed by the private sector) will be reached much faster in Flanders.

15.2.3
Putting It All into Practice: Implementing "FISCH"

A series of recommendations, written down in the FISCH business plan, have regularly been brought to the attention of the Flemish Government and to the ministers concerned, and were thoroughly discussed. In addition, FISCH was presented on several occasions by essenscia flanders to the Flemish Government as being their transition strategy to sustainable development for the chemical industry. In addition, in a separate "IAP" (written in the form of an "Acceptance File") it was also explained in more detail how exactly the SIA, OIC, and SCKC (as defined in the business plan) were to be implemented and described what FISCH and its stakeholders need to do to facilitate this process, and how to organize the different activities and actions. All these activities ultimately led to the acceptance of FISCH by the Flemish Government as being the Flemish (sub) innovation hub for sustainable chemistry. At the end of 2011, the FISCH initiative (shaped as a "competence pool") was awarded a government grant of approximately €5 million per year for a period of (at least) four years. Through this "competence pool", FISCH aims to accelerate the creation of new value chains in Flanders and develop the necessary skills/competences to achieve this. The "FISCH competence pool", organized as a separate legal entity, officially started its activities in March 2012. Although next steps will need to unfold to further shape the Flemish innovation hub for sustainable chemistry, putting in place and launching the "FISCH competence pool" has been the first (and a major) step in the right direction.

15.3
Concluding Remarks and Lessons Learned

Developing the FISCH initiative proved a process of learning by doing. Various obstacles and hurdles had to be overcome. In doing so, some lessons were learned:

- It takes (the persuasion of) individuals (the so-called believers or die-hards), who are willing to work hard to get support for the initiative, who are able to transcend their own organizational barriers, and who are willing to work together toward a common goal and toward the realization of a sector transition. These individuals also build trust and give the initiative a human face.
- It is of the utmost importance to bind the actors around a single vision. However, this is not as straightforward as it seems, as innovation is a complex process, involving many players. It is influenced by the interaction between science, industrial activities, and policy, and closely connected to societal trends. Innovation is also more than just the development of new knowledge and new technologies. Sustainable development in chemistry requires that the socioeconomic and regulatory drivers and barriers that determine the speed and extent of innovation are known to policy makers, industry, and other stakeholders, so that they can act on them with a clear view. This requires that the dialog between academia, industry, authorities, and societal stakeholders is managed in a structured and constructive manner.
- Each actor (industry, government, research organizations, and other societal stakeholders) can act within its own sphere of influence and according to his responsibilities, thereby creating the basis for joint actions that strongly contribute to an overall strategy toward sustainability. Often, it is believed that government must take the initiative for such concerted actions. However, government action alone surely is not enough; although policy can indeed provide the incentives

for the other actors (e.g., the development of human capital and the facilitation of certain kinds of technology development and technology transfer), it is the industry that holds the key to open up the innovation paradox.
- Building consortia to work together during several, subsequent phases in the innovation process is not easy and, therefore, not to be underestimated. As a matter of fact, it is all about building trust and establishing working relationships that last longer than one project to provide the continuity needed at consortium level to build up experience and operational excellence.
- In order to get a sector initiative (similar to FISCH) going, trying to acquire *ad hoc* financing for individual projects (because of a lack of initial government funding for the entire initiative) will ultimately fall short. First, this is due to the fact that such a process is labor intensive and time consuming. Second, there is no option for securing continuity at the consortium level beyond each separate project, which hampers the development of more stable, innovative consortia in the long run.

With sustainable development being one of the most decisive trends of global socioeconomic development, the generation and implementation of innovations leading to cleaner technologies will be of crucial importance for the Flemish chemical and chemistry-using industries in order to maintain or regain their leading worldwide position. We strongly believe that FISCH will provide the momentum needed to increase the amount of innovation so that technology and innovative breakthroughs will become more frequent, and thereby bring about more sustainable development in the Flemish chemical and chemistry-using industries.

Acknowledgments

We thank the more than 400 dedicated individuals (working in more than 200 chemistry-related organizations), who gave input to the ambitious business plan of FISCH for their invaluable contribution!

References

Anastas, P.T. and Warner, J.C. (1998) *Green Chemistry: Theory and Practice*, Oxford University Press, New York.

Cefic (2010) Facts and Figures 2010: The European Chemical Industry in a Worldwide Perspective, http://cms.cefic.be/files/Publications/Facts-and-Figures-12102010-report-rev3.pdf (accessed December 2011).

Chynoweth, E. (2008) A growing band, in *Innovating for a Sustainable Future*, An ICIS Chemical Business Supplement, ICIS Chemical Business, Surrey.

Eder, P. and Sotoudeh, M. (2000) Innovation and Cleaner Technologies as a Key to Sustainable Development: the Case of the Chemical Industry, Institute for Prospective Technological Studies Seville (IPTS), Sevilla, ftp://ftp.jrc.es/pub/EURdoc/eur19055en.pdf (accessed December 2011).

essenscia (2008) Facts & Figures–The Chemical and Life Sciences Industry in Belgium, http://www.essenscia.be/01/MyDocuments/essenscia_F&F_2008_HR.pdf (accessed December 2011).

essenscia (2010) Chemie, Kunststoffen en Life Sciences in Vlaanderen–Kerncijfers 2010 (in Dutch), http://www.essenscia.be/01/MyDocuments/Kerncijfers_vlaanderen_2010.pdf (accessed December 2011).

Kim, W.C. and Mauborgne, R. (2005) *Blue Ocean Strategy: How to Create Uncontested Market Space and Make the Competition Irrelevant*, Harvard Business School Publishing, Boston, MA.

SusChem (2005) The Vision for 2025 and Beyond: A European Technology Platform for SustainableChemistry, European Technology Platform for Sustainable Chemistry (SusChem), Brussels.

SusChem (2006) Innovating for a Better Future: Putting Sustainable Chemistry into Action – Implementation Action Plan 2006, European Technology Platform for Sustainable Chemistry (SusChem), Brussels.

Verbeeck, A., Debackere, K., and Wouters, R. (2004) De Chemische Industrie in Vlaanderen – Op weg Naar 2010, Vlaamse Raad voor Wetenschapsbeleid (VRWB), Brussels (in Dutch).

16
The Transition to a Bio-Based Chemical Industry: Transition Management from a Geographical Point of View
Nele D'Haese

16.1
Introduction

In 2010, the European chemicals industry posted sales of €578 billion, accounting for one-fourth of the world chemical sales in value terms and making the sector one of the most important players worldwide (Cefic, 2012). The future, however, looks less bright as consensus is growing that this frontrunner position is at risk: a number of developments, for example, the construction of vast chemical complexes in the Middle East, the increasing scarcity of fossil resources and the low economic growth in Europe relative to the increase in consumption in China and other emerging economies, threaten the competitiveness of the European chemicals sector (Cefic, 2004; SusChem, 2005; European Commission, 2009; KPMG International, 2010).

In response to these threats, policy makers and sector organizations almost unanimously say that the European chemicals sector should become more sustainable, hoping that by doing so its current frontrunner position can be secured. They point at emerging new technologies, for example, nanotechnology and industrial biotechnology, as innovation areas critical to achieve this desired future state. In addition, the use of renewable energy sources, a switch to bio-based feedstocks, improved logistical systems, and process intensification (the design of synthesis routes with less environmental impact) are also often mentioned as pathways leading to a more sustainable European chemicals industry (Cefic, 2004; SusChem, 2005; European Commission, 2009; KPMG International, 2010). However, what remains rather vague in the reports and action plans referred to here, is how "sustainability" should be understood exactly in this context. The question "What is precisely meant by a sustainable European chemical industry?" remains unanswered.

Although one would think a shared and unambiguous goal could be helpful to revitalize the European chemicals industry, the lack of a clear answer to this question may not be so remarkable. Sustainable development, following the well-known Brundtland interpretation (UN, 1987), is defined as development that meets the needs of the present without compromising the ability of future generations to meet their own needs. It is, therefore, a normative, subjective, and ambiguous concept (Matthews *et al.*, 1997; Rotmans, 2002). Bringing this concept into practice implies making trade-offs between different values and beliefs. As a result, sustainable development should be regarded as a continuous process in which each time new societal challenges arise these values and interests are discussed and negotiated anew (Matthews *et al.*, 1997; Meadowcroft, 1997). For this reason, the concept is more process than outcome oriented and a clear-cut long-term industrial development goal cannot be defined.

Over the past decade, this insight was translated into several new management approaches that are more adapted to the complex nature of processes of sustainable development than conventional goal-oriented forms of planning and policy. One of these new approaches is transition management, developed in the Netherlands and already implemented by the Dutch ministry of economic affairs to initiate the transition to a more sustainable energy supply. In contrast to traditional policies, transition management is characterized by a systemic and evolutionary perspective, is explicitly oriented toward sustainability and has a strong focus on innovation (Loorbach, 2007; Rotmans and Loorbach, 2010).

An important role in the transition management framework is assigned to transition experiments, which are understood as innovation projects aimed at exploring radically new ways to meet societal needs. Transition experiments have four characteristic features that separate them from traditional industrial innovation: (i) they take a societal challenge as the starting point; (ii) they are conducted in a real-life societal context; (iii) they focus on social learning, not on a preset goal or endpoint; and (iv) they are conducted by a multiactor alliance involving industry, governments, knowledge institutes, and other stakeholders (van den Bosch, 2010).

In this chapter we scrutinize innovation projects set up to accelerate and amplify the transition to a bio-based chemicals industry in four port regions in the Rhine–Scheldt Delta, that is, Antwerp, Ghent, Rotterdam, and Terneuzen, and draw lessons from this to improve the management of experiments aimed at strengthening this transition. The Rhine–Scheldt delta is argued to be the main gateway to the Northwest European continent, accommodating the two biggest ports of Europe, that is, Rotterdam and Antwerp. Both these ports house a vast petrochemicals complex, of which the importance can be illustrated with the following figures: 27% of Europe's refining capacity is located in the petrochemicals cluster of Rotterdam (Ministerie van Economische Zaken, 2009). In addition to this, the port of Rotterdam has a market share of more than 50% for crude oil and derivates in the Hamburg–Le Havre range (Havenbedrijf Rotterdam, 2010). The petrochemicals cluster situated in Antwerp, on the other hand, is said to be the largest integrated cluster in the world after Houston, Texas (Vandecasteele, Van Caneghem, and Block, 2007). The distance by road between these two large port areas and the ports of Ghent and Terneuzen is 150 km at most. Terneuzen and Ghent accommodate some petrochemicals companies as well. But these companies are far less in number compared to the number of companies situated in Antwerp and Rotterdam and include less producers of bulk chemicals. Overall, this makes the Rhine–Scheldt delta an interesting case as there is the possibility of comparing the nature and success of innovation projects aimed at accelerating the use of bio-based feedstocks in port regions, which accommodate different types of petrochemicals complexes, simultaneously as these complexes operate in the same macroeconomic context.

As a consequence, we examine in this chapter, more specifically, the regional factors that affected the success of innovation projects contributing to the transition to a bio-based chemicals industry in the chemicals complexes of Antwerp, Ghent, Rotterdam, and Terneuzen and whether these factors differ in the respective port regions owing to regional characteristics of the vested petrochemicals clusters. An additional aim is to picture the successive stages that are distinguishable in this transition process. Finally, we will also clarify what can be learned from this research for the management of transition experiments aimed at increasing the use of bio-based resources in this kind of port regions.

Analytically, we distinguish in our research between, on the one hand, the use of bio-based resources for unconventional, innovative purposes and, on the other hand, the use of bio-based resources in established production processes, such as oils of animal or vegetable origin used as feedstock in the oleochemical industry. Only this first category is taken into account. The analysis is based on publicly available literature and semistructured interviews representing knowledge institutes, port authorities, governments, the first-generation biofuel industry, and local and regional organizations (see Appendix 16.A).

The next section describes the composition of the petrochemicals clusters in the ports of Antwerp, Ghent, Rotterdam, and Terneuzen. In the section thereafter, we present the different projects that were initiated in these regions to accelerate and amplify the transition to a bio-based chemicals industry, arranged in order of the successive stages we could discern in this transition process. In the final section, we discuss the influence of regional characteristics on the presented innovation projects and draw some lessons from this discussion to improve the management of experiments set up to accelerate the transition to a bio-based chemicals industry.

16.2 Composition of the Chemical Clusters in Antwerp, Ghent, Rotterdam, and Terneuzen

16.2.1 The Rhine–Scheldt Delta

The ports of Antwerp, Ghent, Rotterdam, and Terneuzen are all located in the Rhine–Scheldt Delta, which stretches from the southern part of the Netherlands to the northern part of Belgium. This delta is unique as it is the only place in the world where three rivers, that is, the Rhine, Meuse, and Scheldt, flow together and discharge into the sea. Together, the basins of these rivers cover most of Northwest Europe. Because of this, and because of its central location within Europe, the delta has always been an important gateway to Europe for people coming from overseas (Rijn-Schelde Delta Samenwerkingsorganisatie, 2006).

Originally, the delta consisted of many small islands surrounded by a myriad of waterways. As a result of inpoldering, most of these waterways disappeared over the past centuries. In combination with the Delta works, as a result of which all but one of the remaining tidal inlets were closed after the big flood of 1953, this caused a *de facto* split in the delta: for seagoing vessels it is impossible nowadays to travel via inland waterways between Rotterdam, located in the Rhine–Meuse basin in the north, and Terneuzen, Ghent, and Antwerp, the three ports located along the Scheldt in the south.

Figure 16.1 illustrates the Rhine–Scheldt Delta region.

16.2.2
Past and Present of the Petrochemical Industry in the Ports of Antwerp, Ghent, Rotterdam, and Terneuzen

The Rhine–Scheldt Delta is crossed by a large number of pipelines connecting the petrochemicals companies in the area and supplying them with crude oil, naphtha, ethylene, propylene, and industrial gasses. Apart from this tangible connection, the petrochemicals industry in the delta is also interlinked through numerous commonalities and complementarities. These links are vertical (buying

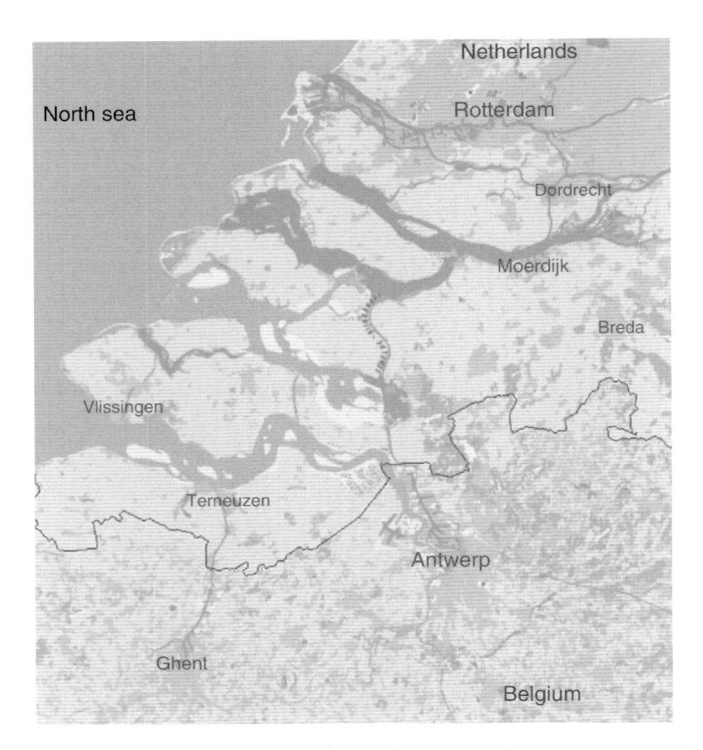

Figure 16.1 Map of Rhine–Scheldt Delta. Source: http://www.mainportdelta.eu/.

and selling chains) as well as horizontal (complementary products and services, the use of similar feedstocks, technologies, etc.). The petrochemicals complexes of Antwerp, Ghent, Rotterdam, and Terneuzen can therefore be considered part of a larger cluster stretching over the whole delta, which is referred to as the world's largest and most diversified petrochemicals cluster (Van den Bulcke, Verbeke, and Wenlong, 2009).

The growth of this cluster goes back half a century, to the years after World War II when port infrastructures were rebuilt with financial support of the Marshall plan. In Belgium, part of these Marshall funds were used to construct a new petroleum port in Antwerp, which became operational in 1951 (Van Hooydonk and Verhoeven, 2007). With these new port facilities, Antwerp tried to take advantage of a new opportunity generated by the switch in the chemicals industry from coal-based resources to petroleum-based feedstocks in the 1940s and 1950s.

Because of this switch, American refineries had been forced to shift their focus back to their home markets and to reduce their export of petroleum derivates to Europe. Producers from the Middle East had taken over the European supply, but did not export refined products, only crude oil. Therefore, new refining capacity was needed in Europe (Stichting Historie der Techniek, 2000). Major ports with associated logistic infrastructure, such as Antwerp and Rotterdam, were ideal locations to build these refineries (Ketels, 2007).

This new refining capacity created a large flow of crude oil to Europe. Although Antwerp's petrochemicals complex had grown substantially in the post-war years, Rotterdam was the only port in Europe at that time capable of handling these large quantities. In 1951, for example, Rotterdam already handled four times more crude than Antwerp (12 versus 3 million tonnes) (De Goey, 2004). This was due to the fact that Rotterdam had been able to take advantage of its technological lead over competing ports: in contrast to Antwerp, which continued to work to a large extent with infrastructure and equipment predating the war, Rotterdam had suffered such destruction that it had to be rebuilt from scratch. Hence, when ultra-large crude carriers were taken into use, Rotterdam was one of the few ports able to accommodate these vessels. The Suez crisis in 1956 further strengthened Rotterdam's leading position as a deep-sea oil port, as did the founding of the European Economic Community (1957) (Van Hooydonk and Verhoeven, 2007).

During the following decades, the port of Rotterdam could further build on and secure this strong market position. In the period 1947–1995, on average 49.4% of its throughput consisted of petroleum and petroleum products (De Goey, 2004). In 2009, this was still 43.6% or 168 608 000 tonnes, which constituted the largest market share in the Hamburg–Le Havre range (Havenbedrijf Rotterdam, 2010).

In contrast to Rotterdam, Antwerp focused less on the import of crude oil and the acquisition of activities directly related to this, such as oil refining. The main reason for that was that over time, the Scheldt had become too shallow for large seagoing tankers. Belgium had been insisting on a further deepening of the Scheldt in view of the growing dimensions of seagoing ships, but could not reach an agreement about this with the Dutch authorities responsible for the part of the river upstream from Antwerp. A solution came in 1971 by means of the Rotterdam Antwerp pipeline (RAPL), which connected the Antwerp refineries with the more easily accessible oil port of Rotterdam. In 2009, this pipeline transported 28 725 344 tonnes of crude oil from Rotterdam to Antwerp, compared to 3 977 717 tonnes that arrived by ship (Gemeentelijk Havenbedrijf Antwerpen, 2010). Hence, most of its maritime traffic in oil the port of Antwerp lost to Rotterdam.

As a consequence, Antwerp's petrochemicals cluster differentiated into a production center for bulk chemicals (De Goey, 2004). From the 1960s onwards, this bulk production demonstrated a very large increase in scale, forming the basis for the highly integrated cluster the petrochemicals complex in Antwerp still is. The main driving factor behind this integration followed from the substitution of batch processes, which were until then the preferred way of production in the chemicals industry, with flow processes. In contrast to batch processes, which are operated intermittently, flow processes made it possible to run a series of chemical reactions in a continuously flowing stream. They can be automated and enable a more precise control of reaction parameters, which in turn allowed the construction of bigger production units (Freeman and Soete, 1997).

These volume increases subsequently led to a reduction in the price of bulk chemicals. Together with the creation of global markets, this made "price" the most distinguishing competitive advantage among producers of bulk chemicals. As a result, companies continually tried to improve their process efficiency. A way of doing this was by also selling side products. Another effect of these volume increases was that bulk companies became more eager to sign long-term contracts with their customers, as these could provide them some guarantee of sales volumes (Stichting Historie der Techniek, 2000).

The outcome of both these effects can be seen, among other things, in the high geographical and functional integration of clusters of bulk producers, such as the ones in Antwerp and Rotterdam: companies built production plants in close proximity to each other and connected them through a dense network of pipelines, resulting in a chemical cluster where production processes are highly interdependent.

The geographical pattern of bulk streams throughout Europe that emerged as a result of these developments in the 1960s and 1970s almost did not change over the past decades. This is due to the capital-intensive character of the installations for the production of bulk chemicals: new facilities and radical technical changes in existing facilities call for investments with long pay-off times (De Goey, 2004; Ketels, 2007). The choice of Rotterdam to focus on the throughput of crude oil and derivates can therefore be seen in the composition of its petrochemicals complex, which is still mainly oriented toward oil refining (common capacity of approximately 72 million tonnes per year) and the production of bulk chemicals such as aromatics and olefins (De Goey, 2004). As can be seen in Table 16.1, Antwerp has a cluster profile similar to Rotterdam, but is smaller in terms of production companies located in the port region and has less refining capacity (approximately 38 tonnes per year).

16.2 Composition of the Chemical Clusters in Antwerp, Ghent, Rotterdam, and Terneuzen

Table 16.1 Cluster profiles.

Port	Size of petrochemical cluster	Refineries	Producers of bulk chemicals	Producers of fine chemicals	Parachemical companies	Logistical companies	Utility providers
Antwerp	57	4 (6.3%)	18 (28.6%)	10 (15.9%)	3 (4.8%)	33 (52.4%)	7 (11.1%)
Ghent	28	0	3 (10.7%)	7 (25%)	5 (17.9%)	13 (46.4%)	2 (7.1%)
Rotterdam	61	5 (6.2%)	21 (25.9%)	13 (16%)	7 (8.6%)	34 (42%)	11 (13.6%)
Terneuzen	13	1 (7.7%)	3 (23.1%)	1 (7.7%)	1 (7.7%)	7 (53.8%)	3 (23.1%)

The figures in each column show the number and percentage of cluster companies that can be classified as a refinery, a bulk petrochemical company, a fine petrochemical company, a parachemical company, a logistical company, or a utility provider (for criteria used to perform this classification see Polders and Huybrechts, 2009). A company can appear in more than one category.

This table also clearly shows that the petrochemicals clusters in the ports of Ghent and Terneuzen are rather small compared to the ones in Rotterdam and Antwerp. The cluster in Terneuzen accommodates only two production companies and one company specialized in custom processing. One of these production companies, Dow Chemical Company, has a refinery as well as production plants for bulk and fine chemicals in Terneuzen, as a result of which the company is responsible for the majority of the activities in the cluster.

The petrochemicals cluster in Ghent, on the other hand, consists of more companies than the cluster in Terneuzen, but is still much smaller than the clusters in Rotterdam and Antwerp. Just as in Terneuzen, the cluster in Ghent is not integrated: none of the companies located in the port of Ghent uses an end or side product of another chemical company as feedstock in petrochemical processes (Van Dyck, 2008). In addition to this, it is also noteworthy that the share of bulk production is relatively small in Ghent.

16.3
Regional Innovation Projects to Strengthen the Transition to a Bio-Based Chemical Industry

In the port regions of Antwerp, Ghent, Rotterdam, and Terneuzen, several innovation projects were set up over the past decade aimed at increasing the use of bio-based feedstocks for the production of chemicals. In this section, we present the projects that were still running at the end of our study period (end of 2009). The focus hereby lies on regional characteristics that facilitated or inhibited project development.

Interviews with the leaders of these projects revealed that the arguments used to illustrate the projects' societal value differed from project to project. Furthermore, the cited arguments appeared to have clear links with the evolution in common knowledge on the use of biomass feedstocks. On the basis of this observation and the specific nature of the projects, we were able to discern three broad steps in the transition process toward a more bio-based chemicals industry unfolding in the port regions of Antwerp, Ghent, Rotterdam, and Terneuzen. These three steps are used here to structure the presentation of the innovation projects.

16.3.1
First Step: Substitution of Fossil Resources by Bio-Based Feedstocks Making Use of Vested Technologies

The most important driver for this first step was the approval of the so-called European biofuel directive[1] in 2003, which obliged member states to ensure that a minimum proportion of biofuels and other renewable fuels is placed on their

---

1) Directive 2003/30/EC of the European parliament and of the council of 8 May 2003 on the promotion of the use of biofuels or other renewable fuels for transport.

markets. This obligation, which provided a legal assurance that a trading market for renewable fuels would be created, made investors more eager to invest in infrastructures for the production of these renewable fuels.

Although the directive did not exclusively apply to first-generation biofuels, that is, biofuels made from vegetable oils, wheat, and other agricultural products originally intended for food or feed, its most visible outcome in the Rhine–Scheldt Delta was the emergence of a first-generation biofuel industry. (In Table 16.2, an overview is given of all first-generation production plants that were operational or under construction in the four investigated port regions by the end of 2009.)

Apart from the production of first-generation biofuels, one other project initiated in the investigated regions also met the above-mentioned criteria of the first step in the transition process: BASF Antwerp substituted fossil resources with bioethanol and Ricinus oil as bio-based feedstocks in the production processes of ethylamines and polyestherol, respectively (Vandecasteele, Van Caneghem, and Block, 2007).

The stakeholders interviewed for this study used three main arguments to explain the societal relevance of these industrial projects. Two of these arguments are similar to the ones cited by the European authorities in the text of the biofuel directive to legitimize the decision to support renewable fuels, namely, the negative impact of fossil fuels on global warming and the drawbacks of Europe's increasing dependency on oil and gas imports. Biofuels were seen as a good equivalent for fossil motor fuels because vehicles currently in circulation are capable of using a low biofuel blend and fuels from bio-based feedstocks were thought to decrease greenhouse gas concentrations in the atmosphere. In addition to this, a majority of these stakeholders also argued that an increased use of bio-based resources would strengthen the European agricultural sector.

During these interviews, we also asked the representatives of biofuel companies about the motives of their company to choose the ports of Rotterdam or Ghent as a preferred location. In the case of Rotterdam, these motives referred, first of all, to the closeness of the chemicals cluster, which includes the clients of first-generation biofuel producers, namely, petrochemical companies such as Shell or Total that blend biofuels into petrol and diesel supplies. These clients can be served more easily and profitably by pipeline because of this geographical proximity. The second reason they gave was that the port facilities in Rotterdam allowed for direct unloading of large bulk quantities of grains, vegetable oils, and other bio-based resources. As a result, the transport costs per unit are relatively low for companies located in the port of Rotterdam.

The first-generation biofuel producers who opted for Ghent, on the other hand, saw an opportunity in the specialization of this port in the transshipment of agricultural products. (In 2009, for example, 22.1% of the throughput consisted of agricultural products and derivates (Vlaamse Havencommissie, 2010).) Ghent also is a cheaper port to establish and operate a new production plant than big ports such as Antwerp and Rotterdam.

Table 16.2 First-generation biofuel companies.

Port	Company name	Type of biofuel produced	Main feedstocks	Production capacity (tonnes yr^{-1})	Year production started
Ghent	Oleon	Biodiesel	Palm and rapeseed oil	100 000	2007
	Alco Bio Fuel	Bioethanol	Wheat	120 000	2008
	Bioro	Biodiesel	Soy and rapeseed oil	250 000	2008
Rotterdam	Abengoa Bioenergy	Bioethanol	Wheat and maize	400 000	2010
	Neste Oil	Biodiesel	Palm oil	800 000	2010
	Biopetrol Industries	Biodiesel	Rapeseed oil and palm oil	400 000	2010
	Dutch Biodiesel	Biodiesel	Rapeseed oil	250 000	2009
	Lyondell Chemie	Bio-ETBE	Bioethanol	650 000	2007

16.3.2
Second Step: Development of a New Technological Paradigm for the Production of Second-Generation Bio-Based Products

The take-off of the second step in the transition toward a more bio-based chemicals industry was triggered by the global food crisis that erupted in the second half of 2007. Food prices had been increasing considerably during the years preceding, leading to malnutrition and food riots around the world. By the middle of 2008, the prices for wheat and maize, for example, were tripled compared to January 2005 (FAO, 2009).

This food crisis fueled the "food versus fuel" discussion, in which first-generation producers were accused of endangering the global food supply (IATP, 2011). Together with other unfavorable announcements penetrating the public debate at that time, such as the message that the beneficial effects of most first-generation biofuels on global warming are much less than once thought (see, for instance, Langeveld and van de Ven, 2010; Fargione et al., 2008), this resulted in a call for alternative products.

Fuels and chemicals made from lignocellulosic biomass, which are called *second-generation bioproducts* here, are such an alternative. They are produced from the inedible parts of plants, such as leaves, straw, or short-rotation coppice wood, and therefore do not put pressure on the global food supply. Two organizations in the regions we investigated, that is, Ghent Bio-Energy Valley, a nonprofit organization operational in the port of Ghent that is governed by the University of Ghent, the Port of Ghent, the City of Ghent, and the development agency of the province of East-Flanders, and Nedalco,[2] a company having an ethanol plant in the port of Terneuzen, are involved in biotechnological research on this kind of second-generation biofuels. At the time of the "food versus fuel" discussion, both organizations announced plans to build a test facility that could simulate industrial fuel production.

Ultimately, Ghent Bio-Energy Valley established a multipurpose pilot plant in the port of Ghent in which agricultural waste products can be converted by means of second-generation industrial biotechnology into biofuels, bioplastics, and other bioproducts (fully operational since 2011). The organization chose to build this pilot plant in the direct vicinity of the production facilities of the first-generation biofuel producers Bioro and Alco Bio Fuel and the bio-based industrial cluster they are part of.

The realization of this cluster, consisting – apart from these two biofuel producers – of an energy company (Electrabel), two storage and warehousing companies storing raw materials and end products (Euro-Silo and Oiltanking), and a crusher of oilseeds and refiner of vegetable oils (Cargill), is the outcome of an intense negotiation process between representatives of Ghent Bio-Energy Valley, the port of Ghent, and the companies mentioned here. Owing to the fact that the negotiations in which all these parties were involved (for the pilot plant on

---

[2] Since 2011 Nedalco is an integral part of Cargill's starches and sweeteners business in Europe.

the one hand and the biofuel companies on the other hand) partly overlapped in time, these could be tuned toward a common goal, namely, the development of a fully integrated bio-based cluster. To facilitate achieving this objective, the port authorities agreed to reserve the remaining free industrial space in the vicinity of Cargill, Euro-Silo and Oiltanking for Bioro, Alco Bio Fuel, and other industrial initiatives that would further strengthen this bio-based cluster. They also granted certain advantages to this kind of initiatives in order to reduce their locational costs and increase the attractiveness of the port. Ghent Bio-Energy Valley, for example, received the site where the pilot plant is established for free.

These advantageous conditions certainly influenced Ghent Bio-Energy Valley in choosing a location in the port of Ghent. Another regional factor that affected their decision was the geographical proximity of the university's laboratories for microbiological and biotechnological research. A final decisive factor was that the pilot plant was financed through European Interreg-subsidy requested by Bio Base Europe, a partnership between Ghent Bio-Energy Valley and BioPark Terneuzen, a sister organization at the other side of the border, that seeks to join forces to develop the region between Ghent and Terneuzen into a major industrial hub for the European bio-based economy.

Nedalco, on the other hand, developed, in cooperation with Delft University and Bird Engineering, a yeast that can convert xylose into bioethanol. The company planned to establish a demonstration plant next to its ethanol plant in Terneuzen. However, Nedalco could not find the necessary funding in the Netherlands in order to realize this plan. Instead, they licensed their technology to Mascoma Company, a market leader in the United States in the field of lignocellulosic ethanol commercialization. Mascoma agreed to further develop the technology in their pilot plant at Dartmouth Regional Technology Centre, close to Dartmouth College where the founders of Mascoma work on research into ethanol engineering.

16.3.3
Third Step: Closing Material Loops

The third step in the transition toward a more bio-based economy consists of organizing closed loops where bio-based products are reborn through reclamation and recycling. This step originates from the increasing awareness about the limits of bio-based resources and the detrimental effect of environmental pollution caused by all kinds of nondegradable chemical substances. Although we found only one concrete project in the port regions that matched the criterion of this third step, we see this step as an obvious one in view of the preceding two.

This one project is, the *F*landers strategic *I*nitiative for *S*ustainable *CH*emistry (FISCH), which has been described in Chapter 15. One of the objectives of this initiative is to increase the use of bio-based resources in the Flemish chemical industry in order to close carbon cycles. Unfortunately, this initiative was still in a rather immature state by the end of 2009. It had already been elaborated conceptually, but no concrete actions had been set up. As a consequence, this third step will be left out of consideration in the following concluding section.

16.4 Conclusions

An overall conclusion we draw from this research is that regional characteristics do affect the outcome of innovative projects that are meant to strengthen the transition to a bio-based chemicals industry. In the cases examined here, namely, the port regions of Antwerp, Ghent, Rotterdam, and Terneuzen, some of these regional characteristics are clearly related to the vested chemicals industry in these ports. We therefore advise innovators who have plans to set up transition experiments aimed at strengthening the transition toward a bio-based chemical industry in these port regions to make the "right" locational choice. What "right" means in this context, will depend on whether the characteristic features of a region are in favor of the experiment's goals.

We reached this conclusion through a thorough comparison of the regional circumstances under which the projects presented in the previous section became successful (see Table 16.3 for an overview of the main regional characteristics affecting the development of an innovative bio-based industry in the four ports). We looked for similarities, but also tried to explain how different developments occur in sometimes seemingly same regional circumstances. The latter can be illustrated, for instance, by the locational choice of first-generation biofuel producers, which stand for the majority of the innovative projects of the first step in the transition toward a bio-based chemical industry we outlined earlier.

First-generation biofuel companies were able to start up a viable business in the ports of Ghent and Rotterdam. In the case of Rotterdam, this success can be explained by the similarities between the economic regimes where first-generation

Table 16.3 Overview of the main regional characteristics affecting the development of an innovative bio-based industry in the ports of Antwerp, Ghent, Rotterdam, and Terneuzen.

	Port of Antwerp	Port of Ghent	Port of Rotterdam	Port of Terneuzen
Locational costs	High	Lower	High	Lower
Strong supportive network	No	Yes (Ghent Bio-Energy Valley)	No	Yes (BioPark Terneuzen)
Large, integrated petrochemical cluster	Yes	No	Yes	No
Prevalence of chemical bulk production	Yes	No	Yes	No
Port infrastructure allows for easy handling of bulk quantities of agricultural commodities	Yes	Yes	Yes	No
Region with research on industrial biotechnology	No	Yes	No	Yes

biofuel producers and bulk petrochemicals companies are operating under and the industrial and port infrastructures in which these are physically translated. Both industries produce commodities that are traded by chemical name and compete on price. The cheaper their products are, the more profitable their business is. Therefore, these companies built large production plants, where they enjoy substantial economies of scale, in port regions where these plants can be supplied by large bulk carriers. Other cost reductions are obtained from, for example, cheap pipeline transport and selling side products. As a result, the port of Rotterdam, accommodating a large integrated petrochemical cluster, including the customers of first-generation biofuel producers, was a logical locational choice for these biofuel producers.

As illustrated in Section 16.2, Antwerp's petrochemicals cluster has a profile similar to the cluster in Rotterdam. However, by the end of 2009 not a single first-generation biofuel producer operated in Antwerp, which raised the question why there was no first-generation biofuel industry developing in this port as well. The answer appears quite straightforward at first sight: In Belgium, the production of biodiesel and bioethanol is regulated by a quota system, which prescribes that only biofuels subject to this system are exempted from taxes, and can therefore compete with fossil fuels on the Belgian market. None of the producers located in Antwerp received such a quota.

However, an underlying reason, presumably, is that the network in Antwerp lobbying for the production quota was not powerful enough. Actors related to Ghent Bio-Energy Valley, on the other hand, managed to obtain a production quota for all three biofuel companies located in the port of Ghent. Our evidence suggests that an important element to understand this realization is the critical mass and expertise Ghent Bio-Energy Valley has built up over time: the major institutions in the region responsible for industrial development (the regional development agency, the university, the city, and the port of Ghent) lead the organization and serve the interests of its members, that is, all bio-based companies in the region.

Thanks to this quota, the first-generation biofuel industry in Ghent currently operates in a protected national market. Other biofuels, such as the ones produced in Rotterdam, cannot be sold in Belgium. Unfortunately, this means that, based on our data, it is impossible at present to conclude about the value of other regional characteristics, such as the strong position of the port of Ghent in agricultural trade, for Ghent's first-generation biofuel industry.

First-generation biofuel producers located in the port of Terneuzen, finally, could benefit from neither the competitive advantages of a port such as Rotterdam nor from the benefits of a supportive network such as Ghent Bio-Energy Valley. Roosendaal Energy, the only producer of first-generation biofuels in Terneuzen, did not survive the turmoil during the "food versus fuel" crisis.

In relation to the second transition step, our analysis shows that the regional characteristics that mattered in explaining the differences in industrial development in step one, lost their relevance. Instead, three totally different regional characteristics were found to be important. First, this is due to the presence of a university in the region carrying out research on industrial biotechnology.

This finding followed from the observation that in both the Terneuzen (Nedalco) and the Ghent case upscaling of second-generation biotechnology takes place in pilot plants located in the direct vicinity of a university. As has been demonstrated in geographical literature, this spatial proximity between facilities for fundamental and applied research facilitates biotechnological innovation, as problems encountered on a pilot scale often urge scientists to start up new tests at the laboratory scale (see, for instance, Murray, 2002). Moreover, universities also play an important role in the creation of biotechnological spin-offs (Zucker, Darby, and Armstrong, 1998; McMillan, Narin, and Deeds, 2000). The absence of a biotechnological research institute can therefore probably partly explain the lack of second-generation biotechnological projects in the ports of Antwerp and Rotterdam. Nedalco, on the other hand, is located in the vicinity of Ghent and could have used the Bio Base Europe pilot plant, but decided instead to license its technology to Mascoma because of competitive reasons.

A second regional characteristic that helps to explain this absence of biotechnological innovative projects in Antwerp and Rotterdam is the innovation-adverse climate in both ports. Because of their strategically favorable position, there is a high demand for industrial areas in these port regions, which makes them relatively expensive. Consequently, most vested petrochemicals companies in Antwerp and Rotterdam are production units of multinational companies with R&D departments located elsewhere. As a result, the innovation rate in the ports is relatively low. Bioport Rotterdam, an initiative in support of bio-based industrial development, for example, tried to set up several demonstration projects on innovative technologies. However, it failed in its efforts because none of the companies in the port wanted to participate. It has also been demonstrated that the innovation rate in Rotterdam's petrochemical industry is very low (Nijdam and de Langen, 2006).

These high locational costs, however, do not only hamper innovation by vested companies but they are also a barrier to innovative second-generation enterprises. Because these enterprises face a great deal of risks (they make use of immature technologies, have to operate in an emerging market, etc), they prefer to settle down in a region with low locational costs.

Finally, a third regional characteristic that can help explain the different second-generation bio-based innovative developments in the four investigated ports is the dominance of bulk production processes in the petrochemicals clusters of Antwerp and Rotterdam. As we explained in Section 16.2, this bulk production is characterized by large production volumes, flow processes, and a high degree of integration between production processes. Innovative biotechnological production of second-generation bio-based products, on the other hand, still tends to be in small volumes in batch processes. Furthermore, in biotechnology, most production processes take place in aqueous media and the end- and side products of these processes usually differ from the set of reaction products of the corresponding fossil-based chemical reactions. As a result, fossil bulk production in the highly integrated clusters of Antwerp and Rotterdam cannot easily be substituted by bio-based biotechnological production (Cherubini, 2010; Scott, van Haveren, and Sanders 2010).

This analysis makes clear that both the first and second steps in the transition to a bio-based chemicals industry in the ports of Antwerp, Ghent, Rotterdam, and Terneuzen are given direction by, among other regional factors, the petrochemicals industry located in these ports: regional characteristics related to these fossil chemicals clusters did affect the success of innovative industrial projects aimed at developing a bio-based chemicals industry in the four port regions. Our empirical observations suggest that this happens in two ways: through the "hardware" of these chemical regions, for example, the extensive industrial infrastructure built up over decades in the clusters of Antwerp and Rotterdam, and through their "software," namely, soft structures such as business and innovation strategies, port development plans, and so on. These hard and soft structures induce path dependent development (Urry, 2005). Bio-based innovative projects in line with this development path, such as the production of first-generation biofuels in the petrochemicals cluster of Rotterdam and the substitution of ethanol with bioethanol in existing production processes in Antwerp, are more likely to be successful than more radical innovation (Levinthal, 1998) deviating from this path, for example, the production of second-generation chemicals in Rotterdam. A notable insight gained from this study is that, unless specific supportive measures are taken, it is better to set up experiments aimed at strengthening the transition to a bio-based chemicals industry where regional characteristics are in favor of the experiment's goals.

16.A Appendix

The following actors have been interviewed (in alphabetical order): Becquart Dirk (port authority Ghent), Norbert Denninghof (Neste Oil Netherlands), Diana De Peuter (Monument Chemicals), MaikkiHuurdeman (Van de Bunt consultancy), Chris Jordan (Deltalinqs), MarijkeMahieu (city of Ghent), David Moolenbergh (port authority Terneuzen), HenkMorelissen (Province of Zeeland), MichielNijdam (Erasmus University Rotterdam), Sandra Prenger (port authority Rotterdam), Bart Rosendaal (Rosendaal Energy), WijnandSchonewille (port authority Rotterdam), WimSoetaert (Ghent University), Brigitte Troost (port authority Terneuzen), Jan van der Zande (port authority Rotterdam), Luc Van Ginneken (VITO), Xavier Vanrolleghem (port authority Antwerp), Adhemar van Waes (city of Terneuzen), Linda Verdonck (development agency of the province of Oost-Vlaanderen), RafVerdonck (Oleon), Frank Vereecken (EWI), RienkWiersma (Biopetrol Rotterdam) and Paulus Woets (province of Zeeland).

References

Cefic (2004) Horizon 2015: Perspectives for the European Chemical Industry.

Cefic (2012) Facts and Figures 2011: The European Chemical Industry in a Worldwide Perspective.

Cherubini, F. (2010) The biorefinery concept: using biomass instead of oil for producing energy and chemicals. *Energy Convers. Manage.*, **51**, 1412–1421.

De Goey, F. (2004) *Comparative Port History of Rotterdam and Antwerp (1880–2000): Competition, Cargo and Costs*, Aksant Academic Publishers, Amsterdam.

European Commission (2009) High Level Group on the Competitiveness of the European Chemicals Industry – Final Report.

FAO (2009) The 2007–2008 Food Price Swing: Impact and Policies in Eastern and Southern Africa. FAO Commodities and Trade Technical paper.

Fargione, J., Hill, J., Tilman, D. et al. (2008) Land clearing and the biofuel carbon debt. *Science*, **319**, 1235–1238.

Freeman, C. and Soete, S. (1997) *The Economics of Industrial Innovation*, Pinter, London.

Gemeentelijk Havenbedrijf Antwerpen (2010) Statistisch Jaarboek 2009.

Havenbedrijf Rotterdam (2010) Haven in Cijfers.

IATP (2011) Excessive Speculation in Agriculture Commodities: Selected Writings from 2008–2011.

Ketels, C. (2007) The Role of Clusters in the Chemical Industry. Study of Harvard Business School supported by the European Petrochemical Association.

KPMG International (2010) The Future of the European Chemical Industry.

Langeveld, J. and van de Ven, G. (2010) Principles of plant production, in *The Biobased Economy: Biofuels, Materials and Chemicals in the Post-oil era* (eds J. Langeveld, J. Sanders, and M. Meeusen), Earthscan, London and Washington, DC.

Levinthal, D. (1998) The slow pace of rapid technological change: gradualism and punctuation in technological change. *Ind. Corp. Change*, **7** (2), 217–247.

Loorbach, D. (2007) Transition management: new mode of governance for sustainable development. PhD thesis. Erasmus University, Rotterdam.

Matthews, E., Rotmans, J., Ruffing, K., Waller-Hunter, J., and Zhu, J. (1997) Global Change and Sustainable Development: Critical Trends, Department for Policy Coordination and Sustainable Development, New York.

McMillan, G., Narin, F., and Deeds, D. (2000) An analysis of the critical role of public science in innovation: the case of biotechnology. *Res. Pol.*, **29**, 1–8.

Meadowcroft, J. (1997) Planning for sustainable development: insights from the literatures of political science. *Eur. J. Polit. Res.*, **31**, 427–454.

Ministerie van Economische Zaken (2009) Economische Visie op de Langetermijnontwikkeling van Mainport Rotterdam: op Weg Naar een Mainport Netwerk Nederland.

Murray, F. (2002) Innovation as co-evolution of scientific and technological networks: exploring tissue engineering. *Res. Pol.*, **31**, 1389–1403.

Nijdam, M. and de Langen, P. (2006) *Leader Firms en Innovatie in de Rotterdamse Haven-Industriële Cluster*, Dutch University Press, Amsterdam.

Polders, C. and Huybrechts, D. (2009) Best Beschikbare Technieken Voor Beperking en Behandeling van Afvalwater van de Sector Organische Bulkchemie. Studie uitgevoerd door het Vlaams Kenniscentrum voor Best Beschikbare Technieken (VITO) in opdracht van het Vlaams Gewest.

Rijn-Schelde Delta Samenwerkingsorganisatie (2006) De Delta Atlas: Landschapsportret van de Rijn-Schelde Delta, NPN Drukkers, Breda.

Rotmans, J. (2002) Duurzame ontwikkeling: Al lerende doen en al doende leren, in (eds B. Wijffels, H., Blanken, M., van Stalborgh, and R., Van Raaij), De Kroon op het Werk, NCDO, Amsterdam, pp. 42–51.

Rotmans, J. and Loorbach, D. (2010) Towards a better understanding of transitions and their governance: a systemic and reflexive approach, in *Transitions to Sustainable Development* (eds J. Grin, J. Rotmans, and J. Schot), Routledge, New York and London, pp. 103–220.

Scott, E., van Haveren, J., and Sanders, J. (2010) The production of chemicals in a biobased economy, in *The Biobased Economy: Biofuels, Materials and Chemicals in the Post-oil Era* (eds H. Langeveld, J. Sanders, and M. Meeusen), Earthscan, London and Washington, DC.

Stichting Historie der Techniek (2000) *Techniek in Nederland in de Twintigste Eeuw. Deel II: Delfstoffen, Energie en Chemie*, Walburg Press, Zutphen.

SusChem (2005) The Visionfor 2025 Andbeyond.

UN (1987) Our Common Future: Report of the World Commission on Environment and Development.

Urry, J. (2005) The complexity turn. *Theory Cult. Soc.*, **22** (5), 1–14.

Van den Bosch, S. (2010) Transition experiments: exploring societal changes towards sustainability. PhD thesis. Erasmus University, Rotterdam.

Van den Bulcke, D., Verbeke, A., and Wenlong, Y. (2009) *Handbook on Small Nations in the Global Economy: The Contribution of Multinational Enterprises to National Economic Success*, Edward Elgar Publishing Limited, Cheltenham.

Van Dyck, B. (2008) Reststrome n in de Gentse Kanaalzone: Onderzoek Naar Mogelijkheden Voor Uitwisseling en Valorisatie. Rapport uitgegeven door het Gents Milieufront.

Van Hooydonk, E. and Verhoeven, P. (2007) *The Ports Portable*, Pandora Publishers, Antwerpen.

Vandecasteele, C., Van Caneghem, J., and Block, C. (2007) Cleaner production in the Flemish chemical industry. *Clean Technol. Environ. Policy*, **9** (1), 37–42.

Vlaamse Havencommissie (2010) Jaaroverzicht Vlaamse Havens 2009.

Zucker, L., Darby, M., and Armstrong, J. (1998) Geographically localized knowledge: spillovers or markets? *Econ. Inq.*, **36**, 65–86.

Part VI
Conclusions and Recommendations

17
Conclusions and Recommendations

Genserik L.L. Reniers, Kenneth Sörensen, and Karl Vrancken

Sustainability is high on the agenda of decision makers in the chemical industry across the globe. This is not surprising: the chemical industry is one of the most important industrial sectors in the world, and consumes enormous quantities of natural resources, while producing similar quantities of greenhouse gases, and other unwanted by-products. The chemical sector is also one of the largest employers, and many thousands of people work inside chemical plants, or live near them, creating a constant challenge for the chemical industry to create a safe environment in and around the chemical plant. At the same time, the chemical industry is very much a part of the global economy. Base chemicals being at the very beginning of the production process, the bullwhip effect caused by the global economic crisis has hit the chemical industry hard. Balancing people, planet, and profit has never posed a greater challenge for the chemical industry.

Chemical engineers in universities, research laboratories, and companies around the world are working hard to reduce the environmental footprint of the chemical industry by improving the efficiency and effectiveness of chemical processes. This book has attempted to demonstrate that, notwithstanding the importance of these efforts, sustainability in the chemical industry has more aspects than those that are directly related to the chemical processes used. An effective *management* is equally important to truly move toward a sustainable operation of the chemical industry.

This book, titled "Management Principles of Sustainable Industrial Chemistry," has attempted to provide a comprehensive coverage of all nontechnical aspects of sustainability in the chemical sector. It has provided different theories, concepts, and industrial examples that should help academics as well as practitioners achieve a sustainable chemical industry. The book has six sections, the last of which you are reading now.

The first section dealt with general topics related to sustainability in the chemical industry. Chapter 2 discussed the transition to sustainable chemistry from a policy perspective. Chapter 3 treated the history of sustainability in the chemical industry and identified several drivers that force the chemical industry to act in a more sustainable way. A general overview of sustainable industrial chemistry from a nontechnological viewpoint is found in Chapter 4.

Management Principles of Sustainable Industrial Chemistry: Theories, Concepts and Industrial Examples for Achieving Sustainable Chemical Products and Processes from a Non-Technological Viewpoint, First Edition. Edited by Genserik L.L. Reniers, Kenneth Sörensen, and Karl Vrancken.
© 2013 Wiley-VCH Verlag GmbH & Co. KGaA. Published 2013 by Wiley-VCH Verlag GmbH & Co. KGaA.

In Part II, the aspects of sustainability *within* a single chemical company were treated. In Chapter 5, a sustainability management system framework was developed for corporate social responsibility. Chapter 6 discussed different methods and tools to assess sustainability. Integrated business- and SHESE (Safety, Health, Environment, Security, and Ethics) management systems were discussed in Chapter 7.

Self-evidently, sustainability stretches across company borders. In Part III, sustainability was widened to other companies on the same level in the chemical supply chain, which we have called *horizontal interorganizational sustainability*. Chapter 9 discussed the difficult balance between exploration and exploitation in the relationship between chemical companies. How cluster management can be a driver to improve safety and security in chemical industrial areas was discussed in Chapter 10.

In Part IV, the *vertical* aspects of interorganizational sustainability were treated. How chemical logistics could become more sustainable was discussed in Chapter 11, while the novel concept of Chemical Leasing® was discussed in Chapter 12.

The chemical industry does not work in isolation, which is why Part V discussed (the transition to) sustainable chemistry in a societal context. In Chapter 14, the continuous strive for sustainability in the chemical industry, looked at from a transition viewpoint, is framed and discussed. In Chapter 15, the Flemish chemical industry transition toward sustainability (known under the acronym FISCH – *F*landers strategic *I*nitiative for *S*ustainable *CH*emistry) was discussed. Geographical aspects of transition management in the transition toward a bio-based chemical industry were treated in Chapter 16.

Repeating the conclusions from each of the chapters in this book would be quite impossible in a single chapter such as this one. Nonetheless, some very general conclusions can be drawn from the contributions in this book. The surprising broadness of topics demonstrates irrefutably that there are many more aspects to sustainable chemistry than simply an improvement of the base processes used in the chemical industry. In each subdomain – intraorganizational, horizontal and vertical interorganizational, and societal – the chemical industry needs to study its sustainability and put in place adequate management systems to measure and control the different aspects involved. Although each situation is different, many companies have already gone a long way toward working in a more sustainable way. By following these leaders and adopting their best practices, other companies can achieve similar advantages. Finally, sustainability essentially boils down to finding a balance between people, planet, and profit. Achieving such a balance is a truly daunting task in a global economy that is in worldwide recession. On the other hand, it is a task that is more needed than ever.

To achieve a truly sustainable chemical industry, many more years of hard work are still necessary. One step at a time, the chemical industry in industrialized countries is showing the way, while the rest of the world is fast catching up. The road is long, but interesting. We hope this book will prove good reading along the way.

Index

a
Accenture survey   17
ADR dangerous goods   202–204
advanced Planning and scheduling (APS)   170
aggregation   106
American Chemical Society (ACS)   28
American Institute of Chemical Engineers (AIChE)   28
Amoco Cadiz oil spill   9
analytic hierarchy process (AHP)   61
annuity factor   76
Antwerp, Ghent, Rotterdam, and Terneuzen ports   254
– closing of material loops   258
– fossil resources substitution by bio-based feedstocks using vested technologies   254–257
– new technological paradigm for second-generation bioproducts production   257–258
– petrochemical industry past and present   250–254
assessment methods and tools   55–56
– framework   56–59
– impact indicators and assessment methodologies   59–62
– – economic impact assessment   75–77
– – environmental impact assessment   62–74
– – interpretation   81
– – multidimensional assessment   79–81
– – social impact assessment   77–79
Australian/New Zealand norm AS/NZS 4360 (2004)   96
average yearly cost (AYC)   76

b
BASF   13, 16, 107
Belgian and Flemish chemical and life sciences industry, in global context   233–234
benefit–cost ratio (BCR)   76
Best Available Technique Reference Documents (BREFs)   23–25
best available techniques (BAT) and BREFs   23–24
biofuel companies, first generation   256, 260
BioPark Terneuzen   258
Bioport Rotterdam   261
bioproducts production technological paradigm, second-generation   257–258
bullwhip effect   267
business as usual (BAU)   242, 243
business management system   89
by-product exchange   133–136

c
Canadian Chemical Producers' Association (CCPA)   10
Canadian Integrated Risk Management (IRM) framework   98–100
cause-and-effect diagram   204, 207
chemical leasing (ChL)   6, 39, 164, 181–182, 211–212
– basic principles   182–186
– chemical management services   186–187
– classical leasing   186
– economic, technical, and juridical aspects   191
– – barriers to model   191–193
– – example   191–192
– – legal requirements analysis impacting projects   193
– legal aspects   196–197

Management Principles of Sustainable Industrial Chemistry: Theories, Concepts and Industrial Examples for Achieving Sustainable Chemical Products and Processes from a Non-Technological Viewpoint,
First Edition. Edited by Genserik L.L. Reniers, Kenneth Sörensen, and Karl Vrancken.
© 2013 Wiley-VCH Verlag GmbH & Co. KGaA. Published 2013 by Wiley-VCH Verlag GmbH & Co. KGaA.

chemical leasing (ChL) (*contd.*)
– outsourcing   187
– practical implications   187–189
– – strengths and opportunities for customer   190
– – supplier strengths and opportunities   189–190
– and traditional business models   183
chemical logistics   166–167
– improvement
– – coordinated supply chain management   170–171
– – horizontal collaboration   171–174
– – multimodal, intermodal and co-modal transportation   174–178
– – optimization   167–170
– and transportation   165–166
chemical management services   186–187
Chemical Manufacturers Association   11
chemical warehousing   199
– risk management
– – control and documentation   213
– – hazard identification   200–204
– – risk acceptance   213
– – risk minimization strategies   205–211
– – risk quantification   205–209
– – risk transfer strategies   211–213
circular resource economy   229
class compatibility matrix   206
classical leasing   186
Clean Air Act, Occupational Safety and Health Act   8
Clean Water Act   8
cluster management   36, 147–148
– cross-organizational learning on safety and security   150
– – knowledge transfer   150–151
– – model   152–154
– – Multiplant Council (MPC)   151–152
– discussion   157
– significance   148–150
community advisory panels (CAPs)   11
co-modality   174
competence pool   244
competition law and chemical leasing
– Comprehensive Environmental Response Compensation and Liability Act. See Superfund Act (1980)
Consumer Product Safety Act   8
contracts, importance of   193–194
Convention on Long-Range Transboundary Air Pollution (CLRTAP)   22
Cook Composites and Polymer Company   13

3M Corporation   12
coordinated supply chain management   170–171
corporate social responsibility (CSR)   3, 36, 45–47
– system framework development   47–49
– – knowledge and commitment from workforce (soft factor)   50–51
– – management knowledge and commitment (soft factor)   49
– – operational planning, execution, and monitoring (hard factor)   51–52
– – stakeholder knowledge and commitment (soft sector)   49–50
– – strategic planning (hard factor)   50
corporate sustainability strategies   14–15
cumulative energy demand (CED)   69
cumulative energy requirement analysis (CERA)   69
cumulative exergy consumption (CExC)   70
cumulative exergy extracted from the natural environment (CEENE)   70

d

Deeming loop. See plan do check act (PDCA) cycle
Deming cycle   46
deposits, methods based on   71
distance-to-target approach   61
Dow Chemical Company   13–16
drayage operators   176
DuPont   14–16

e

eco-efficiency   12–13, 107
economic impact assessment   75–76
– indicators   76
– methodologies   76–79
EcoScale   74
ecosphere   57, 58, 68, 70, 81
emergy   71
end-haulage   175
environmental impact assessment   62
– assessment methodologies   72
– – gate-to-gate methodologies   72–74
– – life-cycle methodologies   72
– – methodologies using technology indicators   74
– – shortcut tool kits   74
– emission impact indicators
– – endpoint indicators   62–67
– – midpoint indicators   62
– resource impact indicators   68–69
– – endpoint indicators   71

– – midpoint indicators 69–71
– technology indicators 71–72
environmental movement and regulations 8
European Commission 22, 27
European countries and environmental legislations 8
European Eco-Management and Audit Scheme, (EMAS) 89
exergetic life-cycle assessment (ELCA) 70
exergy consumption and entropy production 70–71
extended exergy accounting (EEA) 70

f

Federation of European Risk Management Associations. See FERMA risk management standards (2003)
FERMA risk management standards (2003) 95–96
Flemish Initiative on Sustainable Chemistry (FISCH) 6, 40, 258
– added value, and spillover effects 242–243
– Belgian and Flemish chemical and life sciences industry in global context 233–234
– feasibility study 236–237
– in Flemish and European context 241–242
– implementation 243–244
– lessons 244–245
– societal chemistry 233
– sustainable development challenge in Flanders 234–235
– vision, mission and setup 237–238
– – open innovation infrastructure cluster (OIC) 240
– – portoon 241
– – strategic innovation agenda 238–240
– – sustainable chemistry knowledge center (SCKC) 241
fossil resources substitution by bio-based feedstocks, using vested technologies 254–256
Framework Directives for Water and Waste 22
functionality 39

g

Gabi methodology 79
gate-to-gate methodologies 72–74
Ghent Bio-Energy Valley 257–258, 260
governance 27–28
Green Alternatives Wizard 74

green chemistry 12, 33, 219
– twelve principles 13
green degree 73–74
– of substance 73–74
– value of production of unit 73
green economy, transition to 27
green engineering 13
greenwashing 45

h

heuristics 168–169
history, of chemical industry 7
– industry response 10
– – corporate sustainability strategies 14–15
– – Responsible Care® program 10–11
– – technology development 12–13
– new issues and regulations 15–16
– recent industry trends 16–17
– rise of public pressure 7–8
– – environmental movement 8
– – public trust 9–10
– and sustainability as opportunity 16
– and sustainability drivers 18
horizontal collaboration 171–174

i

IChemE 107
industrial ecology 13
Industrial Emissions Directive (IED) 22
Industrial Emissions Directive (IED) and voluntary systems 25–26
industrial symbiosis 133–135
– leading to decreased ecological impact 135
– and regional cluster 136–137
– requiring highly developed social network 136
– resourcefulness 137–138
– – Moerdijk industrial park 141–142
– – petrochemical cluster in Rotterdam harbor area 138–139
– – Terneuzen port 139, 141
Institute for Sustainability at the American Institute of Chemical Engineers (AIChE) 16–17
– Sustainability Index™ 16–17
integrated management systems 89–90
– models 95
– – Australian/New Zealand norm AS/NZS 4360(2004) 96
– – Canadian IRM framework 98–100
– – FERMA risk management standards (2003) 95–96
– – ISO 31000 (2009) 97
– obstacles and advantages 92–95

integrated management systems (*contd.*)
– in practice, and characteristics   100–102
– requirements   90–92
Integrated Pollution Prevention and Control (IPPC)   21
– best available techniques (BAT) and BREFs   23–24
– in chemical sector   24–25
– environmental policy for industrial emissions   22
intermodal operators   177
intermodal transportation   174–176
internal rate of return (IRR)   76
International Council of Chemical Associations (ICCA)   11
interorganizational sustainability management
– horizontal   5, 37–38
– vertical   6, 36
intraorganizational sustainability management   5, 34, 36–37
Ishikawa diagram. See cause and-effect diagram
ISO 31000 (2009)   97
iSUSTAIN™   74

k
knowledge transfer   150–151

l
landscape and transitions   28
land use   70
life-cycle assessment (LCA)   70, 72, 106, 163
– and life-cycle design   12
life-cycle costing (LCC)   76, 77
linkages, symbiotic   135–136, 138, 139, 142
lock-in   224, 225, 227, 231
logistics   38–39
long haul   175
Love Canal controversy   9

m
Mascoma Company   258
mass and energy   69–70
material input per service (MIPS)   69
material safety data sheet (MSDS)   200–201
metaheuristics   169
Moerdijk industrial park   141–142
– symbiotic exchanges   142
monetization approach   61
Monte Carlo simulation   126
multicriteria decision analysis (MCDA)   61
multimodal, intermodal and co-modal transportation   174–178

multiobjective optimization   170
multiplant management. See cluster management

n
National Environmental Policy Act (NEPA)   8
Nedalco   257–258, 261
net present value (NPV)   76, 117
network operators   177
niche and transitions   29
nontechnological viewpoint   33–36
– interorganizational sustainability management
– – horizontal   34, 37–38
– – vertical   36, 38–39
– intraorganizational sustainability management   36–37
– sustainable chemistry in societal context   36, 39–40
normalization   61
– of KPI   111
NP hardness   168, 169

o
Occupational Safety & Health Administration (OSHA)   200
open innovation infrastructure cluster (OIC)   240
Operations Research   167
optimization   34, 36–38, 40, 167
– cycle   168
outsourcing   187

p
panel approach   61
payback time   76
plan do check act (PDCA) cycle   91–92
– application to integrated management   92
– basic philosophy of   91
policy perspective   4–5, 21
– from Industrial Emissions Directive (IED) to voluntary systems   25–26
– Integrated Pollution Prevention and Control (IPPC)   21
– – best available techniques (BAT) and BREFs   23–24
– – in chemical sector   24–25
– – environmental policy for industrial emissions   22–25
– sustainability challenges for industry   26–27

– – and drivers for sustainable chemistry 27–28
– – transition management approach 28–30
"Pollution Prevention Pays" program 12
potential impact factor (PIF) 117
potential impact indicators 117
prehaulage 175
primary impact category 114
– structured matrix example 114–117
process, definition of 91
process flow diagrams (PFDs) 107, 108
Procter and Gamble (P&G) 13
product service systems (PSSs) 182
Profit, People, and Planet 45, 55, 59, 90, 225
public trust, problem of 9

q
Q-RES 46
qualified working time (QWT) approach 79
quality-adjusted life years (QALYs) 79

r
regime and transitions 28
regional innovation projects, to strength transition to bio-based chemical industry 254
– closing of material loops 258
– fossil resources substitution by bio-based feedstocks using vested technologies 254–256
– new technological paradigm for second-generation bioproducts production 257–258
Registration, Evaluation, Authorization, and Restriction of Chemicals (REACH) 190, 199
– and chemical leasing 195–196
– Directive 15–16
Resource Conservation and Recovery Act 8
resource extractions, future consequences of 71
resource sustainability 229
Responsible Care® program 4, 10–11
Restriction of Hazardous Substances (RoHS) Directive 15
reverse logistics 164
Rhine-Scheldt delta 248–250
– Antwerp, Ghent, Rotterdam, and Terneuzen ports 254
– – closing of material loops 258
– – fossil resources substitution by bio-based feedstocks using vested technologies 254–256

– – new technological paradigm for second-generation bioproducts production 257–258
– – petrochemical industry past and present 250–254
risk classification matrix 206
– general 208
Rotterdam Antwerp pipeline (RAPL) 252
Rotterdam harbor area 138–141
– industrial symbiosis development characteristics 139

s
Safe Drinking Water Act 8
safer chemical processes, inherent 13
Safety, Heath Environment, Security and Ethics (SHESE). See integrated management systems
service-based chemical supply relationship. See chemical leasing (ChL)
Sevilla Process 21
SHESQ 34, 37, 38
Sigma Guidelines 46
social impact assessment 77
– indicators 78–79
– methodologies 79
social system 55, 57–59
societal context 6, 36, 39–40
steady-state cost (SSC) 76
Superfund Act (1980) 9
supply chain management 199, 210, 212. See also chemical logistics
– coordinated 170–171
supply chain management 199 210, 212. See also chemical logistics
"SusChem" 242
Sustainability External Advisory Council (SEAC) 14
Sustainability Key Performance Indicator methodology 105–107
– conceptual diagram 108
– customization and sensitivity analysis in early assessment 123–128
– input data summary 108–109
– normalization and aggregation criteria 121–123
– primary 114–120
– quantitative assessment, in process design activities 107–111
– tree of impacts 111–120
sustainable chemistry knowledge center (SCKC) 241
sustainable development, definition of 33

sustainable direction 49
SWOT analysis 157–158
– of chemical leasing legal aspects 197
synthetic chemicals 8
systems thinking 90

t

target area 121
– definition 110, 121
– examples, to KPIs normalization 121–122
technosphere 57, 68, 70, 81
terminal operators 176–177
Terneuzen port 139, 141
– exchange patterns 141
The Center for Waste Reduction Technologies (CWRTs) 77
Three Mile Island nuclear incident 9
total cost assessment (TCA) 77
Total Material Requirement (TMR) 69
Toxic Substances Control Act 8
Tragedy of Commons 157
train derailment incident, Canada 9
transition perspective 217–218
– on chemical industry 219–221
– geographical point of view. See Rhine–Scheldt delta
– governance strategies 227–230
– management perspective 28–30, 39–40
– – critical issues 225–226
– and pathways 223–225
transparency 10, 11, 15
tree of impacts 110–120
– basic structure 113
– conceptual diagram 112
triple bottom line 45, 55

u

uni-objective optimization 169
Union Carbide incident (Bhopal), India 9–10
United States Environment Protection Agency (USEPA) 8, 13, 106
– Tool for the Reduction and Assessment of Chemical and Other Environmental Impacts (TRACI) 106

v

vendor-managed inventory (VMI) 172, 213
vertical integration. See coordinated supply chain management

w

Water Framework Directive 15
water pollution 7
weight factors 122–123, 126–128
whole-life costs (WLCs) 76